HEMEON'S
PLANT &
PROCESS
VENTILATION

Third Edition

HEMEON'S
PLANT &
PROCESS
VENTILATION

Third Edition

Edited by
D. Jeff Burton

LEWIS PUBLISHERS
Boca Raton Boston London New York Washington, D.C.

Library of Congress Cataloging-in-Publication Data

Hemeon, W. C. L. (Wesley Chester Lincoln)
 Hemeon's plant and process ventilation : new edition of the
classic industrial ventilation text / revised and edited by D. Jeff
Burton.
  p.   cm.
 Rev. ed. of Plant and process ventilation.
 Includes bibliographical references and index.
 ISBN 1-56670-347-6 (alk. paper)
 1. Factories--Heating and ventilation. 2. Industrial buildings-
-Heating and ventilation. I. Burton, D. J. II. Hemeon, W. C. L.
(Wesley Chester Lincoln). Plant and process ventilation.
 III. Title.
 TH7684.F2H396 1999
 697.9′2—dc21   98-15349
                                                               CIP

© 1999 by D. Jeff Burton
Lewis Publishers is an imprint of CRC Press LLC

No claim to original U.S. Government works
International Standard Book Number 1-56670-347-6
Library of Congress Card Number 98-15349
Printed in the United States of America 1 2 3 4 5 6 7 8 9 0
Printed on acid-free paper

Preface to the New Edition
by D. Jeff Burton

This edition of the classic textbook by W.C.L. Hemeon—first published in 1955—retains the great practical and historical aspects of that book. I have modified the text as necessary for clarity, to add new information, or to eliminate outdated or unnecessary information.

Where changes and updates could be made seamlessly, I incorporated them in the text without notation. Where it seemed important to identify new information [or to clarify retained text], I used brackets.

Hemeon was both a theoretician and a successful practitioner. He actively sought the fundamental principles underlying industrial ventilation practices and then distilled his theoretical approaches into practical "handbook data."

During the past forty years many of Hemeon's unique approaches and principles have found their way into the ACGIH *Ventilation Manual* and other textbooks (e.g., the contour area approach to hood design and the VP method of duct design). He coined many words widely used today ("receiving hood") and developed new ways of characterizing subjects that are still being studied by others (e.g., the behavior of stratified warm air in large, open buildings).

Hemeon wrote in the style of his time: long sentences written with flare and distinction. I like his style and have tried to maintain the flavor. You will appreciate the brilliant practicality of the man in his own words.

Hemeon was the principle engineer in Hemeon Associates, an air pollution firm located in Pittsburgh during the 1950s and 1960s.

Before that he was an Associate Professor at the Graduate School of Public Health at the University of Pittsburgh. Most of his work was completed and published in the late 1940s and 1950s after Hatch, McElroy, Tuve, Alden, DallaValle, Silverman, and others had mostly completed their groundbreaking work.

Hemeon assumed that readers would be somewhat familiar with the subject. (He thought most would be mechanical or industrial hygiene engineers.) If you are new to the subject, it may be useful to review a primer like my *Industrial Ventilation Workbook.*

The text assumes standard conditions (STP: 70° F, 29.92" Hg, dry air) except where noted. See recent references (in References and Bibliography) for ways of dealing with non-standard air conditions.

Because I am not an expert in every facet of ventilation, and because I have not had access to every research report published since 1955, I cannot vouch for everything that is retained in the text. (Hemeon's development of formulas related to air ejectors, for example, leaves me a little nervous; yet it seems sound.) Even after forty years, some of Hemeon's theoretical approaches are yet to be tested and verified experimentally. I have moved a few of these to the Appendix where they will continue to patiently await the careful scrutiny of graduate students.

Because the book has been revised and retypeset, errors and typos are inevitable. Please contact me and we will make corrections for the second printing. (My address is in the References and Bibliography section.)

Finally, remember this is a *textbook*. When designing and installing industrial ventilation systems, always involve competent licensed engineers and follow current codes and standards of good practice.

Prefaces to the Original Editions
by W.C.L. Hemeon
[Edited for clarity and length]

Design of a ventilation system for industrial spaces consists essentially of three problems: (1) determination of the airflow rate and arranging airflow patterns in the space; (2) design of the duct system; and (3) selection of the fan. Of these three problems, the characteristics of the first largely distinguish *industrial* ventilation from others.

Principles for design of ductwork and selection of the fan were developed long ago and are well understood. It sometimes seems that engineers experience such sheer satisfaction in their ability to handle the design of duct work on a neatly quantitative basis that they are led to slight the initial problem of selecting suitable exhaust or ventilation rates as though it were a minor detail to be covered as quickly as possible so the job of ductwork design and selection can be completed, whereas, in fact, the part they overlook is the essential ingredient of industrial ventilation.

Skill in the design of mechanical arrangements is also of great practical importance, sometimes overriding other considerations in the practical worth of an industrial ventilation system. But this aspect belongs in the "department of mechanical ingenuity," rather than in the field of understanding principles; i.e., the basic principles of mechanical design belong to a broader engineering category.

Any branch of engineering may pass through three stages during its development. First, it is practiced as an art, where success is dependent mainly on experience and empirical data, often ill-defined. Extension of these data to new situations is a haphazard business.

In the second stage, experience has become crystallized into a body of principles and design becomes the practice of applying them. Attainment of this stage makes it possible to communicate a large body of information by reference to concisely stated rules of analysis and design.

As more experience develops, a third stage occurs in which there is an extensive reduction of design principles to handbook data forms.

Industrial ventilation engineering as currently [1955] practiced is largely in the first stage among practitioners. The exposition of principles by DallaValle and Hatch (1932) describing the nature of airflow adjacent of an exhaust opening and ways of exploiting these principles for the design of one type of local exhaust hood was an important first contribution to the development of Stage 2.

The author [Hemeon] has attempted to contribute to a maturing of this subject by developing certain principles in undeveloped areas pertaining to air motions of various processes. The first half of this book is concerned with methods for analyzing a ventilation problem and the dynamics of the air polluting process to determine what ventilating air quantities are needed, local or general exhaust, and in what manner the air is to be channeled through the space

The principles developed are based on theoretical considerations so that in some cases experimental verification is [still] necessary.

Having in mind the needs of the user for "handbook" data, we have, wherever feasible, attempted reduction of the various principles to tabular or graphical summaries.

In the course of developing principles, the author has taken the liberty of coining some words and phrases, e.g., pulvation, inertials, exterior hood, receiving hood, null point, and loss-and-recapture.

A primary acknowledgement is due to some early contributions of Professor Theodore F. Hatch, a former colleague of the author. Perhaps most important is his proof (1936) that fine particles of dust and fume of interest to occupational health have no power or motility independent of the air in which they are suspended. He was associated with J. M. DallaValle (1930-1933) whose work led to development of empirical equations describing airflow adjacent to local exhaust hoods. Others include C. E. Lapple and C. B. Shepherd (particle kinetics in air), R. T. Pring (air induction by falling materials), G. L. Tuve and G. E. McElroy (air jet behavior and duct loss factors), and J. Alden (airflow measurement using static pressure).

Table of Contents

Chapter 1

Objectives and General Considerations

This chapter provides an introduction and an overview of materials presented in subsequent chapters.

If one had a giant's eye perspective of a typical industrial plant, one would see various processes inside the plant shielded by thin walls from the large-scale weather conditions outside. One could watch the continual struggle of management to warm (or cool) the plant, while at the same time trying to maintain adequate air quality.

Industrial ventilation (IV) is concerned largely with engineering techniques for controlling emissions, exposures, air movement, and for introducing outdoor air in a pattern and on a scale that is adequate to maintain satisfactory air quality without excessive exhaust of tempered air.

The objectives of a ventilation system for offices, conference rooms, and commercial spaces are to promote comfort and to suppress odors due to human occupancy, building equipment, and building materials. Additional important objectives include temperature control and effective arrangement of air supply points in relation to the space dimensions, its shape, and nature of its occupancy.

In contrast, the objectives of an industrial ventilation system are to control airborne dusts and fumes and to control adverse thermal conditions, or to do both. The purpose may be mainly to eliminate a hazard to health or to remove a merely disagreeable atmospheric contamination. The ventilation process may consist of gentle flushing of the interior space with clean outdoor air at *calculated* rates just sufficient to dilute contaminants to predetermined, acceptable concentrations, or it may take the form of local exhaust ventilation in which contaminants are withdrawn *at their point of origin* into duct systems for discharge outside the building or to collection units.

1

Local Exhaust Ventilation (LEV). By removing the contaminant at its point of origin before it can escape into the general atmosphere of the space, satisfactory air purity can be maintained with relatively small quantities of outside air. This is in contrast to a system of natural or general ventilation in which large air flows sweep through a building effectively removing the contaminant but also making it impossible to conserve tempered air.

Design of a successful local exhaust system depends on the correct estimation of the rate of airflow into an exhaust opening, the hood. The flowrate, in turn, depends on the character of the contaminating process, on the type and dimensions of the hood, its placement relative to the process, the amount of contaminant generated, and its degree of toxicity.

General Exhaust Ventilation (GEV). The approach associated with the term *ventilation,* in the minds of many people, consists of the flushing action of a total space and has been more specifically referred to as *general ventilation.* Since the action of such an approach is primarily a matter of diluting some contaminant with clean air, it might more descriptively be termed *dilution ventilation.* This method is especially useful where a large number of sources are widely dispersed in a room and the total quantity of contaminant is small enough to be diluted effectively by practicable rates of ventilation. In that case, general ventilation may provide a more economical system than one of local exhaust.

In the design of a ventilation system in which control of contaminants will be effected by dilution, the overall ventilation estimate is made as simply as in this example: if a process generates 0.1 acfm of vapor and the concentration in the air must not exceed 10 parts in 1,000,000 parts of air, then the required dilution ventilation airflow rate is:

$$Q = q/C = 0.1 \text{ acfm}/(10 \text{ parts}/1{,}000{,}000 \text{ parts}) = 10{,}000 \text{ acfm}$$

where acfm is *actual cubic feet per minute;* the flowrates are q and Q.

In plants where there are no air contaminating processes, general ventilation systems are also needed to remove the excesses of either solar heat or heat generated within the space by operation of lighting and machinery. In the past, air for cooling has been generally estimated on the illogical basis of space volume, as a certain number of air changes per hour. In today's world, ASHRAE 55 (standard on thermal comfort, latest version) is generally applied to working situations and required airflow is based on the requirements to meet that standard.

Air Distribution. The turbulent process whereby an air stream enters and mixes itself with the air of a room is of importance in the design of either local or general ventilation systems. The process is related to the problem of avoiding worker discomfort due to excessive local air velocities, and to effective dilution of contaminants. It is possible to analyze the behavior of such air streams or jets, but this certainly cannot be done merely by drawing curved lines on a diagram connecting the air inlet of a room to the air outlet—a practice which is all too common.

Origins of Design Information

Plant engineers and those charged with the design of industrial buildings often design ventilation systems without the assistance of a ventilation engineer. Likewise, occupational health and safety professionals are often called to assist in design. In these cases, all should rely on concise standards of good practice, as available. These include the references and bibliography included at the back of this book.

Historical Background: Woodworking Machinery; Leather Working; Metal Grinding and Buffing

The reason for an exhaust system in a woodworking plant is quite different from reasons applying to the host of modern air contaminating processes. The tremendous volume of wood waste (shavings, sawdust, wood chips, and like) produced in a woodworking factory is so great that, in the absence of automatic conveying equipment for its continuous removal, the manufacturing operation would literally choke to death in its own offal within a few hours. Pneumatic conveying is more nearly suited to the task than any other, particularly since it simultaneously removes the fine wood dust produced in such operations as sanding.

Industrial activity just before and after the turn of the century saw a phenomenal growth in the manufacture of articles of commerce made of wood, and this, in turn, led to a correspondingly flourishing activity for the contractors who designed and installed exhaust systems for the handling of wood waste and similar problems.

Because of the fire hazards in such plants, safety regulations were developed that had a profound influence on exhaust ventilation design for many years. The best of these regulations specified the diameter of branch duct to be attached to a given machine and the static suction to be maintained at the throat of the hood. In this approach, the code writers succeeded in specifying the two elements most essential to the satisfactory operation of the system, i.e., rate of

air exhaust, which together with a suitable hood design, determines the effectiveness of waste removal, and air velocity in the connecting duct which assured adequate transport of the waste to the central cyclone dust collector. Thus a specification of two inches hood static suction was equivalent to fixing the branch duct velocity at some magnitude between 4,000 to 5,000 feet per minute, values known from experience to be adequate to ensure continuous self-cleaning of the ducts. Specifications in these terms were simple to understand, easy to check by measurement and—as regulations—simple to administer.

Two circumstances contributed to the development of standard specifications for practically every type of woodworking machine. One of these was the fact that since the machines were, in a sense, standard units, data once obtained would be applicable to practically all other installations. Second, unlike most industrial ventilation and dust control problems, the prime criterion upon which performance was based, and correctly so, was the effectiveness of the installation in removing the wood waste produced by the machine, a visual test based on simple common-sense observation. In some operations, control of atmospheric wood dust was a factor of equal importance, but here also visual observation served as an adequate criterion of exhaust efficacy.

The early development of standards for metal grinding exhaust systems is also interesting. The bulk of solid wastes resulting from these operations is not great. The prime objective of grinding wheel exhaust is the removal of fine dust. In the early part of the 1900s, it was thought that all dust from abrasive grinding was harmful to health; that the hard sharp particles of abrasive dust caused silicosis or led to tuberculosis. Until the advent of synthetic abrasive wheels of carborundum (silicon carbide) and alundum (fused alumina) on a large scale, grinding wheels were of natural sandstone, the dust from which did cause silicosis.

The factors described above led to adoption of today's standard grinding wheel hoods and standard rates of exhaust. As in the case of wood waste and wood dust removal, visual evidence sufficed as an adequate criterion of exhaust hood performance, and all these exhaust standards were so based.

Standard rates of exhaust serve well enough for the control of metal grinding dust. They may be entirely inadequate, however, where the grinding results in dust which is toxic or highly objectionable in small concentrations. Ground lead or other heavy metals, or irritant dusts such as those from incompletely cured phenol-formaldehyde resins, in concentrations that could be regarded as minor, can constitute a significant health hazard or a major nuisance.

With the rapid development of new industrial processes requiring some form of ventilation for control of dust or fumes, it was natural for engineers to turn to state and federal agencies for standards of good practice. Unfortunately, there was insufficient experience with which to meet this demand. The hazards and exhaust requirements of many new processes could not be appraised by visual means as had been done in the case of woodworking and metal grinding, and so the analysis and tabulation of experience lagged.

Not until the tools of air analysis that implement the techniques of industrial hygiene came into wide use in the middle 1930s was it possible to gain the data necessary for precise knowledge of the subject and for subsequent crystallization of that knowledge into concise form. The pressure for the development of this kind of information has been great, and consequently increasing effort has been expended to codify existing knowledge (e.g., 29CFR1910.94). Most of this effort, however, has been confined to specific industrial operations rather than to broad classifications. See, for example, Chapter 10 of the ACGIH *Ventilation Manual*, which lists specific requirements for over 200 unique operations.

Classification of Processes

Even though there will always and inevitably be many instances of misapplication of such "coded" requirements, more and more codification has been nevertheless an objective earnestly sought and developed. It will be clear from the reading of this text, however, that an effective summary of ventilation knowledge cannot be based practically on individual industrial operations nor upon individual industries (with occasional exceptions). Rather it must be accomplished by classification of situations, to each of which some general rules of procedure can be formulated.

The broad classification of hoods and processes employed in this book is simple and adequate to bracket the entire field as shown in Table 1-1 (next page).

Inasmuch as the foregoing tabulation anticipates the subject matter discussed in succeeding chapters, a complete definition of terms employed and their significance must be obtained in those chapters. A broad discussion is given in Chapter 4.

Design Objectives

Before one can come to grips with a ventilation design problem it is necessary to obtain a quantitative conception of the objectives. Unlike the problems of commercial ventilation in which the objectives are confined to suppression of body and building-materials

odors while maintaining a comfortable distribution of temperature, industrial ventilation is concerned with the control of a host of physical contaminants—dust, fumes, vapors and gases—and with the relief of excessive industrial heat.

Table 1-1
Classification of Processes and Hoods

Hood Type	Type of Process
Exterior	a. Cold multidirectional dispersion
	b. Heat generating
Booths	a. Cold; null point of dispersion entirely within booth hood
	b. Cold, and null point outside hood
	c. Heat generating
Complete enclosure	a. Cold, no air induction into enclosure
	b. Cold, with air induction into enclosure
	c. Cold, with "splash" impingement on wall openings
Receiving	a. Heat generating
(e.g., canopy)	b. Mechanical projection of particles toward exhaust opening

Toxicological Factors. The effect of exposure to excessive concentrations of air contaminants may be one of severe poisoning occurring acutely after very brief exposure (e.g., carbon monoxide poisoning) or from chronic daily exposures over a period of weeks, months, or years (e.g., lead poisoning or benzol poisoning). On the other hand, exposure may produce only a mild illness, symptoms of which are obscure. The great majority of air contaminants encountered in industry are not toxic but are, nonetheless, often objectionable. For example, they may be irritating or malodorous if present in certain concentrations.

A fundamental concept, of primary importance and utility in the study of air purity in relation to health of exposed workers, is that the human organism can tolerate small amounts of most toxic substances without injury, because the physiological flushing, excretion, and repair processes operate at a rate greater than the destructive effects of those amounts below the tolerance quantity.

Threshold Limit Values (TLV®). Tolerance levels vary over a wide range in accordance with the properties of the substance. A human, in the course of an average working day, may inhale some 300 to 600 cubic feet of air. A figure of 350 cubic feet is conventionally referred to as average, mainly because it is numerically equivalent to 10 cubic meters, a convenient decimal figure in the SI system. A daily dosage figure for a particular air contaminant is therefore reducible to an atmospheric concentration basis. For example, a permissible daily intake of 10 milligrams can signify that the threshold limit

value, previously called maximum allowable concentration (MAC), is about one milligram per cubic meter.

In the first half of the 1900s, conceptions and techniques were developed for measuring atmospheric concentrations of dusts, fumes, and the like, and appraising the degree of hazard to health from the results. Originally, attention was confined to substances recognized to be definitely toxic. Later, lists were extended to include many substances, toxic properties of which were uncertain, and, finally, a still larger number were added, the limit values for which served simply to define objectionable concentrations from the standpoint of sensory response without any allegation of toxicity. In the latter case, air purity standards are not designed for complete elimination of odor or irritant. (This, again, is in sharp contrast with the practice in nonindustrial ventilation.)

Acceptable exposure concentrations are given in the latest ACGIH TLV® list and OSHA's PEL tables 29CFR1910.1000. (See References.)

[Editor's note: 2nd Edition Tables 1-2 to 1-7 showing older exposure limits have been deleted from this edition.]

Metal Fumes and Dusts. Fume is that particulate matter formed from parent metal at high temperature, usually molten, by a process of volatilization and oxidation. If the same material is condensed to a powder and is involved in some handling process whereby the particles become airborne, it is considered to be *dust*. Lead oxide in the atmosphere around a high temperature pot of melted lead is a fume; the individual particle size is a fraction of one micrometer (or micron). Lead oxide in the form of litharge, however, on becoming airborne is properly termed dust; the individual particle size is at least several microns due to aggregation of individual particles that are much smaller.

Metallic fumes and dusts having a TLV® or PEL of 0.1 milligram per cubic meter are "high toxicity." Control of such compounds by ventilation often presents problems of more than average difficulty which is aggravated by the fact that atmospheric concentrations in this range are invisible in ordinary circumstances. On the other hand, it is relatively simple to meet the specifications for manganese and zinc oxide which have higher acceptable exposure levels.

The distinction between toxic metal fumes and nuisance fumes is indicated partly by the magnitude of the limits, except that in the case of zinc oxide, a transitory systemic toxic action known as metal fume fever may occur at concentrations above the PEL.

Mild steel welding fume, a common example of a "nuisance fume," contains a high proportion of iron oxide plus the constituents of the welding rod coating. Visual criteria in this case has been a

traditional basis for judging objectionable concentrations. The threshold limit value has been variously defined for electric welding fumes as from 1-5 milligrams per cubic meter. These concentration figures merely describe atmospheric conditions that have been found, by experience, to be objectionable. Classified as toxic rather than nuisance are those fumes generated by welding of certain metals which introduce compounds like lead, cadmium, chromium, etc.

Irritants. Irritant substances that cause discomfort to the nose or eyes are found among gases, vapors, fumes, mists, and dusts. While their physical properties vary widely, for the present it is sufficient to group together all those substances having in common this one irritant characteristic. There is no need, of course, for a measurement of concentration to inform exposed people that concentrations of some contaminant are present in irritating concentration. However, it has been found useful to convert irritation experience into terms of atmospheric concentrations which then often become the limiting value. Such limits do not, in general, represent complete freedom from irritation for all persons. Rather they describe a compromise concentration, to which workers find it possible to become accustomed.

Toxic Gases. These compounds differ from the toxic fumes, dusts, and mists only because of their gaseous state. While some are characterized by faint odor discernible to a specialist in environmental hygiene, many of these substances at low but hazardous concentrations may be regarded, for practical purposes, as lacking in any sensory warning properties and their presence is therefore not perceptible to the average person. In this case, the identification and evaluation of contaminant sources is often the principal problem in ventilation design and will require the guidance of an industrial hygiene engineer.

Solvent Vapors. Solvents differ from other substances because, as used in industry, they are often intended to (or perform as a part of the process) vaporize completely at some stage of the process. The quantities evaporating are usually known, and it is therefore possible to compute ventilation requirements on the basis of air flows required to dilute them.

Nuisance Dusts are, by definition, "nontoxic" materials. The criterion, therefore, of a satisfactory exhaust system is logically based on the same considerations that determine the objectionable situation in the first place. This is sometimes a visual one. The principal consideration applying to this group of contaminants is that few special problems exist as to identification of sources and there is

little need for extraordinary care in providing for complete control of all potential sources, as is sometimes true of toxic materials. Visual observations are often adequate for appraisal of the magnitude of the design problem, and also for appraisal of the completed unit.

Common examples of this type of dust are seen in wood sanding, gypsum handling, and the synthetic abrasive powders: silicon carbide and fused aluminum oxide.

Ventilation Design for Low Concentrations

Reference has been made to the scale of difficulty in the design of ventilation control systems to attain low atmospheric concentrations, such as 0.1 or 0.05 milligrams per cubic meter. It is true that the low concentration specification is an important indication of the magnitude of the problem but it does not in itself completely define it. Difficulty may stem from the existence of a large number of sources of contamination in a given area, and also from the problem of locating all of them. Ordinary visual methods of identifying such sources are of limited utility. Understanding of the physical mechanism whereby fumes or dust become airborne and detailed observations of the various operations provide a sounder basis for inferences preliminary to satisfactory design.

Still another factor of importance is the rate of contamination of the atmosphere. Although data are usually lacking on this rate, it nevertheless has a bearing on the problem equal in importance to the value of the maximum allowable concentration. This point is effectively illustrated in the following problem.

Example 1. A new process, which is to be installed in a plant, consists essentially of a rectangular tank 8 ft long by 3 ft wide, containing a solution which generates mists that are described as extremely toxic to humans. The PEL is given as 0.001 milligrams per cubic meter, equal to 0.00003 milligrams per cubic foot. (This is about 50 times as toxic as lead, at 0.05 milligrams per cubic meter.) It is known that the process generates the dangerous mist at a maximum rate of 0.72 mg per hr from the tank in question. The tank is to be exhausted by slots along each long side. What is a suitable *minimum* safe rate of exhaust in these circumstances?

Solution. Conventional exhaust ventilation rates for tanks might call for 100 to 250 cfm per sq ft of tank area. In the absence of data, one might expect a ventilation specification of, say, 3 x 8 x 200 = 4,800 cfm. However, the emissions data permit one to calculate the airflow rate required to *dilute* the contaminant to a safe level, say 25% of the PEL. The escape rate is 0.72 mg per hr. To

dilute this flow to 0.00003 mg per cu ft requires 0.72/(0.00003 x 60 x 0.25) = 1,600 cfm. Even though extremely toxic as indicated by the PEL, the quantity evaporating is quite small. In any case one can conclude that an exhaust at some level *below* 4,800 cfm would be entirely adequate *if the minimum exhaust rate was important.*

Appearance of Various Dust Concentrations

Dust and fume concentrations can be roughly described in terms of sight perception. Some dusts may be present in potentially harmful concentrations though not visible in ordinary interior illumination and visa versa. For example, quartz dust or lead oxide dust can be present in concentration well above the allowable limit yet not be apparent visually. Harmful concentration of manganese oxide dust on the other hand, would nearly always be apparent.

Table 1-2 provides a rough scale of dust concentration values in terms of the sense of sight. These figures are based on the author's impressions and should not be regarded in any way as other than an approximate description. It is clear by comparing this scale of values with threshold concentrations of free silica dust (typically 1-5 million particles per cubic foot) that hazardous concentrations are apparent visually only under the most favorable conditions of illumination.

Concentrations of metals like lead, with maximum permissible concentrations of 0.01 to 0.2 milligram per cubic meter, are seldom visually apparent until very toxic concentrations are attained.

Table 1-2
Sight-Perception Dust Scale
(Approximate Visible Dust Concentrations in General Air)

	Perspective	
	Short Distances < 10-15 feet	Long Distances > 50-200 feet
Lighting	Concentration, Million Particles/cu foot	
Beam of sunlight, background dark	2	2
Beam of sunlight, background dark	10-20	2-5
Bright daylight, no direct sun	10-20	5-10
Low intensity daylight	20-40	10-20
Dim artificial light (night)	100-200	75-100

Chapter 2

Dynamic Properties of Airborne Contaminants

The subject matter covered in this chapter is fundamental to the field of industrial ventilation. It deals with the motion of gaseous and particulate contaminants in relation to the air with which they are mixed; in other words, with their segregation tendencies.

For this discussion, it suits our purpose to present a classification based on physical rather than hygienic properties of the various airborne substances. While the previous chapter emphasized human reactions to such substances, they are now divided simply into physical categories: dust, fumes, mists, vapors, and gases.

Dust. The term *dust* has the ordinary connotation. Dusts are finely divided solids that may become airborne from the original state without any chemical or physical change, other than fracture. Dispersion of a dust in air may result from disintegration of lumps of parent material, as in crushing and grinding of rock, or from disturbance of a deposit of already pulverized material. Dust in atmospheric suspension generally consists of particles ranging in size, for the most part, from 0.25 to 20 micrometers, with numerically most-common sizes in the zone between 0.5 and 5 micrometers. On a weight basis, however, the majority of dust is composed of particles larger than 10 to 20 micrometers.

Fumes. Fumes are solid, air-borne particles that have resulted from some chemical or physical process that involved a change of state—usually a thermal process of oxidation, sublimation, or evaporation and condensation. Consequently, the size of the particles is very much less than that of dusts, even though their respective ranges overlap. In general, the weight proportion of fume particles larger than 0.5 to 0.75 micrometers is negligible. The electron microscope reveals that there are vast numbers in the range between 0.01 and 0.1 micrometers (the ordinary light microscope is incapable of revealing particles smaller than about 0.25 micrometers). Table 2-1 gives a scale of common objects in microns (an older term for micrometers).

11

Table 2-1
Micrometer (micron) Dimensions of Common Objects

Object	Size, Micrometers (microns)
325 mesh screen	43
Red blood cells	8
Common pollens	15-25
Human hair, coarse	75
Human hair, fine	50
Cotton fiber	15-30

The rapid volatilization at elevated temperatures of molten metals such as lead or zinc, accompanied by oxidation to solid particles of lead oxide or zinc oxide, illustrates a common type of fume production:

$$2Zn \text{ (plus heat)} ---> 2Zn + O_2 ---> 2ZnO$$
$$\text{metal} \qquad\qquad \text{vapor} \quad \text{air} \quad \text{fume}$$

Some salts such as ammonium chloride experience sublimation when the temperature is raised sufficiently. In this process, the salt does not pass through a molten state. The particles resulting from recondensation by cooling in the surrounding atmosphere are very small and are often classed as fumes.

Waxes may also be volatilized at temperatures above their melting point and when the vapors recondense as reduction of their temperature permits, a finely divided airborne suspension of wax particles results. In some cases chemical changes, partial thermal decomposition, may also occur. The formation of an oil smoke is the same process, but, because it results in minute droplets of liquid oil rather than solid wax, we prefer to consider it under "*Mists.*"

Boundary Between Dusts and Fumes. Examples can be cited of processes in which the distinction between dusts and fumes may not be clear cut. The combustion of pulverized coal produces fly ash which we prefer to classify as dust because of its coarse particle size characteristics, even though it results from a drastic chemical and physical process.

Melted lead surfaces rapidly accumulate a film of lead oxide, dross, which becomes airborne when the metal surface is agitated. It can be argued that this material is either dust or fume. However, these cases need not concern one if the nature of the particles is understood. Our objective in classifying contaminants is to direct attention to their significant physical and chemical properties, i.e., size, rather than for the sake of classification system per se.

Mists. Atomization of liquid to form an airborne suspension of tiny liquid droplets—a mist—may take place in different ways. Pneumatic atomization of paint particles in a spray gun results from shear forces on liquid filaments.

Of significance are those mists formed by the collapse of an air or gas bubble. This may be seen in the fracture of a large soap bubble. Small bubbles of froth that accumulate on the surface of an electroplating solution create a mist of the plating solution. A liquid stream falling into a tank of liquid or onto the floor also creates bubbles and mist. Sea salt may become airborne by this same mechanism where conditions are favorable to the formation of foam.

If the liquid is a water solution of a solid, evaporation of the water takes place after mist formation and a system of dry particles in atmospheric suspension, having the same composition as the material dissolved in the mother liquor, results. Particle size of the dry suspension is a function of (a) the concentration of solids in the original solution, and of (b) the size of the original liquid droplets comprising the mist. The particle size of the dry material in the exhaust air from a lacquer spray booth is relatively coarse, because they are formed from a mist of which the droplets are large—several hundred microns in diameter—and the concentration of solids in the original liquid lacquer is high.

As is the case of dusts and fumes, one can readily cite systems showing how the mist classification frequently overlaps with the others. A dry salt suspension resulting from the misting process might be duplicated by a dusty operation involving the same material in dry powdered form. An oil mist can be formed by the atomizing action of a high speed mechanical process, or it could result from thermal volatilization and subsequent recondensation by natural cooling to temperatures below its dew point. (The particles from the latter process would, nevertheless, be much smaller than from the former.) It may also be noted that volatilization and condensation of an oil is the same process as that cited for formation of a wax fume. The difference is only in respective forms, liquid vs. solid, that are normal for each at the final temperatures.

Gases and Vapors. There is seldom any confusion in the meaning of the terms gases and vapors. They are molecular dispersions, in intimate mixture with molecules of the air. Gases are those which cannot exist as liquids at ordinary temperatures except at high pressures, like carbon dioxide, carbon monoxide, nitrogen oxides, etc. Vapors, on the other hand, are those which are commonly liquid at ordinary temperatures with vaporization characteristics determined by the vapor pressure of the liquid and the concentration of its vapor in the surrounding air. A vast number of industrial solvents illustrate

this group, as well as aniline, nitrobenzene, nitroglycerine, and mercury, to cite a few. It is of interest to note, in passing, that mercury differs from the molten metals zinc and lead, which produce fumes, in that it does not readily oxidize to form its solid oxide. It vaporizes at normal temperatures to form uncombined mercury vapor.

Density and Inertia — Some Historical Fallacies

It is sometimes mistakenly assumed that the specific gravity of dusts or other particulate matter play an important role in the action of ventilating or exhaust systems; that the density of vapors and gases of high molecular weight dictates the necessity for designing an exhaust system to withdraw a stream of air from the lowest point of a room or enclosure. Let's examine a few commonly asked questions.

1. In a room where, for example, carbon tetrachloride evaporates from numerous pieces of equipment, is it correct to assume that, since the molecular weight of this solvent is more than 5 times that of air, the vapor will accumulate near the floor; therefore, several suction openings placed to withdraw air a few inches from the floor will prevent the accumulation of excessive concentrations of vapor in the air at breathing level?

2. The specific gravity of lead dust is greater than that of other common dusts. Does this mean that its settling rate is so great that this factor must be taken into account in the design of exhaust for operations involving dusts of this character?

3. In local exhaust design for dust control, where the solid particles are projected through the air at high velocity, what is the significance of the advice that the hood opening be placed so that the material will be projected into the throat of the hood? Does this signify that satisfactory dust control is impossible in those instances where those directions cannot be followed? If that design procedure is sound, how critical is the positioning of the hood throat relative to the trajectory?

In the following discussion, the role of density and of particle inertia will be considered in relation to these questions. First, it will be shown that airborne contaminants (whether gaseous or particulate and in concentrations or sizes that are of practical significance in occupational hygiene) have little real significant weight or inertia, and that those factors, therefore, do not normally enter into calculations for design of ventilation.

Vapors and Gases

The action of gravity on a volume of gas or vapor is determined not by the weight of the gas molecules alone, but on the average

weight of all molecules of whatever nature contained in the particular mixture. Also, molecular mixtures (of health concern concentrations) are permanent; there can be no spontaneous separation of the heavier molecules from the lighter ones. The density ratio (specific gravity of significance in any settling rate) therefore, is

$$\frac{\text{Density of mixture of air and vapor}}{\text{Density of air}}$$

rather than, as is sometimes mistakenly suggested

$$\frac{\text{Density of pure vapor}}{\text{Density of air}} = \frac{\text{MW of pure vapor}}{\text{MW of air}}$$

where MW is the molecular weight.

The maximum possible concentration that may be attained by any vapor relative to air, as in the air space above a covered container of the liquid, or in a film very close to the surface of evaporating solvent, is limited by the vapor pressure of the solvent at the temperature in question. If the characteristic vapor pressure of a solvent is 76 millimeters Hg at room temperature, then the maximum possible concentration of vapor attainable in the air space above the liquid is the ratio of this pressure to that of the atmosphere—760 millimeters, i.e., 76/760, or 10%. Expressed generally, if c is maximum percent concentration, and p_f is vapor pressure of the liquid in millimeters of mercury, then

$$c = \frac{p_f}{760} \times 100 \qquad \text{in percent at standard conditions}$$

The average molecular weight of such a mixture is the average, weighted according to the percentage. The "molecular weight of air" is 29; therefore, the average molecular weight of the mixture

$$MW_{average} = \frac{c \bullet (MW_{solvent}) + (100-c) \bullet 29}{100}$$

This average molecular weight may be expressed as a ratio with air's molecular weight (MW = 29) to describe the specific gravity.

In Table 2-2 the specific gravities (SG) of some common organic solvents are compared with the molecular weight ratio. The latter is frequently tabulated, erroneously, as the specific gravity of the

vapor. The discrepancy is especially misleading in the case of solvents of low vapor pressure, like toluene.

A common industrial exception to the preceding remarks is typified by the case of boiling degreaser solvents in open contact with air. Thus, trichlorethylene is 3.7 times as heavy as air in a boiling degreaser. Even here, however, the density is not represented by the ratio for molecular weights because the solvent vapor is typically 85 to 90°C and the air at 20°C, and a temperature correction must be applied.

Table 2-2
Density of Air-Solvent Vapor Mixtures

Substance	Vapor Pressure at 20°C, mm Hg	SG of Saturated air mixture	Ratio of $MW_{solvent}$ to MW_{air}
Acetone	184.8	1.24	2.0
Benzene	74.2	1.16	2.7
n–Butanol	6.3	1.01	2.6
Ethyl ether	430.0	1.88	2.6
Methyl alcohol	92.0	1.01	1.1
Toluene	22.0	1.06	3.2
Trichlorethylene	60.0	1.28	4.5
Xylene	10.1	1.04	3.7

Industrial Values

Concentrations approaching the maxima cited are scarcely ever encountered in industry except as accidental occurrences. Measured concentrations of benzol vapor above a table surface with a source in the center of the table are shown in Figure 2-1.

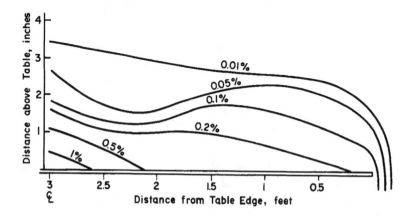

Figure 2-1. Concentrations of benzene vapor above table top; from evaporation at center at a rate of 2.5 cc per minute. (After Gunner, Transactions, 19th Annual Safety Congress.)

Experiments of Fig. 2-1 were conducted in a small, closed room and precautions taken to avoid drafts of air over the table top. It is significant that even in these circumstances representing minimum air disturbance, there was appreciable air circulation as evidenced by the rapidity with which concentrations decreased within two to three inches above the source. The vapor pressure of benzol at ordinary temperatures is 74 mm Hg; therefore, the maximum concentration in a film above the vapor source would be a little under 10%. In a vertical distance of 2 inches, therefore, concentrations had decreased by two orders of magnitude. Note also the "spill" from the edge of the table. These hydrocarbon vapors (concentrations of 0.1-0.01%) will quickly mix with room air and settle only a few inches.

Concentrations of 0.1 to 0.5% by volume may be taken as the maximum likely to be encountered in occupational hygiene-related ventilation work. (Higher concentrations may be found within the ventilated enclosures attached to vapor recovery systems.) The specific gravity of such low air/vapor mixtures is practically unity, since they typically consist of 99.5 to 99.9999% air.

Air mixing. Random air currents of more than 15 to 20 fpm are found in the most carefully draft-protected spaces. Most important is the ease with which mixing for dilution occurs. Actual conditions one encounters in ventilation problems are highly favorable for mixing. One almost never deals with large masses of vapor but, rather, with slender streams of vapor-air mixtures comparable with streams of cigarette smoke observable in random motion in a smoking lounge.

Another analogy can be given to illustrate these principles. Into a large tank of water which is being gently heated and, therefore, has noticeable convection currents, a very dilute solution of salt is allowed to trickle from a pipette. Rather effective mixing would be found to occur, with no accumulation of higher salt concentrations at the bottom of the tank than elsewhere; a permanent molecular mixture occurs in mixing and no separation can occur.

A word of explanation is required regarding some circumstances in fire and explosion protection which might seem to refute some of the conclusions of previous paragraphs. If a large vessel of solvent were to be heated accidentally or otherwise to effect rapid evaporation, massive quantities of vapor would pour over the rim and move toward the floor with velocities that are predictable by the methods illustrated above. In such large quantities, dilution with distance of fall would be relatively slow, and fire or explosion by ignition at a lower level is not an uncommon occurrence.

The travel of ether vapors in a channel can be demonstrated on a laboratory scale in which the stream can be readily ignited after traveling several feet. The channel serves to minimize mixing with

air. A similar effect is seen without a channel where the ether is heated with resulting rapid evaporation.

All these circumstances, of vital concern in fire protection, are of an accidental nature, a sudden, unexpected chain of events. They do not affect the conclusions applicable to the design of industrial ventilation for occupational hygiene purposes.

Inertials. We shall have frequent occasion in later discussions, as in the present, to distinguish between very fine dust and coarser particles. The distinction is of such importance as to warrant a formal designation. The term *inertials* will be used in this text to describe the larger particles having relatively high inertia, a term which will serve as a reminder of special properties and will assume more meaning as the subject matter develops. For the present, it suffices to think of inertials as particles of such weight that they fall readily by gravity as contrasted to fine dust which tends to remain in air suspension. Inertials may be considered to have a minimum size of around 50 to 100 microns, and fine dust a maximum of 10 to 20 microns.

Fine dust particles have important health significance, partly because they account for practically all of the particles in the atmospheric suspension. They are the particles able to penetrate the respiratory tract, and they are of the size having the greatest chemical reactivity (i.e., potential toxicity).

Aerodynamics of Dust Particles

When a particle falls under the influence of gravity it reaches a constant terminal velocity, very rapidly in the case of fine dust. The magnitude of this velocity for spherical fine dust particles is given by the expression (derived and discussed in detail later in this chapter)

$$u_t = 1.37 \times 10^6 \rho_s D^2 \tag{1}$$

where u_t is terminal velocity in feet per second, ρ_s is density of the particle in pounds per cubic feet, D is diameter of the particle in feet (3.05×10^5 microns equals 1 foot)

This equation permits calculation of the settling velocity for a 10 micron particle, of 2.7 specific gravity (water = 1), to be about $u_t = 1.5$ fpm. This is a low velocity in comparison with air velocities above 20 fpm characteristic of most occupied industrial spaces. Even particles of 30 to 40 microns of the same specific gravity will attain settling velocities of only 15 to 20 fpm. On the basis of gravitational forces alone, it is plain that very small particles have little power of motion independent of the air in which they are suspended.

The inertial properties of dust particles may also be examined by considering the behavior of a particle that has been projected at high initial velocity by some mechanical force. The frictional resistance offered by the air decreases the speed of the particle in accordance with known physical principles.

Using these principles, one can estimate the distance of travel, s, for a spherical particle of any size and density to decrease the velocity in any given ratio. This was done for particles of specific gravity = 2.7 (re: water) which had an initial velocity of 1,000 fpm. The results are shown in Table 2-3 and Figure 2-2.

Table 2-3
Distances Traveled by Spherical Particles of 2.7 SG (re: water)
with Initial Velocity of 1,000 feet per minute

Final Velocity, fpm	Distance, feet		
	$^1/8$ inch	25 microns	2.5 microns
250	95	0.8	0.08
50	200	1.7	-
25	260	2.2	-

The formulas employed for these results provide the only single value for the 2.5 micron particle. The expression also assumes that gravitational effects are negligible in comparison with the force of air friction and particle momentum.

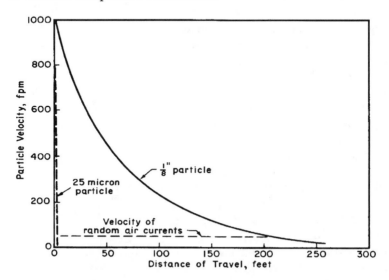

Fig. 2-2. Distance reached and velocity of spherical particles projected horizontally with initial velocity of 1000 fpm (SG = 2.7 re: water) against air resistance. Figure illustrates distinction between "inertials" and "fine" dust.

The results illustrate much more strikingly the difference between inertials and fine dust than the previous demonstration involving gravity alone. Obviously the eighth-inch particles are inertials and the 2.5 micron particles are fine dust, in accordance with our definition; and the 25 micron particles may be given either term depending on which size it is being compared with and the physical circumstances of the particular problem. In general, however, one would class their behavior as that of fine dust.

It is strikingly clear, also, from these results that fine dust particles, even in the conditions of violence postulated in this example, have almost no power of motion independent of the air in which they are suspended. From these demonstrations springs the extremely important principle applicable to problems in dust control, that since fine dusts of occupational hygienic significance may be considered as lacking any weight or power of independent motion through air, controlling their motion reduces to the problem of controlling the motion of the air. In other words, the problem is one of coping with the characteristics of air motion, and one's attention is therefore directed entirely to air motion characteristic of a particular process, in the knowledge that control of the air automatically cares for its dust burden. This important principle was first set forth by Hatch in the 1930s.

It has been shown, however, that small particles can be dragged along with inertials when they are accelerated to high velocities. So, in mixtures of particle sizes which include inertials, smaller particles can travel greater distances.

Dynamics of Particles. The motion of solid or liquid particles in air is retarded by air friction. The relationship between velocity and distance is outlined for particles falling freely starting from rest, and for particles projected horizontally with initial velocity where effect of gravitational force is negligible.

Types of Motion. In the flow of fluids through ducts, two types of motion are recognized and discussed in elementary texts on hydraulics. They are streamline (or laminar) and turbulent flow, and the

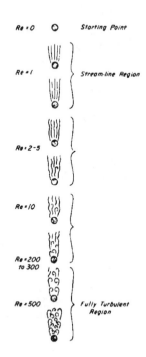

Figure 2-3. Changing character of the wake turbulence in a falling particle.

frictional resistance between fluid and duct wall follows different laws in each case. The type of flow is largely determined by the magnitude of the *Reynolds number*, Re.

When air flows past a particle, the magnitude of the drag or resistance is also determined by the character of relative airflow which includes streamline flow, turbulent flow, and an extensive region intermediate between these two. The magnitude of the Reynolds number, Re, defines the type of flow to be expected for a given particle shape. In this application of Reynolds number, the diameter of the particle, D, is used (assumes a spherical particle).

The transition between streamline flow and intermediate flow is not sharply defined, nor is there a sharp break from intermediate flow to turbulent flow. The gradual change in character of flow, i.e., air resistance, is depicted in Figure 2-3, suggesting the changes in relative air turbulence in the wake of falling particles as the Reynolds number increases due to increasing velocity. In the case of particles of occupational hygiene interest, Re is almost always less than 1 and, as such, particles are constrained to fall in the streamlined stage. Large particles, such as those thrown from a grinding wheel, may reach intermediate or turbulent stages.

Stokes Law and Terminal Velocity. The magnitude of the force tending to retard the motion of a particle in air, regardless of the type of relative motion prevailing, includes a term, C, the drag coefficient, or coefficient of resistance

$$F_r = \frac{C \rho u^2 A}{2g} \qquad (2)$$

where F_r is the resisting force, pounds; ρ is the density of the air (0.075 lb per cu ft at standard temperature and pressure); u is the particle velocity relative to the air, feet per second; A is the projected area of the particle, square feet; and C is a dimensionless number.

The coefficient, C, has a value variable with the Reynolds number. Some values are given in Table 2-4.

Table 2-4
C for Equation 2-2

Re	C		Re	C
0.1	240		1,000	0.46
1.0	26.5		5,000	0.38
10	4.1		50,000	0.49
100	1.1		100,000	0.48
500	0.5		1,000,000	0.13

Streamline Motion in Fall by Gravity. The slope is constant for values of Re < 1; and that in this region

$$C = \frac{24}{Re} \tag{3}$$

As mentioned before, this is the region of streamline flow. When this value of C is substituted in general equation (2), noting

$$Re = Du\rho/v \quad \text{and} \quad A = \pi D^2/4$$

where D is particle diameter in feet and v is viscosity, one obtains the well-known expression, Stokes Law:

$$F_r = \frac{3\pi vuD}{g} \tag{4}$$

Terminal Velocity. The terminal velocity, u_t, of a particle subject to the force of gravity is attained when that force is in equilibrium with the resistance force. For a falling body, the force is that of gravity, F_g.

When F_g is equated to the resistance force F_r, one again obtains

$$u_t = 1.37 \times 10^6 \rho_s D^2 \tag{1}$$

Projection of Particles. Where large particles are projected, bullet-like, from a source with a high initial velocity, analysis of the motion may be assumed to have negligible gravitational effect.

A large particle may start its travel in the turbulent flow region because of high velocity (i.e., high Reynolds number), pass through the next zone, intermediate motion, and finally slow to streamline flow. By the time its velocity has attained low enough velocities for the last stage, however, it is unlikely that gravity affects could be neglected.

More coverage on inertials is presented in Chapter 3.

[Editor's note: The original text provided additional pages of formula derivation and approximations related to particle flow in intermediate and turbulent regions (which are briefly summarized in this chapter). Because of their complexity, the absence of corroborating data (even after 40 years), and the lack of simple applications to occupational health engineering, they have been moved to the Appendix.]

Chapter 3

Dispersion Mechanisms:
Pulvation, Entrainment, Induction

It was suggested in the preceding chapter that most airborne contaminants of health significance have very little independent mobility. They tend to move about because of the motion of the air in which they are suspended. It follows that local exhaust design is concerned merely with the means for controlling the motion of contaminated air. It will be shown in later chapters how the airflow characteristics of various suction openings can be readily described quantitatively, and, therefore, the problem of local exhaust design reduces fundamentally to a matter of designing the airflow pattern of the exhaust opening to meet the requirement of the air motion characteristics of a particular contamination mechanism. It is, therefore, necessary first to understand the character of air motion in various situations.

Evaporation and Pulvation

Evaporation of liquids to a vapor state is well understood; vapor pressure relations determine what will occur, and molecular theory provides an explanation of the actual mechanism of evaporation. There is need, for the sake of convenience and descriptive clarity, of a word corresponding to "evaporation" that is applicable to dust. We have therefore coined the word "pulvation" which is defined as the non-molecular mechanical or physical process whereby dust or mist becomes suspended in air from a previous state of rest.

Mist formation from liquids is not a molecular process and therefore our term pulvation should logically apply to it. However, the principal need for the term is in consideration of dust processes and our use of it will be largely confined to that and will imply dust formation.

Pulvation

Clouds of dust are formed as the result of two events in sequence: (a) the primary action of pulvation, a mechanical or pneumatic action that projects fine particles at high velocity from a coagulated state into the air of immediate vicinity as individual particles, and (b) secondary air currents which transport the localized dusty air, formed in the prior action, away from the site of formation.

The pulvation action may merely disperse previously pulverized material from a state of rest to an air-borne state, or it may involve simultaneous reduction from a solid mass, like rock, to powder. It may be regarded as essentially an "explosive" process with a high rate of acceleration of the particles from the state of rest to the air-borne state. It must be distinguished from the dispersion action of secondary air current, which may be violent or gentle like the random convection currents characteristic of all spaces.

Forces Causing Air Motion

There are at least two fundamental forces that cause primary air motion integral with pulvation: (1) the drag of air behind inertials moving at high velocity, and (2) the shearing action of air suddenly expelled from between the pores of a loose bulk of material.

Air Motion by Inertials. Any solid body moving through the air is resisted by the force of friction, and the interaction of the two forces results in a part of the energy going to the creation of a wake behind the body (Figure 3-1). Highly turbulent, a portion of the wake is a linear air motion in the direction of motion of the body. This phenomenon is illustrated by the dusty air currents which are seen in motion after passage of an automobile traveling at high speed along a dry, dirt road (Figure 3-1a) . The motion is small when only one vehicle is involved, but would be multiplied markedly were there a procession of them, each contributing added momentum to the air that is already in motion from preceding vehicles.

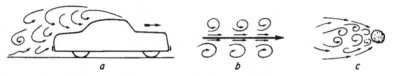

Figure 3-1. Turbulent streams of air induced as a wake behind high speed objects. a. Motor vehicle. b. Plane, moving parallel to its length. c. A single solid particle.

This action occurs where inertials are created in any dusty process. It is of greater interest and importance to note that inertials and fine dust are almost always formed simultaneously because, if the inertials are numerous and have appreciable initial velocity imparted to them in the process, they will give mobility to the fine dust particles which otherwise would move only with the random direction of miscellaneous air currents unrelated to the process.

Unidirectional Pulvation. If, for example, a grinding wheel is in heavy contact with its work so as to produce liberal quantities of inertials continuously, at high velocity, the latter will induce a unidirectional stream of air which will serve as the transport medium for the fine dust formed in the same process. Thus, both classes of particles travel together in the same stream but by different mechanisms.

One might liken the stream of inertials to a series of microscopically small fans in fixed positions, where any one of these small fans represents one particle at any single instant, but is immediately replaced in that position by another particle, then another, and so on. (Figure 3-2).

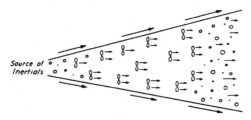

Figure 3-2. A stream of inertials inducing a continuous flow of air suggests the action of numerous fixed miniature fans.

In the case of a grinding wheel, the air currents generated by the wheel itself have no perceptible effect when inertials are formed at a high rate; the fan action of the inertials overwhelms the former. If, however, the grinding contact pressure is lessened or the material is changed, the layer of air carrying the fine dust will travel part way around the periphery of the wheel. This phenomenon is never seen in the operation of a cut-off wheel because its lack of thickness provides no similar surface for angular drag of air.

Pulvation at a Swing-Frame Grinder

The principles are neatly illustrated in the operation of swing-frame grinding wheels, equipped with a 180 degree top guard (Figure 3-3). In conditions favorable for a high rate of formation of inertials,

full contact of the whole width of the wheel, and application of high pressure on the work, the fine dust travels in the stream of inertials in a fairly narrow cone. When, however, less than the full width of the wheel is in contact, as when a small elevation is being ground, the air following the wheel periphery may carry fine dust with it. It will travel only 80 to 90 degrees before it reaches the guard housing.

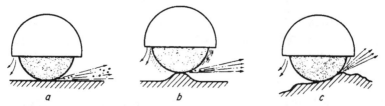

Figure 3-3. Pulvation at a grinding wheel. a. Heavy grinding produces large quantities of inertials and a strong resulting air stream. b. Incomplete grinding contact forms fewer inertials, permitting peripheral air stream to carry some of the fine dust. c. Grinding contact disposed with inertials trajectory toward housing.

At that point any tendency to leave the wheel surface will not matter because the air now becomes a part of the circular stream traveling between the wheel and the guard by drag of the grinding wheel surface whence it is discharged at the opposite diameter into the open air.

The above analysis, it will be noted, has involved reference to three air streams: (1) that induced by inertials, (2) that following the surface of the wheel due to its high velocity, and (3) that traveling under the guard, differing from the second, by its size and by being channeled by the guard.

Another situation is illustrated in which irregularity of the surface to be ground results in deflection upward of the stream of inertials toward the guard opening where it is entrained in part by the under-guard air stream. (Figure 3-3c).

Multidirectional Inertials. Another illustration of the air-drag action of inertials is seen in the impact of a steel chisel point, or of a sledge hammer head, on a block of stone (Figure 3-4). The dust clouds observed from a single blow simulate, in miniature, the smoke clouds formed in the explosion of a bomb. Since it has been proven that the fine dust particles could not travel the observed distances in the absence of air motion, one can readily deduce that the pulvation action includes integral air currents induced by the drag of inertials that are projected at high velocity from the point of rock fracture. The action is essentially the same as grinding wheel pulvation except that, in the present instance, inertials and integral air motion move

in all directions. It is convenient for later treatment to distinguish these two classes of action by the terms unidirectional and multi-directional respectively.

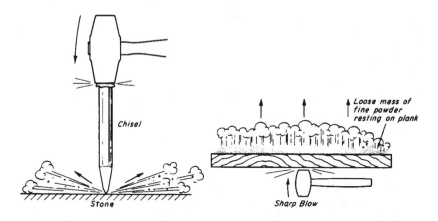

Figure 3-4 (left). Multidirectional pulvation from an impact of steel chisel on stone. Stone fracture forms inertials and fine dust simultaneously. High velocity trajectory of inertials drags air (and fine dust) as shown.

Figure 3-5 (right). Pulvation in which sudden compaction of a bulk of fine powder expels air from pores of powder mass. No inertials are involved in this action.

Shear in Air Expulsion

Airborne dust will be created when finely powdered material resting on a board is disturbed by the blow of a hammer on the board. The pulvating action is probably due, mainly, to sudden expulsion of air in the pore space between the particles of the mass as the force of the blow causes an instantaneous compaction of the loose mass. The escaping air, in a fluid shearing action, lifts some fine dust particles into suspension in air.

The *pulvation distance*, i.e., the length of high-velocity travel of the integral air motion, would be very low in this particular action.

An action in which fine powder at rest is lifted into suspension by passage of exterior high velocity air currents is another illustration belonging to this class.

Similarly, when a mass of pulverized material is allowed to fall, its motion relative to the surrounding air induces a shearing action and results in fine suspension in air of a portion of the falling mass.

An interesting demonstration of some of these principles may be seen in a comparison of the fall of fine powder and coarse material containing a fraction of fine powder.

Figure 3-6 (left). In gravity fall of fine powder, rapid disintegration of the falling column occurs with dispersion of powder into air.

Figure 3-7 (right). In a column of falling material having a liberal proportion of large inertials, air inducing power retains dust within the column.

The fine powder, free of inertials, will become well dispersed in air suspension by the fluid shearing action and, when this dispersion is completed, the downward fall is arrested and the powder is more or less completely airborne. In contrast, the falling material shown in Figure 3-7, consisting mostly of inertials, will continue its fall. While the fines become dispersed into air suspension as in the former case, the air in which they are suspended will be carried downward by the induction action of the inertials, except for a fraction nearest the periphery which finds it possible to become separated from the main column.

Splash. When a mass of falling dusty material strikes the floor (see Figure 3-8), sudden compacting of its bulk occurs with a resulting violent escape of air which carries fine dust with it by the shearing action characteristic of this category. Pulvation is multidirectional.

Secondary Air Motion—Entrainment

Secondary air currents are those that are not an integral part of the pulvation action, but, impinging on concentrated air suspension near its source, cause further dispersion. The random air currents of the work area are obviously of the secondary classification.

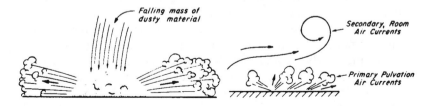

Figure 3-8 (left). Splash, in which a sudden compaction of dusty material occurs when its fall is arrested. Pulvation action is fundamentally the same as Figure 3-5.

Figure 3-9 (right). Primary air currents are those that are integral with pulvation. Secondary air currents, or, separate origin, transport dusty air away from pulvation sites.

Less obvious are cases where the machine responsible for the primary air currents of pulvation also creates secondary air streams which may impinge on the primary pulvation zone. A criterion aiding in the distinction between these air currents is that it is usually physically possible to interpose a baffle between the source of secondary streams and the pulvation zone. The emphasis on the high velocity, explosive character of the primary pulvation should not lead to the error of identifying high velocity air currents with primary pulvation, because secondary air currents are often seen with high velocities as in Figure 3-10.

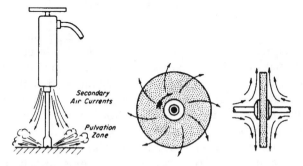

Figure 3-10 (left). Secondary air currents from compressed air leakage in a pneumatic chisel impinging on the zone of pulvation.

Figure 3-11 (right). Secondary air currents induced by rotation of a disc.

It is also in order to warn against drawing any conclusions of significance from the fact that machinery elements create high air velocities in a dusty atmosphere. One may see such actions visibly churning the dusty air, but if they do not interfere with the pulvation action by impingement on the primary dust streams, they have no practical significance.

The grinding wheel of the preceding discussion will also serve to illustrate secondary air streams. The air film which follows the periphery of the wheel could be classified as secondary, although almost unique in its intimate relation to the primary pulvation.

Grinding wheels and cut-off wheels also induce air currents at their sides somewhat in the fashion of centrifugal fan impellers. The ambient air moves toward the hub of the wheel, then close to the rotating surface, and is thrown outward radially (Figure 3-11). Such centrifugal currents are not very obvious near small grinding wheels having a bulky hub and usually do not impinge on the pulvation zone.

One of the best examples of the interaction and interference of secondary air currents with primary pulvation in the same piece of machinery as in the operation of pneumatic chisels used in carving stone (see Figure 3-10). Pulvation occurs at the surface of the stone and, if leakage of compressed air occurs, as is common, around the chisel receptacle in the barrel of the pneumatic hammer, it will impinge at high velocity on the pulvation zone and aggravate dispersion markedly.

Rising hot air streams sometimes may occur in such a manner as to disperse the primary pulvation cloud, although such streams are usually integral with a contaminating action which is considered separately under the following subject of air induction.

Air Induction

Where granular material falls through a chute, for example into a bin, each particle imparts some momentum to the surrounding air and the aggregate effect is the induction of a stream of air along with the material in the identical manner already described under "uni-directional pulvation." On reaching the bin, the air streams outward through all available openings, carrying with it the fine dust that has become suspended in the process of falling. If the bin is air tight, a slight pressure will be created by the action (Figure 3-12).

Air induction assumes, of course, that there is a source of air that the falling material can entrain. In some closed transport systems —for example in screw conveyor systems attached directly to a chute—there may not be a source of air. In this case, a negative pressure will build within the chute and its source.

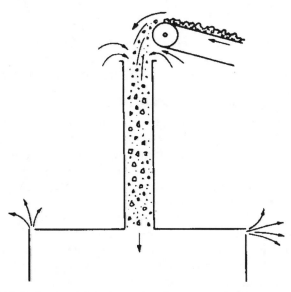

Figure 3-12. Induction of air through a chute into a bin by action of falling materials.

This phenomenon occurs in varying degree wherever granular material is allowed to fall by gravity or is thrown. It is most commonly demonstrated in material-handling systems composed of crushers, belt and bucket conveyors and elevators, vibrating screens, chutes, and bins.

The action may be observed, with the aid of smoke, in a water spray from a garden hose or in the shower bath where water droplets supplant solid particles. The same action is spectacular around water falls like Niagara.

Heat

Another form of air induction by gravitational force is in the draft created by hot gases which are unenclosed. Air receiving heat from a hot process becomes lighter than the cooler air, hence buoyant, and rises. When the hot process is also responsible for pulvation, or produces a fume, that contaminant is carried upward in the rising stream of hot air or gases.

One sees this problem in an obvious form in hot metal casting shake-out, brass melting furnaces, electric welding, lead dross handling, electric furnaces, zinc galvanizing tanks, hot water tanks, pouring of molten metal, and innumerable others.

Enclosed Machines

Certain kinds of machinery act like fans and induce a flow of air through their casings. When this operation is associated with the handling of materials that become airborne, by evaporation or pulvation, the phenomenon is of importance in the design of ventilation.

High speed hammer mills have marked fan action which causes strong inward flow of air with the feed and an outward flow with the material, the latter heavily laden with dust from the pulverizing action of the mill. Some other high speed pulverizing machinery and disintegrators have a similar action of varying degrees.

Basket centrifuges strongly resemble centrifugal fans with closed outlets. That is, the common centrifuge has no opening corresponding to the fan outlet but does have openings corresponding to a double inlet fan. The air motion around a centrifuge, therefore, is very similar to that observed in a fan with a blocked outlet. In addition to the internal circulation of air which does not concern us, there is a turbulent spillage at both openings, top and bottom.

Centrifugal air separators are designed for internal circulation of air by means of the built-in fan, but there is commonly a net flow of air through it, inward with material and outward with tailings or product.

A classification of pulvation, dispersing, and induction processes appears in Table 3-1, while a tabulation of secondary air motion types is given in Table 3-2. (See next two pages.)

Table 3-1
Classification of Pulvation, Dispersing, and Induction Processes
(Processes not in a housing)

Primary pulvation process	Examples	Characteristic integral air motion	Typical secondary air motion
1. Dispersion shock	Blow on dust-covered plank, dispersing material already pulverized	None by primary process	Possibly, by striking mechanism
2. Disintegration shock (shattering)	a. Smashing rock by impact	Currents induced by flying materials	With pneumatic tools, compressed air escapes
	b. Crushing rock by pressure	No primary air motion	In crushers, gravity discharge to and from machine
3. Mechanical shear	a. Abrasive cut-off wheel	Currents induced by flying inertials (unidirectional)	Radial flow of rotating discs, but not dust entraining
	b. Abrasive grinding wheel	Currents induced by flying inertials (unidirectional)	Periphery of wheel may carry air part way around
	c. Disc grinder	Practically multidirectional	Same as (a), except they are dust entraining
4. Aerodynamic shear	a. Air streams of high velocity lifting settled dust	The causative air streams themselves	
	b. Liquid atomization		
5. Splash (related to but distinguished from air induction)	Dropping of dusty material from a height, on reaching floor or pile	Lateral, disc-shaped air streams by compaction	
6. Film collapse (liquids)	Breaking of bubbles on surface of a solution	None by primary process	

Table 3-2
Secondary Air Motion Induced by Machinery
(Processes not in a housing)

Secondary air moving process	Examples	Directional tendency	Magnitude of air currents
1. Rotation of discs, fly wheels, etc.	a. Abrasive cutting wheels	Inward at hub and radially outward	Considerable, but may not disperse dust
	b. Abrasive grinding wheels	a. Same as above	a. Considerable
	-	b. Air carried around periphery of wheel	b. Variable
2. Linear motion, belts, etc.	a. Conveyor belt	Longitudinal, modified by convention currents	May not be great, especially if no pockets are formed; no greater than belt velocity
	b. Open bucket elevator	Same as buckets	At least the volume of moving buckets
3. Reciprocating elements	-	-	Depends on size and speed of moving part
4. Stirrers, mixers	-	-	-
5. Pneumatic tools	Air leakage at shank of pneumatic chisels	Directly downward into pulvation zone	A function of degree of wear, and may be a major consideration in specification of exhaust

Chapter 4

General Principles of Local Exhaust

Local exhaust ventilation (LEV) describes an important approach for preventing contamination of workroom air. The LEV system withdraws the contaminant at its source into an exhaust system for discharge to the air cleaner or building exterior. The contaminant is thus "captured" before it escapes into the general air of the space.

This approach is contrasted with general exhaust ventilation (GEV), in which the contaminant is emitted into the surrounding air and its concentration is subsequently prevented from rising above some acceptable level by continuous flushing or dilution of the total space with clean air. GEV systems have their place (sometimes a superior one), but in many cases are incapable of accomplishing the degree of control possible by local exhaust. In some cases, adequate emission or exposure control can be achieved only by a local exhaust system. We will examine these various cases in upcoming chapters.

The local exhaust hood is the point of air entry into the duct system and the term "hood" is used in a broad sense to include all suction openings regardless of their shape or their physical disposition. The term is used here to embrace suspended canopy-type hoods, booths, exhaust through grille work in the floor or a bench top, slots along the edge of a tank or table, wet-sink hoods, exhausted enclosures, the simple end of a duct or pipe and, in the present discussion, the exhaust from enclosures such as bins.

Enclosing Hoods

Enclosing hoods and booths are typified by the common laboratory hood or spray painting booth in which one face of an otherwise complete enclosure is open for access. Air-contaminating action takes place inside the enclosed space, and air is exhausted from it at a rate that induces an average velocity through the front

opening (the face) of from 50 to 400 fpm, which is usually sufficient to overcome escape tendencies of the contaminated air within (see Figure 4-1). This type of exhaust arrangement is the most positive, requires the smallest rate of exhaust, and should be the aim of hood design wherever conditions permit. The walls of the enclosure not only reduce the requirements as to rate of exhaust but also serve as baffles against the adverse effect of extraneous room air currents.

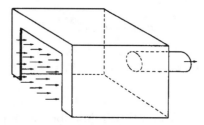

Figure 4-1. A booth is a partial enclosure with an open face for manipulative access to the interior.

The exhaust connection may be located at the rear wall opposite the opening or at the top. Except in the case of booths which enclose heat releasing processes, it does not matter greatly where the exhaust connection is located provided the dimensions of the hood, relative to the exhaust opening, result in reasonably adequate distribution of the airflow over the face. If the booth were five feet by four feet at the face, and only two feet deep, poor distribution would result regardless of where the exhaust connection was made (Figure 4-2).

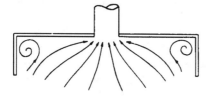

Figure 4-2. Shallow booth in which distribution of velocity at face is not characteristic of deep booths. Velocities are high nearest exhaust openings and low at either end. Velocities may be analyzed for prediction of distribution.

Where there are units inside the booth that release appreciable quantities of heat, upward convection currents of air carrying the contamination will be set in motion. In this case, it is best that the exhaust opening be located at the top of the booth opening (Figure 4-

3). Such a combination of process air currents in a booth having the exhaust outlet at the top is actually a quite different apparatus and it could logically be argued, therefore, it should not be classified as a booth.

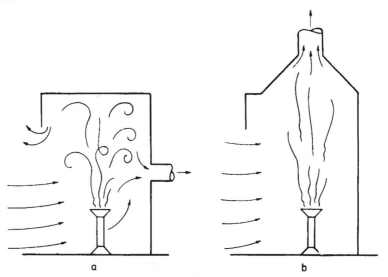

Figure 4-3. Exhausted booth enclosing a heat releasing process. a. With exhaust connection low, spill of hot air occurs at top edge of opening. b. Exhaust connection at top may avoid spill.

The air velocity to be employed at the face together with the area determines the exhaust capacity required (Q = VA) and is, therefore, an important element in the design of a booth. The area is obvious and the required air velocity is dependent on the same considerations that apply to all cold processes whether enclosed by a booth or served by an exterior hood as discussed below.

Exterior Hoods

This class of hood includes numerous types of suction openings located adjacent to sources of contamination which are not enclosed (Figure 4-4). These include exhaust slots on the edges of tanks or flanking a work bench; exhaust duct ends located close to a small area source; large exhaust hoods arranged for lateral exhaust across an adjacent area; exhaust grilles in the floor or work bench below the contaminating action; small canopy hoods, except those above hot processes (see below); and large propeller exhaust fans in outer walls, which is adjacent to the zone of contamination. All these have one common characteristic: the evaporation or pulvation zone is exterior to the boundaries of the physical hood, nearly always because the

manipulative or mechanical requirements of the process will not permit the obstructions that partial enclosure of that area would entail.

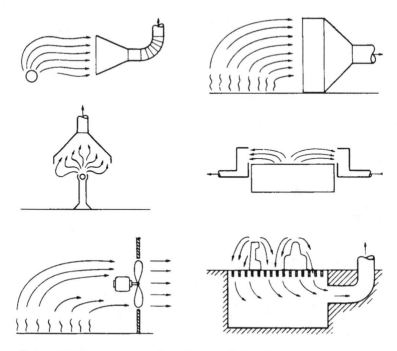

Figure 4-4. Various types of exterior hoods. In all cases, the source of contamination is beyond the boundaries of the hood structure, hence control action is effected by inducing velocities in the adjacent space.

Exterior hoods effect their local exhaust action by creating directional air currents adjacent to the suction opening. They must "reach" beyond their own physical limits by inducing adequate air velocities at those points of the contamination zone most distant from their opening to draw the contaminated air into the duct system.

The design of an exterior hood involves specifying its shape and dimensions, its position relative to the evaporation or pulvation zone, and the rate of air exhaust.

The rate of exhaust, as in the case of booths, is dependent on the air velocity required and on the size of an imaginary curving area, the contour area, which corresponds to the face opening of the booth; being exterior to the physical limits of the hood itself, its magnitude is not as obvious. The significant area is a portion of either a sphere or a cylinder and its radius defines the size of a protective envelope of air currents. All air contaminants originating within that zone are withdrawn directly into the exhaust opening. Methods for estimating

the surface area of any significant contour are largely matters of geometry and are examined in detail in Chapter 5.

Just as in designing a booth—its dimensions are determined by those of the operations to be controlled—the contour radius of exterior hoods is selected in accordance with the characteristics of the air contaminating process. This is also bound up with the question of air velocity to be induced through that area. Both are discussed in Chapter 6. The product of velocity in feet per minute and area in square feet then gives the required air exhaust rate in cubic feet per minute ($Q = VA$). In the SI system: $m^2 \times m/sec = m^3/sec$.

It will be seen from this or later discussions that, while face velocity is a critical value for booths, the air velocity *at the opening* of an exterior hood has no significance as to its ability to control contamination.

Receiving Hoods

The term *receiving hood* applies in this text to those hoods in which a stream of contaminated air induced by and characteristic of the process is delivered to a hood especially located for that purpose. For example, a canopy hood located above a small furnace from which fumes escape receives and exhausts the fume-laden warm air which rises to the hood level by reason of its own buoyancy. Superficially similar to the exterior hood in that the contaminated air originates beyond the physical boundaries of the hood, it differs fundamentally in the air motion induced in the present case *by the process*, independent of the hood, whereas an exterior hood is required to induce the necessary air velocities at a distance from its opening to attract adjacent contaminated air to it. Exhaust requirements for a receiving hood are determined by the rate at which contaminated air is delivered to it by the process; the considerations entering into the design of an exterior hood, for a cold process are, therefore, of no concern here.

In order to design a receiving hood it is clear that information is necessary on the air pumping characteristics of such processes, in particular the heat-releasing processes cited in the example above, which are the most common ones encountered. Air motion characteristics of hot processes are studied in Chapter 8 and a basis developed for estimating their magnitude.

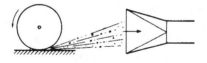

Figure 4-5. Grinding wheel inertials inducing a jet of air into the hood.

Another process to which receiving hoods might be applied is illustrated by a grinding wheel discharging a stream of inertials along a fixed line as was described in Chapter 3. A stream of air is created which carries dust along with it (see Figure 4-5). A receiving hood located on the axis of such a stream could be arranged to exhaust such dust provided the exhaust rate were at least equal to that delivered by the inertial stream. The mathematics of such an airflow induction process are developed in Chapter 7.

Arrangements in which an exterior hood is supplemented by a jet of air blowing toward the hood from the opposite side of a contamination zone are occasionally of some interest. By the step of introducing the air jet, usually in the form of a sheet of air, what was an exterior hood now becomes a receiving hood and its required exhaust capacity is determined by the rate of air motion due to the jet process at the hood location. Development of quantitative information on the characteristics of such jets is in part the reason for much of the subject matter developed in Chapter 9.

Exhausted Enclosures

Exhausted Enclosures (Cold Process). These are some of the most important devices for the control of air contaminants. In the sense of this discussion it is assumed that there are some opening such as slots, cracks, or observation apertures, because if the enclosure were perfectly tight there would be no need for any exhaust. For many enclosures the exhaust requirement is determined, as in the case of booths enclosing cold processes, by the product of the total area of openings in square feet and some velocity, usually between 50 and 400 fpm. However, enclosures frequently house processes in which there may be an important inward flow of air, the magnitude of which may vastly overshadow any exhaust requirements calculated on the above basis. This is illustrated most frequently in materials-handling systems where the fall of crushed material through a chute into an enclosed bin may pump a considerable flow of air into the bin along with the material. In a dust control system, the rate of exhaust from the bin enclosure would need to be based on the rate at which air flows into the enclosure. In order to provide some basis for estimating that airflow induction rate in various circumstances, a theoretical development is offered in Chapter 7.

Exhausted Enclosures (Hot Process). Where heated processes are enclosed and exhausted, the necessary rate of ventilation is based on different considerations than those outlined for cold processes. The existence of a higher temperature inside the enclosure than that outside results in a thermally induced pressure on the ceiling and

upper walls. If there are any openings in the upper part of the enclosure, an outward leakage can result even though, due to the enclosure exhaust, there is inward flow through lower apertures. In order to prevent this effect, it is necessary to exhaust from the interior at a rate sufficient to create a negative pressure inside that is equal to or greater than the pressure at the topmost openings. By use of hydrostatic principles, a basis is presented for computing this rate that is set forth in Chapter 8.

One other criterion sometimes applies to the ventilation of enclosures; that is where an explosive vapor (or dust) exists inside the enclosure. Then it is necessary to compute the airflow rate which, when drawn through the enclosure, is sufficient to dilute the vapor continuously to a concentration below its lower explosive limit. This requires an estimate of the rate of evaporation of the liquid inside the enclosure and data on its lower explosive limit. Results of such a calculation should be considered in conjunction with one based on one of the others previously discussed and the one giving the higher value employed in the design.

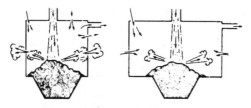

Figure 4-6. Splash-escape from an enclosure. (Left) An enclosure in which dusty air escapes by splash in spite of exhaust through nearby openings. Inward movement of air can occur through other openings on which splash does not impinge. (Right) A similar enclosure where splash escape is avoided because of distance to walls.

Splash-Escape. Enclosures shown in Figure 4-6 illustrate a situation in which contaminated air escapes from the interior in spite of the air exhausted therefrom due solely to a violent action which causes high velocity impingement of interior air on a wall opening. This phenomenon is of importance because it so commonly interferes with otherwise adequate dust control. The term "splash-escape" is intended to aid in distinguishing it as a special phenomenon requiring particular recognition. In Figure 4-6(b) a similar action is illustrated where this condition is avoided because there is enough interior space in which the pulvation air streams can spend their velocity before reaching the openings in the wall.

Splash-escape may occur together with air induction, which illustrates the importance of distinguishing between the two. Escape due to excess air induction can be distinguished from splash-escape by

observing whether air issues from openings in only one section or level of the wall. Excess air induction would cause escape from any orifice regardless of its location. It may result from the fall of dusty material onto a pile or from the action of any high speed mechanical elements inside an enclosure. The obvious corrective measures are to avoid openings in the splash zone, and/or to provide a enclosure of more spacious dimensions.

One may observe splash-escape in which there is immediate return or "recapture" of the dusty air by the action of the exhaust; this is usually a pulsating action in which escape projects dusty air a short distance to the exterior, following which it is withdrawn into the enclosure by the same orifice. It can readily be shown that such leakage from small cracks, as distinguished from the large openings, is more likely to result in permanent loss from the enclosure.

Control of Inertials vs. Fine Dust

The distinction between inertials and the fine dust traveling with it that was made in Chapter 3 is further drawn in the following discussion of some practical factors in the design of hoods for their control. Dust from grinding wheels is usually of little hygienic significance, but it is a machine familiar to all and therefore serves conveniently to illustrate the difference in behavior of inertials and fine dust. Several fanciful arrangements of grinding wheel and suction opening are depicted in Figure 4-7 (next page). Figure 4-7A shows an arrangement which would exhaust all fine dust effectively, but would permit loss of most of the inertials by rebound. It depends for success on the continuous formation of a liberal quantity of inertials to deliver an air stream laden with fine dust to the baffle plate.

By contrast, the conventional hood of Figure 4-7B results in an action whereby the fine dust is dispersed in the entire volume of air traveling in the space between guard and wheel. Since some of this air follows the periphery of the wheel, it escapes at the top opening of the guard. (The effect is usually overcome in practice by the exhaust of more liberal volumes of air which results in highly dilute dust concentration in the escaping air and in "recapture.")

Figure 4-7C is an arrangement in which all inertials, as well as fine dust, are captured. The facts illustrated by the foregoing may be summarized in the following general statements. (See also Table 4-1.)

(1) Inertials can only be controlled by an exhaust hood that is disposed so that its opening coincides with their trajectory. Their high velocity and weight make it utterly impractical to divert them from their natural path with air currents induced by an exhaust hood. (Figure 4-7C)

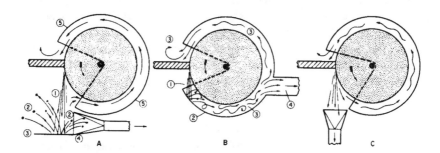

Figure 4-7A. Arrangement of grinding wheel exhaust illustrates distinction between control of inertials and of fine dust. (1) Unidirectional stream of inertials delivers fine dust in an air stream to plate (3) within influence of exhaust hood (4) which withdraws fine dust. Inertials rebound (2) from plate in all directions and are largely free of dust.

Figure 4-7B. Stream of inertials (1) from heavy contact between work and wheel induces a stream of air as in (A), but here delivery point of stream is within wheel enclosure (2). A rotating stream of air (3) within the hood exists as before but now carries a portion of the dusty air past the exhaust outlet (4) which escapes at front.

Figure 4-7C. An arrangement in which inertials and fine dust both are captured by the hood. Air stream flowing under the guard receivers none of the dusty stream. Another form of (A).

(2) A design in which the suction opening is adjacent to a baffle plate and with its axis at right angles to that of the stream will capture only those inertials which chance to rebound at right angles into the hood opening. On the other hand, the fine dust in the same stream is readily exhausted, because the air in which it is suspended is not subject to elastic rebound from the plate as are the inertials and is therefore separated from the latter and drawn into the hood opening.

(3) A hood that captures all inertials will generally control all fine dust, but one can be designed which will capture all fine dust but not capture inertials.

The foregoing discussion shows that design of hoods for control of inertials presents many more mechanical difficulties than those for the control of fine dust. This springs from the necessity for placing the hood opening in one specific position which the machine operation may not be able to tolerate practically. One sees this difficulty in some wood-working machinery and in certain machining operations.

Pros and Cons of Inertials Control

In exhaust systems for woodworking machinery and for leather trimming and cutting in shoe factories one sees examples of hood

systems whose prime objective is the removal of inertials—waste material generated in such large quantities that if not cared for would so clog up machinery and adjoining spaces and seriously impede manufacturing operations. There are other operations where the disposal of waste shares importance with the need for control of fine dust.

Table 4-1
Hood Design Principles Applicable to Inertials and Fine Dust

Item	Inertials	Fine Dust (micron sized)
Definition	Material readily settles	Tends to remain suspended in air
Typical size	40+ microns	Up to 40 microns
Examples	Sawdust, grinding dust, brick cutting	Windborne dust
Coexistence	Varying proportions of fine dust are usually produced when inertials are created.	
Control hood	Receiving (in trajectory)	Capture
Physical principle	Momentum rules; elastic	Motion of air stream rules
	Rebounds from baffles	Particles move with air stream
	Particles move independent of air stream	Particles move with air stream
Relative control	Control of inertials controls fine dust but not visa versa	
Hood design	Easier for fine dust; inertials may require housekeeping types of control.	

In operations where the total bulk of inertials is not great, deliberate consideration should be given to designing the system for removal of inertials, as well as fine dust, as compared with one for control of fine dust alone. The decision affects not only the design of hoods as shown above, but has an equally important bearing on design of duct work as brought out in Chapter 12.

The alternative to removal of inertials by the exhaust system is illustrated in the booth arrangement for control of dust in operation of swing frame grinders. Inertials are permitted to drop to the floor of the booth and are removed manually at suitable intervals. This

arrangement contrasts with the frame-attached hood and flexible hose exhaust system sometimes used, which in ideal circumstances operates to remove both inertials and fine dust. In other cases, it is practical to remove inertials accumulation by suitably arranged mechanical conveyors.

Selection of Hood Type. The operational characteristics of an air contaminating process often determine the type of hood that is permissible. Thus canopy hoods are not practical where the operation is served by an overhead crane; some processes do not permit enclosures or even booths; and some permit neither overhead nor lateral exhaust, forcing one to resort to downdraft exhaust. Where this results in serious conflict between the type of hood best suited to control and that permitted by the process, the next best hood type may serve at a greater cost in air capacity or the conflict may be irreconcilable. In the latter case some modification in the process must be sought or some ingenious mechanical adjunct to the hood arrangement made to resolve the difficulty.

Enclosures are best suited to massive dust-producing operations; indeed, they may sometimes be the only practical means for control. Enclosures or booths are in fact the most effective types of hoods in all cases. But for small-scale operations, the exterior hood, being simpler, may be better even though at a cost of greater exhaust capacity.

Contaminants from large-scale heat-releasing operations cannot be effectively controlled by downdraft exhaust and, where canopy hoods are excluded for operational reasons only, lateral exhaust remains, although enormously greater exhaust capacity is required over that possible by overhead exhaust. Exhaust for hot operations represents one of the most troublesome problems encountered in industry because of the frequently encountered conflict between operational requirements and those of exhaust.

Chapter 5

Local Exhaust and Exterior Hoods

Terms and Units for Chapters 5 and 6

A_c = area of any velocity contour, sq ft
A_f = area of face opening of a hood, sq ft
D = Diameter of hood face, ft
L = length of cylindrical contour
Q = airflow rate by exhaust, cfm
W = width of a slot, ft
X = axial distance (radius) from hood face to significant contour, ft

In a later chapter we consider the essential nature of general or dilution ventilation in which volumes of air are caused to stream through a room and provide beneficial results solely by reason of quantitative diluting effect, rather than by direction of the air flow.

Local exhaust ventilation (LEV) differs fundamentally from general ventilation in that the arrangement of exhaust equipment envelops the source of contamination by positive air currents which immediately remove contaminants before they can escape into the general air. A laboratory hood or paint spray booth are common examples of local exhaust which *entrains the contaminant at its point of origin* for removal to the outside.

The design of a booth is completely described by reference to its total face area (opening area) and to the total airflow rate through it, that is, by face area and average velocity of air through that area. The relation between the two quantities is

$$\text{Airflow rate} = \text{Face area} \bullet \text{Face velocity}$$
$$\text{ft}^3/\text{minute} = \text{ft}^2 \bullet \text{ft}/\text{minute}$$
$$Q = V \bullet A$$

The physical space requirements of the operation determine face dimensions of the booth. Then, if adequate information exists to specify face velocity, the required airflow rate is determined.

It will be noted that the face dimensions of the booth are determined not by ventilation requirements, but by the space requirements of the operations taking place within; also, that only a very small fraction of the total air streaming through the face may be required for entrainment, i.e., the local exhaust action. Hence, even though the fume is adequately diverted by the portion of air streaming through a narrow section of the face area, the same velocity must be maintained through the entire area.

Two-Dimensional Air Flow

Hemi-Cylindrical Contour. A situation may now be imagined in which the process responsible for the contamination requires free access at both sides in addition to the front; and that the two side walls as well as the front of the booth are removed, leaving only the bottom plane (bench top or floor), back (e.g., outer wall of the building), and top. The face area determining airflow requirements may not be so immediately obvious as in the case of the booth. The latter could be termed one-dimensional or unidirectional air flow, whereas, in the present case, the airflow is two-dimensional in that air tends to flow toward the suction opening from all directions with equal facility, that is, air will flow laterally toward the exhaust opening, parallel to the back wall as readily as it does along the extended axis of the exhaust opening. (See Figure 5-1.)

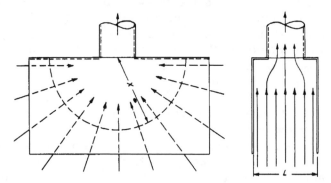

Figure 5-1. Partial exhaust enclosure resulting from removal of two sides from a booth. The locus of all points of equal velocity, i.e., the contour area, is a hemi-cylindrical surface of area equal to πXL.

If the suction opening were a vertical slot extending from the bottom to the top of the space bounded by the two planes, and its

width were insignificant, an ideal velocity shape would result which has the following properties. The locus of all points in space that have a given air velocity, is the surface of a hemi-cylinder. The area of the hemi-cylinder corresponds to the face area of the booth where, ideally, the air velocity is the same at all points.

The principle or equation of continuity reminds one that in any fluid stream of constant flow the flowrate is the product of velocity, V, and area, A, of the surface through which velocities are equal; and that when the area is reduced in another location of the stream, the velocity at that point is increased in direct proportion to the area reduction. Because air flows readily toward the slot equally from all directions, velocities would be roughly equal at all points on the circumference of any semicircle of fixed radius that one might draw in a horizontal plane from any point along the exhaust slot; these circumferences constitute the hemi-cylinder.

Thus, in order to generate a desired velocity at some distance X from the slot, at a point directly in front, it is necessary to exhaust sufficient air to generate that velocity at all points of the cylindrical surface. Even though air streaming toward the exhaust opening laterally may serve no useful purpose, it is a "cost" which must be paid in order to realize the desired velocity at the point of interest.

The area of the contour surface, from geometry, is

$$A_c = \frac{2\pi XL}{2}$$

and by analogy with the booth, the "width" may be said to be πX and the length, L. The rate at which air must be exhausted, Q, is, then, estimated from the product of the above area, and the velocity, V, which is desired at the distance X.

Actually, the performance described above is that of fluid lacking viscosity; in that sense it is ideal and not realized in practice. The distortion, however, mainly occurs close to the exhaust opening at distances of little practical consequence. (This limitation is, however, significant for other shapes that will be discussed later.)

If, in the partial booth of Figure 5-1, the exhaust opening were circular instead of a slot, the distortion from cylindrical contours would, of course, be considerable in the vicinity of the opening.

Quarter Cylindrical Contours. The shape shown in Figure 5-2 is suggested by a situation where requirements for free access to the operation might permit one side wall to remain in position. If, in this case, the exhaust opening is situated close to a dihedral corner, the contour surface of equal velocity will be seen, by application of the above reasoning, to approach that of a quarter cylinder of which the

radius is fixed by the distance to the outermost boundary of the
contaminant source.

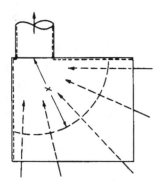

Figure 5-2. Partial exhaust enclosure (plan
view) similar to a booth but where one side
wall is missing, and exhaust opening is
displaced to one corner. The contour area is
one-quarter cylinder, $A = \pi XL/2$.

Other Cylindrical Contours. Similarly one can conceive of an
arrangement of baffles relative to suction openings in which the
significant area is any fraction of a cylindrical surface up to a whole
cylinder. The contour will always be cylindrical where two parallel
planes flank the exhaust opening.

In summary, it is clear that one may define the shape and
location of the significant contour area as *the surface which is the
locus of all points having the same air velocity induced by a source of
suction,* and that to induce a specified air velocity at any location on
that area, it is necessary to exhaust air at a rate sufficient to create
the same velocity throughout the entire contour area. Then the
product of the desired velocity and the contour area gives the
required rate of exhaust.

Example 1. Air is exhausted between two parallel planes
arranged as in Figure 5-1. The distance between planes is 3 ft.
What rate of exhaust would be required to induce an air velocity
of 100 fpm at a distance of 4 ft from the exhaust opening, assuming
the parallel planes are more than 8 ft long by 4 ft wide?

Solution. To induce this velocity at one point 4 ft distant from the
opening, it will be necessary to provide for the same velocity at
all other points at the same radius. The significant contour area
is roughly semi-cylindrical and is πXL, where X is 4 ft, and L is 3
ft, or 38 sq ft. Required exhaust rate, Q, therefore, equals VA_c or
$100 \cdot 38 = 3{,}800$ cfm.

Example 2. A propeller fan in the corner of a *large room* exhausts
air at a rate of 10,000 cfm. The fan is *midway* between floor and
ceiling, the height of which is 8 ft. The two walls at right

angles, the floor and the ceiling are the only structural surfaces bounding the space under consideration (Figure 5-3, left). Calculate the average velocity (near waist level) at the point in space that is found by moving from the corner 12 ft along one wall, thence out from the wall, at right angles, a distance of 9 ft, resulting in an X-distance of 15 ft from the corner.

Solution. The contour shape approximates a quarter cylinder of area equal to $0.5\pi XL$ where $X = 15'$ and $L = 8'$. With $V = Q/A$, velocity is 50 to 55 fpm.

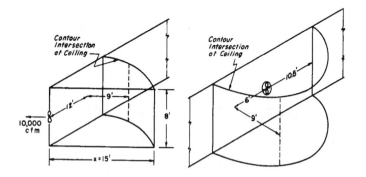

Figure 5-3 (Left) Arrangement in Example 2. (Right) Arrangement in Example 3.

Example 3. Suppose the position of the fan in Example 2 were shifted along one wall a distance of 18 ft from the corner (6 ft beyond the marker point referred to in the Example 2.) (Figure 5-3, right). Calculate the velocity in the space at the same point defined in the Example 2. All conditions are the same except the position of the fan.

Solution. The contour shape approximates a half cylinder. Velocity is about 40 fpm.

Three-Dimensional Flow

Exhaust openings that are not flanked by at least two parallel planes produce an airflow pattern in which air streams converge toward the opening from along the three fundamental axes, in contrast to the two-dimensional and one-dimensional patterns previously discussed. The shape of the contour surface which is the locus of all points of equal velocity is therefore some portion of a spherical surface depending on the number and position of flanking planes; and further assuming that the distances are great enough to

permit one to ignore the distortion of contour shape that is discussed later in connection with the practical limitations to geometric estimation.

One-Eighth Sphere. An exhaust opening at the apex of a trihedron provides a velocity pattern in which equal velocity contours tend to approach the shape of one-eighth of a sphere. That is, velocities in the adjoining space will vary inversely as the square of the distance outward measured along any radius; conversely, the velocities will be roughly equal at any imaginary surface, all points of which are at an equal distance from the opening. This shape may be illustrated by visualizing a suction opening adjacent to the upper corner of a room, bounded by the ceiling and two walls (Figure 5-4, left). The area of such a surface is $4\pi X^2/8$.

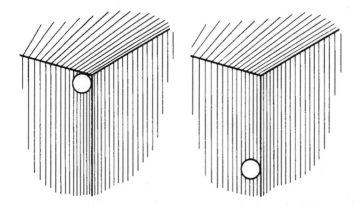

Figure 5-4. (Left) Exhaust opening at the corner formed by 3 planes (trihedron). Equal velocity contours will be one-eighth sphere. (Right) Exhaust opening flanked by two planes at right angles. Equal velocity contours are one-quarter spheres.

One-Quarter Sphere. Where an exhaust opening is flanked by two planes at right angles to each other, the contour would tend to assume the shape of one-quarter sphere of area. (See Figure 5-4, right.)

$$A = 4\pi X^2/4$$

Hemisphere and Sphere. The same geometric concept could be applied to an opening flanged by a single plane and to a simple opening with no flanking planes, except for the distance limitations discussed in the following section. If attained, the area of the contours of equal velocity would $4\pi X^2/2$ and $4\pi X^2$, respectively.

Practical Limitations to Geometric Estimation

Unflanged Plain Hoods. In Figure 5-5 a circular exhaust opening of diameter D is represented. At a distance X equal to the diameter of the opening, D, the shape of equal-velocity contours cannot be spherical, for the geometric concept, previously discussed, requires a true sphere or, at least, that its bulk be negligible compared with the dimensions of the contour surface, i.e., X >D. The spherical contour represented in the Figure 5-5 by the circle of radius where X = D, cannot be valid unless the spherical area cut by the duct diameter is subtracted from the total spherical area (i.e., the area is better represented by $A = 4\pi X^2 - \pi D^2/4$). And when X ≤ 0.5D, the spherical contour area approach is totally invalid.

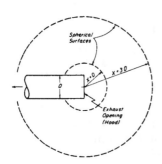

Figure 5-5. Spherical surfaces with center at exhaust opening may be described to estimate area of contour surfaces of equal air velocity—illustrating that spherical contours of small radii become invalid due to relatively large bulk of hood.

The spherical contour X = 3D has dimensions materially larger than that of the opening,. The error in applying geometry to an estimation of contour area would not be as serious in practical use. The contour area, A_c, for this case would likely be approximately $A = 4\pi X^2$.

Zone of Practical Control. How far away can the hood be positioned and still be effective? Contour velocities less than 50 fpm are not of practical significance; also sizes of exhaust openings often result in face velocities of about 3,000 fpm. Taking these values, we can calculate the farthest distance X in terms of D at which a contour velocity of 50 fpm would be found with an exhaust opening having a face velocity of 3,000 fpm:

$$Q = VA = 0.25\pi D^2(3,000) = 50(4\pi X^2 - \pi D^2/4) \qquad \text{and so} \qquad D = 2.0X$$

Analyses of this kind demonstrate that distances of up to two diameters are those of practical interest for plain, unflanged, unobstructed exhaust openings where velocity contours tend ultimately to approach a spherical shape.

Similar distance limits, calculated for the other shapes discussed in previous sections and for V =150 fpm, are given in Table 5-1.

Table 5-1
Limiting Distances of X in Terms of Hood Face Diameter
Significant for Practical Air Velocities
(Duct velocity V = 3,000 fpm)

Nominal contour shape	Q	A_c	X for V = 50 fpm	X for V = 150 fpm
Whole sphere	$0.25\pi D^2(3000)$	$4\pi X^2 - \pi D^2/4$	2.0D	1.1D
Hemisphere	$0.25\pi D^2(3000)$	$2\pi X^2$	2.7D	1.6D
Quarter sphere	$0.25\pi D^2(3000)$	πX^2	3.9D	2.2D
One-eighth sphere	$0.25\pi D^2(3000)$	$0.5\pi X^2$	5.5D	3.2D
Whole cylinder	$LW(3000)$	$2\pi XL$	9.5W	3.2W
Half cylinder	$LW(3000)$	πXL	19.1W	6.4W
Quarter cylinder	$LW(3000)$	$0.5\pi XL$	38.2W	12.7W

Note that where contour surfaces assume the shape of a sphere, the bulk of the hood may be so large that higher capture velocities (≥150 fpm) are encountered only in a zone close to the hood, and not at sufficient distances to permit application of the spherical contour.

Where flanking planes (i.e., large flanges or baffles) reduce the contour surfaces to smaller fractions of a sphere, the geometrical surface tends to provide a more realistic basis for estimating contour areas. Similarly, cylindrical contour areas may be readily estimated on this basis, since practical velocities are seen to extend outward from 10 to 40 slot widths from the suction opening.

[Studies by the editor suggest that all points equidistant from the center of the *vena contracta* in the throat of the duct exhibit similar velocities. This can easily be demonstrated by tying one end of a string to a thin wire affixed at the center of the *vena contracta* (about $1/2$ D inside the duct) and the other end of the string to the measuring tip of a velometer. Measured velocities will be approximately equal at any point within reasonable reach (with the string stretched tight) except near surfaces.]

Experimental Determination of Velocity Contours

DallaValle, noting the difficulty of the contour equation when X≤D, investigated the velocity pattern in the space adjoining circular

and rectangular exhaust openings. (See Figure 5-6.)The results cover X-distances up to one diameter which, as has been shown, is the zone of principal concern from a practical standpoint.

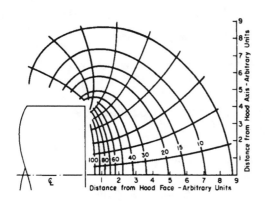

Figure 5-6. DallaValle's experimentally determined velocity contours in front of a circular exhaust opening. Diameter of opening is 8 distance units. Velocities are expressed as percentages of that at opening.

The contour area, expressed as a function of distance, X, measured outward along the unflanged hood axis, in feet, was found to be approximately $(10X^2 + A_f)$, where A_f becomes insignificantly small as X increases, and when X becomes zero, this term alone remains, as should be.

To induce any velocity, V, in the space in front of such a hood and along the hood axis, it follows from the relation $Q = VA_c$, that

$$Q = V_c (10X^2 + A_f)$$

This expression applies to distance measured outward along the extended axis of the hood throat. DallaValle's findings can also be described in terms of the ratio of indicated contour area to the face area, as shown in Table 5-2.

Table 5-2
Relation Between Axial and Face Velocities of Hoods

Distance, measured along along hood axis, in fraction of hood face diameter	Area of contour intercepting this point, expressed as a ratio with face area, i.e., $A_{contour}/A_{face}$
1/4	1.8 to 2.2
1/2	3.5 to 4
3/4	6 to 7
1	10 to 13

Example 4. An industrial operation requires an air velocity of 150 fpm for control of dust it generates and it cannot be enclosed. An unflanged, unobstructed hood can be placed for lateral exhaust at an X-distance of 10 inches (plane of hood opening to source of dust). Using both the DallaValle equation $[Q=V(10X^2+A_f)]$ and spherical area equation $[Q=V(4\pi X^2 - \pi D^2/4)]$:

(a) Calculate necessary rate of air flow, Q, when the duct diameter is 3"; when it is 5"; when it is 10".

(b) Calculate Q for the same three hoods when the X-distance is only 5 inches (instead of 10 inches).

Solution. See the following tabulations:

D Inch	X Inch	X/D	Contour Area (Sphere) Flowrate Q, cfm	DallaValle Area Flowrate Q, cfm	Ratio of Sphere to DallaValle
3	10	3.3	1,300	1,050	1.24
5	10	1.7	1,290	1,070	1.21
10	10	1.0	1,230	1,120	1.10
3	5	1.7	320	270	1.19
5	5	1.0	310	290	1.07
10	5	0.5	250	340	0.74

Results of this example suggest the following conclusions:

• There must be an X-distance where both approaches give the same flow rate. That can be determined by setting both equations equal to each other and solving for X in terms of D:

$$Q = V(10X^2 + A_f) = V(4\pi X^2 - \pi D^2/4) \qquad \text{and}$$

$$10X^2 + \pi D^2/4 = 4\pi X^2 - \pi D^2/4 \qquad \text{and solving for X}$$

$$X = 0.78D$$

From this calculation several points are noted:

— The DallaValle equation "underestimates" Q when

X/D ≥ 0.78. When the X-distance is small, the DallaValle equation is better used.

— The Contour Area equation "underestimates" Q when X/D ≤ 0.78. When the X-distance is large the contour area equation is better used. (Indeed, when X ≤ 0.5D, the spherical contour area approach is invalid.)

• Variations in the velocity at the hood entrance have little effect on the velocity at a distance from the opening. (The magnitude of face velocity *does* help determine the energy loss for the hood entrance as will be seen in a later chapter.)

• The area of the hood opening of itself has relatively little influence on the capturing performance of the hood, all other things being equal.

Example 5. A grille 2 ft by 2 ft is set, centrally, six inches above a bench. Dust generated on the bench originates always at levels no greater than 6 inches above the grille surface.

(a) Using Table 5-2, estimate the area of the contour which coincides with the point 6 inches above the center point of the grille. Note that this describes an unflanged hood, facing upward, and the central point lies along the hood axis.

(b) What exhaust rate is required to induce a velocity of 50 fpm through this contour area?

Solution using Table 5-2. The X-distance, 6 inches, is one fourth of the face diameter and from Table 5-2 it is noted that the contour area at this distance is 1.8 to 2.2 times the face area, i.e., 7.2 to 8.8 sq ft. More directly, obtain the value $(10X^2 + A_f)$, or 6.5 sq ft, which is to be regarded as good agreement with the value 7.2 above. The exhaust rate for a contour velocity of 50 fpm is then about 6.5 x 50 = 325 cfm, or 7.2 x 50 = 360 cfm.

Example 6. An unflanged hood having the face dimensions 4" high by 8" long rests flat on a bench, and exhausts laterally across the bench surface. The plane of the face is perpendicular to the bench top. Calculate the exhaust rate required to induce a velocity of 100 fpm at a distance of 6 inches from the hood, along its axis.

Solution. The bench surface can be regarded as a plane that bisects a hood of twice the size of the real hood, i.e., 6 inches

high instead of 3 inches, and exhausting air at twice the rate to be calculated for the real hood. Therefore, one calculates the result for the double hood and then divides by 2. Note that A_f in the formula is double that of the real hood. (See "Plane of Symmetry," page 61.)

$$Q = \frac{V(10X^2 + 2A_f)}{2} = \frac{100(10[0.5]^2 + 2(4 \times 8/144))}{2} = 375 \text{ cfm}$$

Flanged Plain Hoods. Comparison between flanged and unflanged hoods (see Figure 5-7, DallaValle's work) shows that in the zone that lies along and in the vicinity of the hood axis there is some difference in the magnitude of velocities induced. The improvement due to the flange is equivalent to a reduction of from 10% to 50% in required exhaust rate for given velocity at X, depending on the flange width. For the flange to have the greatest effect (and to allow the use of the hemisphere contour area approach) its width must be $W \geq X - 1/2 D$.

Another effect of the flange is in the extension of a given velocity contour laterally for appreciable distances, a factor which may be of importance in some applications. Its baffling effect against disturbing room air currents is also advantageous.

The ACGIH *Ventilation Manual* (22nd Ed., see References and Bibliography) suggests flanged hoods should modify the unflanged DallaValle area equation by reducing the volume flowrate by 25%, or

$$Q = 0.75V_c (10X^2 + A_f) \qquad \text{(Flanged plain hood)}$$

to account for the extra flow expected in front of a flanged hood. The size of the flange is not specified so the amended formula tends to err on the side of safety as flange widths approach $W = X - 1/2 D$.

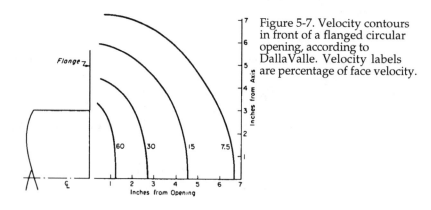

Figure 5-7. Velocity contours in front of a flanged circular opening, according to DallaValle. Velocity labels are percentage of face velocity.

Slot Velocity Contours

Several studies of the velocity pattern induced by slot-shaped exhaust openings have been conducted, all employing the heated thermometer anemometer (Figure 5-8). The data tend to confirm that velocities decrease linearly with distance in accordance with the reasoning applied in earlier sections to be a consideration of cylindrical shaped contour.

Airflow at End Zones

In the earlier discussion of cylindrical contours, airflow was restricted to two dimensions by the existence of baffles at either end perpendicular to the slot's length-wise dimension. In practice, slot exhaust arrangements are sometimes not provided with such baffles and air will flow from the end zones as in a three-dimensional airflow pattern (Figure 5-9).

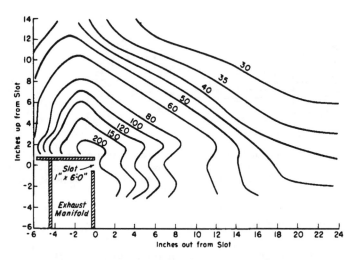

Figure 5-8. Velocity contours for plating tank exhaust slot.

However, if the slot is very long relative to the X-distance of interest, flow from the end zones may be negligible in comparison to the principal two-dimensional flow through the cylinder. Study of the experimental work referred to suggests that cylindrical contours may be applied to axial distances up to about 1/4 to 1/3 of the slot length (and not much less than 2 or 3 slot widths).

[The ACGIH *Vent Manual* suggests using the following equations for flanged and unflanged slots:

Q = 3.7LVX   unflanged, free standing slot

Q = 2.6LVX   flange

These compare to the contour area approach:

Q = 2πLVX   unflanged, free standing slot (full-cylinder), L>>X

Q = πLVX   flanged, flange width > X, (half-cylinder), L>>X

The contour area approach has the potential advantage of being able to more accurately predict velocities at the X-distance when careful evaluations of the contour areas are possible.]

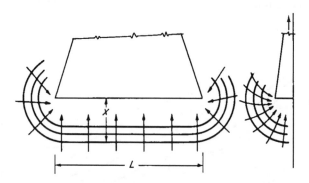

Figure 5-9. Velocity contours around an exhaust slot resting on a flanking plane, illustrating airflow through cylindrical contours in central portion and airflow at ends. If L is great relative to X, end flow may be negligible.

Synthesis of Velocity Contours

With an understanding of the character of airflow in the space around exterior hoods, as brought out in the earlier discussion of this chapter, it becomes possible to estimate airflow requirements to induce velocities in space for baffle arrangements quite different from those already outlined.

For this purpose DallaValle suggested a procedure in which the velocity contours of his different hoods are synthesized by vectors to obtain a new pattern corresponding to the hood shape of interest. While useful as a concept, this method is too laborious and lacks flexibility for practical use and will not be treated further here.

Plane of Symmetry

A very useful device described by the same author involves manipulation of *planes of symmetry*, as in the derivation of the velocity contours for an exterior exhaust hood flanked by a plane parallel to its axis—e.g., a hood resting on a horizontal plane. Insertion of an imaginary plane along the axis of an unobstructed exterior hood bisects the airflow pattern but does not modify it.

If the lower half of the airflow is considered only as the image of the "real" flow above the plane, one notes, referring to Figure 5-10, that the two taken together represent a hood with face area of double the "real" size, A_f, and of double the "real" flow. The contour area of the double opening would be $(10X^2 + 2A_f)$. Therefore, the contour area of the "real" hood is given by DallaValle's "half-hood-formula" $(10X^2 + 2A_f)/2$

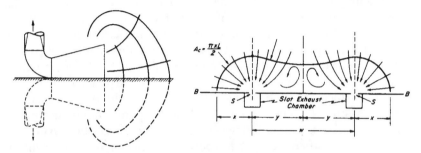

Figure 5-10 (left). DallaValle's half-hood diagram.

Figure 5-11 (right). Double parallel slots, S, in bottom plane B-B, through which air is exhausted. Construction of center plane of symmetry does not alter airflow pattern. Slot bisector planes, shown dotted, are not planes of symmetry, but at point P, inner and outer contours of equal velocity coincide; therefore, both contours shown, at $X = W/\pi$, are equal. Total contour area for this value of X is 2WL.

Double Parallel Slots

Two long slots may be imagined, arranged as in Figure 5-11. The character of the velocity contours is more readily visualized by use of planes of symmetry. Apparently one such plane can be inserted vertically midway between, and parallel to, the slots within certain dimension limits discussed further below. Another could be erected to bisect each slot longitudinally, as shown. Evidently then, half the flow of each slot can be considered outside its bisecting plane, and to create quarter-cylindrical contours of area, $\pi XL/2$. The flow pattern in the rectangular space between symmetry planes can be inferred to be somewhat as drawn—this on the basis that air cannot flow

directly toward the slot *away from a barrier* (this excepts the indirect air motion of vortices in confined corners).

Another condition can now be made apparent by reference to the fact (1) that the flow, Q, on each side of the slot bisector is equal, (2) the position of the inner and outer contours corresponding to any velocity, V, must coincide at the bisecting plane, and (3) it follows from (1) and (2) that the area of the contour, A_c, within the rectangular space, must be equal to that of the outer contour, $\pi XL/2$. These considerations require that the distance between the barrier planes must be large enough to provide equal contour area inside to that outside. Clearly, it cannot be much less than $\pi XL/2$ without requiring an impossible shape of the inner contour, i.e., excessive lowering of the end of the contour near the center plane. If, for simplicity, one considers the inner contour to be sensibly flat, i.e., a horizontal plane, then its area is that of a plane, yL, and it must be equal to $\pi XL/2$. Therefore, $y \approx \pi X/2$ or $X \approx 2y/\pi$.

It will now be clear that the dotted lines representing the slot bisector planes are not planes of symmetry. Comparison of contour areas, $\pi XL/2$ and yL, has shown that they are equal for only one value of X. At greater values, the inner velocities would be greater than those at equal distance to the other side of the bisector plane, and at shorter distances, the reverse would be true. However, a curved surface could be constructed that would be a surface of symmetry, and it would intersect the bisector plane shown at point P, where $X \approx 2y/\pi$. It is not necessary to construct the surface of symmetry; it suffices for our purposes to consider only the contour surface shown.

The total contour area can be calculated as follows: the area of the quarter cylinder is $\pi XL/2$ and equals $wL/2$. The inner contour has the same area, by the definition above and, therefore, the total contour area above plane *B-B* is $2wL$.

The arrangement of planes and slot in the above example suggests the combination of slots along the long edges of an industrial tank, and might be regarded as an idealized form thereof.

Example 7. A rectangular table is 12 ft long and 6 ft wide. If it is required to induce air velocities of 50 to 75 fpm above the surface of the table to prevent escape of room-temperature vapors, estimate the required rate of exhaust using multiple table slots.

Solution. The analysis in preceding paragraphs permits calculation of contour areas for an arrangement of slots facing upward (see Figure 5-11, left) which may be less favorable to the induction of air velocities over the surface than the one to be employed on the table where slots will face each other. Use of

the idealized analysis will, therefore, provide a conservative result and, lacking other information, constitutes a useful and practical basis of computation. Air velocities of 50 to 75 fpm must be induced throughout the contour area, $A_c = 2wL$,

$$Q = VA = 50 \text{ to } 75 \, (2wL) = 100 \text{ to } 150wL$$

or 100 to 150 cfm per square foot of table surface area. This exhaust rate will tend to produce the velocity contours shown in Figure 5-11 which meets the conditions of the example.

The calculation of the preceding example is interesting in that experience has shown that good control of mists from large-surface emission sources (e.g., tables and tanks) can be obtained with exhaust rates of 100 to 150 cfm per square foot, which, combined with the fact that control velocities of 50 to 75 fpm are realistic values for processes of these types, demonstrates that an analysis in this manner leads to practical results.

Sketching Streamlines

Estimation of the contour area, A_c, can be roughly made for any combination of exhaust opening and baffle. In the previous discussion of double parallel slots, the character of airflow in the rectangular central section was inferred and streamlines were drawn. A "sense" can be readily developed with a little practice, permitting reasonably good construction of streamlines of airflow toward suction openings by free-hand sketching. The contours are then drawn to cut the streamlines at right angles. The shape and dimensions of the contour intersecting the point of interest then permit an approximate estimate of contour area. This is illustrated in the following example.

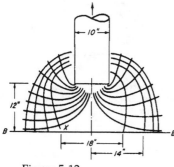

Figure 5-12.

Example 8. A 10-inch duct hangs vertically with its open end a distance of 12 inches from the bench. A circular source of contamination 18 inches in diameter is to be controlled by exhaust through the duct. A velocity of 75 fpm at the outmost boundary will provide control. Estimate the required exhaust rate.

Solution. The arrangement is illustrated in Figure 5-12. Streamlines have been drawn "by

feel," with liberal use of the eraser for correction of lines deemed false; then contours intersect adjacent streamlines at right angles. The contour intersecting the point X which is the boundary of the 18-inch zone of contamination roughly appears to assume the form of a hemisphere with its center on the plane *B-B*, with an area at the apex blocked off by the duct end. The area is obviously greater than a hemisphere of radius 9 inches and less than one of radius 14 inches (each corrected for duct end area). These two contour areas are each calculated from

$$A_c = 4\pi y^2/2 - \pi D^2/4 \ = \text{a range of 3-8 sq ft}$$

A comparison of these areas on the diagram with the actual contour area suggests that A_c = no greater than about 6 to 7 sq ft, so Q = 75 (6 to 7) = 450 to 525 cfm.

Example 9. A circular tank is to be exhausted by means of a continuous exhausted slot at its periphery as indicated in Figure 5-13. Draw approximate airflow streamlines.

Solution. It will be recalled that the significant contour is one which intersects that point of the contamination zone most remote from the exhaust slots; in this instance the intersection must be imaginary.

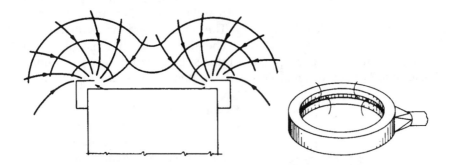

Figure 5-13. Idealized contour lines illustrating general character of airflow induced by a circular exhaust slot disposed at the periphery of a cylindrical tank or pot. Right figure shows a freestanding slot.

Other hood forms are summarized on Figure 5-14.

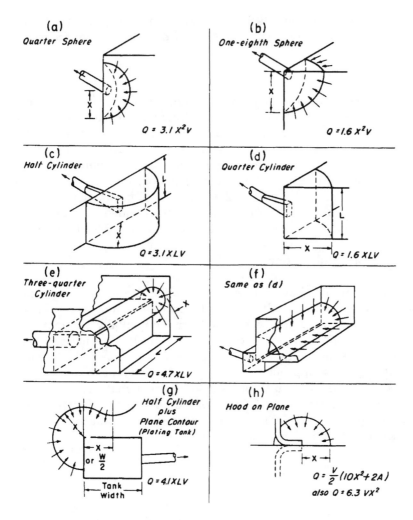

Figure 5-14. Typical exhaust arrangement and formulas for Q.

Comments on Practical Application

The design of exterior hoods is a process in which an airflow pattern (contour) of suitable size, shape, and strength (velocity) is superimposed on the air motion pattern characteristic of the pulvating action to envelop and dominate the latter. This will be further apparent during study of the next chapter.

There are practically never any physical arrangements in practice that fit the ideal shapes and patterns that have been employed in the discussions of contour characteristics of suction

openings, but with just a little imagination the designer will see the appropriate applications and the effect of deviations from the ideal.

Obstructions

More often than not, small obstructions will be present in front of hoods where the various contours are drawn but in most cases their effect is of little practical importance. Air will stream around them without any "splashing" and velocities beyond them will be seen to prevail intact. That is, air beyond the obstruction will part momentarily as it approaches the obstruction, then close in again after passage; so that while a "shadow" is cast by the object on the outside, the "shadow" itself is completely surrounded by positive air currents, hence no contaminant can escape the local exhaust action.

Synthesis and Adaptation

The contour characteristics that have been discussed are adaptable to almost all physical arrangements that will be encountered, particularly when one learns to mentally eliminate the minor obstructions referred to above. It will often be necessary to simplify one's mental description of particular arrangements to make them fit the most appropriate hood type. Moreover, as experience is gained, a sense of proportion is also attained that prevents hair-splitting in the final specification of exhaust rate as between, for example, 600 cfm and 700 cfm; and this sense will aid in the rapid application of simplifying assumptions.

Chapter 6

Capture and Control Velocities and the X-Distance for Exterior Capture Hoods/Booths

Having established a basis in Chapter 5 for estimating the *contour area* for a particular radius (X-distance) *in front of an exterior hood*, it remains for us to fix the X-distance and select a controlling/capturing velocity before we can complete the hood design, i.e., calculate the required exhaust rate, Q.

In the case of a booth or enclosure, it is necessary to select a value only for the controlling velocity for *the area of the booth face* which will exert an adequate force against the escape tendencies of the particular pulvation action, or, if not violent, against the action of random air currents that may impinge on the booth.

Null Point

To facilitate a description of the design procedure used in a cold, multidirectional pulvation action, I have defined a term, *null point*. As I define it, this is the location where the emissions comprising the pulvation "explosion" have expended their initial energy and contaminant velocities have decreased to the magnitude of the random velocities of the surrounding air (Figure 6-1).

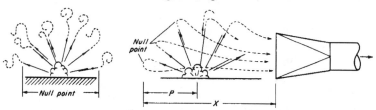

Figure 6-1. Loss and recapture concept for the design of exterior hoods. Null point of a multidirectional pulvation action, and its relation to X-distance. (Left) Pulvation action with no exhaust hood. Solid lines represent high velocity spurts of dust air; dotted curved lines represent the continued motion of the air, but after the initial energy has been spent. (Right) Same action but in the presence of an exterior capture hood which induces velocities at farthest null point adequate to overcome random air currents.

The distance from the origin of pulvation to the null point can be thought of as the "maximum pulvation distance." For a lateral exhaust hood, the X-distance for calculation of contour area is often the distance from the face of the hood to the null point.

Capture

See Figure 6-1 again. With the exhaust flow in operation and in favorable conditions of illumination for visibility, one would observe streamers of dusty air spurt explosively outward away from the hood. Then, with this energy expended, dust emissions would return toward the hood opening under the influence of the velocities induced by the exhaust system. If one could follow the action in slow motion, the initial outward pulsations would appear at first to indicate a loss of the dust to the hood influence, but this turns out to be temporary as it is recaptured, after spending itself, by the exhaust velocities that are still effective at the null point. This "capture" concept provides a useful and practical basis for design in a wide variety of situations.

It is noted that the X-distance from which the exhaust rate is calculated is the most remote from the source of suction, yet adequate velocities are induced at that location. At the areas closer to the hood, velocities are even greater and the contaminant, therefore, diverted with even greater dispatch. Lateral areas are brought under the influence of suitable exhaust velocities by making the hood of suitable breadth.

Velocity

In specifying the design velocity near the location of the null point, one assumes no retarding influence on the escaping air streams until its explosive stage has been spent. This assumption is, of course, not true, since the pulvation violence must be reduced when the hood-induced velocities are superimposed on the pulvation area. However, no useful result is served by introducing this complicating factor into the design.

While the selection of a design velocity might be related to the degree of draftiness of the work space, experience indicates that most completely enclosed work rooms will not require more than 50 to 75 fpm at the null point. If draftiness (secondary air motion) suggests the need for higher values, then baffles are probably required to intercept and prevent them from interfering with the exhaust action.

Some consideration, qualitatively at least, should be given to the rate of generation of contaminant by the process, and also to its toxicity rating in relation to the selection of a control velocity. If process A generates 1 milligram per second of airborne lead dust and

process B, 1,000 milligrams per second, and if one percent of the dust in each case were to escape removal by the hood, the absolute rate of escape would be 0.01 vs. 10 milligrams per second, respectively. The consequence would be altogether different in each case and one would therefore desire an additional factor of safety for process B in the form of a somewhat larger control velocity. If 50 fpm seemed satisfactory for process A then one might well decide on 75 fpm or a little more for process B. Quantitative data on generation rates like those in the example above would hardly ever be available, therefore judgment in this connection must necessarily rest on experience. The principle that has been referred to in this discussion receives more complete and definite treatment in the chapter on solvents.

Table 6-1
Loss and Recapture Design Method
and Required Control Velocity at Null Point

Draft Characteristics of the Space	Lower Safety Factor[a]	Higher Safety Factor[b]
	Controlling velocities required at null point, fpm	
Nearly draftless space, process easily baffled	40-50	50-60
Low drafts in space	50-60	60-70
Moderate drafts, w/baffling	70-80	75-100

(a) Lower Safety Factor: Moderate amounts of nuisance dust or fume; or, only small amounts of toxic dusts or fume.

(b) Higher Safety Factor: Toxic dusts or fume or large amounts of nuisance dusts or fume.

Note: These values are illustrative, minimum guideposts. It is impossible to estimate velocity requirements as closely as implied by the above figures.

Procedure (as applied to exterior hoods)
(1) Establish most suitable location for hood relative to source of contamination; as close as possible.
(2) Observe pulvation distance, add baffles as possible, and thus establish distance, X, from hood to null point (equal to distance from hood to source, plus pulvation distance).
(3) Select category of space to cross drafts, and in relation to baffles.
(4) Select category as to safety factor, and, together with selection (3), pick appropriate controlling velocity from table.
(5) Compute exhaust capacity by appropriate Q-V-X relationship. (Chapter 5.)

Hood Dimensions and Position

It has already been indicated that the face of the hood should be located as close as possible to the source of contamination. If the action takes place on a bench or close to the floor, consideration should be given to exhaust through a grill in the bench top or in the floor because the X-distance in these circumstances is the least possible value and, consequently, effective hood action is attained with a smaller exhaust rate.

The length of hood (regardless of downdraft or lateral exhaust) should roughly correspond to the length of the pulvation zone.

Its height or width may correspond to the other dimensions of the pulvation zone or, if that aspect is of no consequence, it may be made a dimension merely for suitable size (from mechanical considerations), or for suitable face velocity. The various steps in the design method discussed are summarized in Table 6-1.

Experimental Controlling Velocities

The capture basis of design is essentially different from methods outlined in other texts dealing with characteristics of local exhaust hoods in that control velocities cited elsewhere are based on experimentally determined values where the X-distance is measured from the actual source of dust generation to the hood face. In other words, the control velocity has been defined heretofore as that required at the source, regardless of pulvation distance. In application to practical problems of differing nature, that method would require the experimental determination of control velocity for each new operation, or inference of a value by comparison with actions of similar scale and violence. For example, it could be based on a scale such as the following: gentle action—50 fpm; moderate violence—100 fpm; quite violent disturbance—200 fpm, and so on.

A real example, translating the data from an experimental determination of velocity for a certain process into the visual observation basis, is given below and will emphasize the points that have been made.

Example 1. Assume the action of Figure 6-2 (top) has been studied and its controlling velocity is to be 450 fpm at the point of dust origin. A hood type is chosen wherein the required exhaust rate is described by $Q = \pi V X^2$. (That is, the velocity contour tends to assume the shape of a quarter sphere.) What will the control situations look like for a hood placed 6″ and 24′ from the source?

Solution. See Figure 6-2, next page.

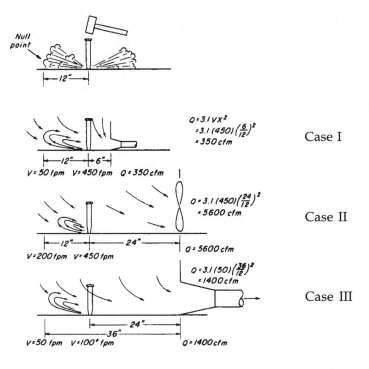

Figure 6-2. Design examples.

Case I. X-distance = 1/2 ft
Then required $Q = 3.1(450)(0.5)^2 = 350$ cfm

Case II. X-distance = 2 ft
Then $Q = 3.1(450)(2)^2 = 5,600$ cfm

Suppose the pulvation distance of this process has been observed to be 12 inches from the origin point. If one calculates the velocity at the null point for each case, a marked inconsistency will be noted. In Case I, a velocity of 50 fpm is calculated at the null point, and in Case II a velocity of 200 fpm. The control velocity specification on which the original calculations were made implies that the two cases are exactly equivalent, whereas it is obvious that the dust control barrier of Case II, represented by the 200 fpm velocity, is very much more powerful than that of Case I.

If the exhaust capacity of the original experimental hood is known to be 300 to 400 cfm, we can infer that a more universal specification for the process can be given in a statement that the process requires a velocity of 50 fpm at a distance 12 inches beyond the point of dust origin. Stated even more simply: the null point of

the process is located at a distance of 12 inches beyond the point of dust origin. This is a universal method for describing the characteristics of any cold multidirectional pulvation process. On this basis, one calculates for Case III, that an exhaust rate of only 1,400 cfm is required instead of 5,600 cfm. This is based on an X-distance of 2 ft plus 1 ft (or 3 ft total) and a capture velocity of 50 fpm.

Table 6-2
Field Observation Questionnaire for Pulvation Action
Dust or Mist Cold Processes
Pertinent Data for Exterior Hood Design

Name or description of operation	
Primary pulvation action	
Site of pulvation action	
Describe inertials	
Secondary air dispersion mechanism	
Likely sites for baffles	
Dimensions of pulvation area	
Pulvation distance	
Fix hood location and record X-distance	
Length, height, and depth of hood	
Notes on fine dust/composition/toxicity	

Field Observations

Careful, detailed field observations of the character of the pulvation action is a key to successful design of local exhaust hoods. The concepts and terminology that have been described are summarized in Table 6-2 as a check list of questions for use in the field. Its use will ensure that no pertinent characteristics of the action are overlooked.

The procedure covered by this questionnaire is applicable only to cold processes. If significant quantities of heat are generated, see Chapter 8. It is also intended mainly for design of exterior hoods; however, if an open face booth, or even enclosure, is found applicable, the field data given by the check list will still be found useful.

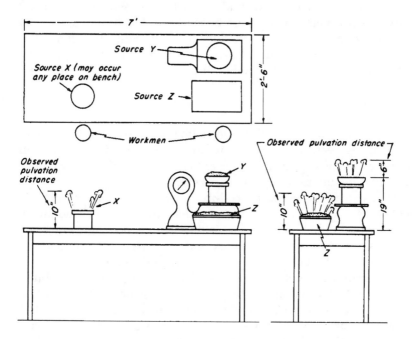

Figure 6-3. Powder handling at a work bench.

Example 2. An operation taking place at a work bench as shown in the sketch (Figure 6-3) involves handling of powdered resin by scoop preparatory to certain molding operations. What types of control approaches come to mind as you examine the drawing?

Solution. There are three primary sources of dust as follows:

Source X--Pulvation distance observed to be not over 10 inches above bench top. (This equipment may be shifted by worker to any position on bench from hour to hour.)

Source Y--Pulvation distance observed not over 6 inches above plate. Results from deposit of powder by scoop onto plate for weighing.

Source Z--Pulvation distance observed not over 10 inches above bench top. Results from dropping excess material from scoop back into tote pan and dropping of scoop on top of powder in pan.

Solution. Dust control design could be based on a hood placed at the rear of bench for lateral exhaust, or based on downdraft exhaust through a perforated bench top. Modifications could be made in the bench construction or equipment arrangement to facilitate more economical exhaust for design. The approaches are numerous; the task is to find the most cost-effective solution.

Salvage Zone

The concept on which all the preceding discussion has been based is that the air velocity induced by an exhaust hood is a complete description of the capturing power at that point. It is necessary, now, to describe the limitations of that concept.

As a practical design procedure, a single velocity contour is selected as the boundary within which all contaminated air is withdrawn into the hood. The fact is that there is no single boundary at all; useful velocity contours extend, continuously, beyond the selected design value, as well as inwardly and, therefore, a *zone* of induced air motion exists beyond the selected contour and contributes importantly to the withdrawal into the exhaust hood. The width of the effective zone, as well as the magnitude of velocities is important to the description of its power to control the contaminant. This is illustrated in the following example.

Example 3. One exterior hood is placed in relation to a contamination zone such that the X-distance is 3.75 inches; in another case, a similar hood is placed so that X-distance is 12 inches. A velocity of 75 fpm is induced at the null point and the velocity contour equation $Q = V(10X^2 + A)$ applies in each case. Show that the hood with X-distance of 12 inches provides a greater safety factor.

Solution. The hood face area, A, can be considered negligible. The exhaust rate for V = 75, and X = 3.75 inches and = 12 inches, respectively, are 75 cfm and 750 cfm, using $Q = V(10X^2)$.

If distance X is increased 1 inch in each case, the X-distances become 4.75 inches and 13 inches, respectively. For the fixed rates of exhaust, above, the velocity for the small hood will have decreased to 47 fpm from 75 fpm. For the large hood, the velocity will have decreased only to 64 fpm. It is clear, then, that if the

violence of the contaminating process were to shift the null point outward one additional inch, the control forces would be reduced to a much greater extent for the 75 cfm hood, as compared with the 750 cfm hood.

Contour velocities for other distance increments are shown in Figure 6-4.

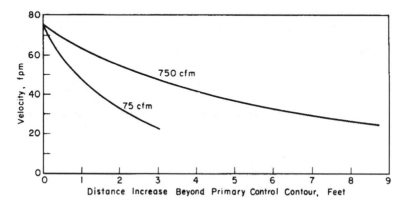

Figure 6-4. Comparison of velocity decrease for two plain exterior hoods.

Capture velocities as low as 25 fpm are unreliable as a primary control, but they do have some "salvage" value. It is a convenient boundary velocity to employ in the present connection to describe the width of the *salvage zone*, which is defined as the zone extending from the primary control contour to the 25 fpm contour. In the last example the primary control contour was 75 fpm. The width of the salvage zone for the lower exhaust capacity is, then, about 2.75", and that for the larger capacity, 8.5 additional inches.

A simple expression for width of salvage zone can be derived by assigning V = 25 fpm in the appropriate contour-velocity formula, and solving for X; from the value, the null point X-distance is subtracted. Thus, in Q = V(10X^2 + A), let V = 25, let A = 0 since it is usually negligibly small, and obtain

Salvage zone width = 0.064(Q)$^{0.5}$ - X   (width, feet)

This and corresponding expressions for the common types of hood are summarized in Table 6-3.

Table 6-3

Expressions of Width of Salvage Zone for Common Hood Shapes

(For the 25 fpm contour line; "A" deleted from salvage equation)

Hood Type	Contour Formula	Salvage Zone width, ft
Unobstructed or flanged opening	$Q = V(10X^2 + A)$	$0.064\sqrt{Q} - X$
Flanking plane parallel to hood axis	$Q = V(5X^2 + A)$	$0.088\sqrt{Q} - X$
1/4 sphere	$Q = V(\pi X^2)$	$0.113\sqrt{Q} - X$
1/8 sphere	$Q = V(\pi X^2)/2$	$0.16\sqrt{Q} - X$
1/2 cylinder	$Q = V(\pi XL)$	$0.013Q / L - X$
1/4 cylinder	$Q = V(\pi XL)/2$	$0.025Q / L - X$

Table 6-4A

Rates of Exhaust Required for Cold Multidirectional Processes with Common Types of Exterior Capture Hoods

X-distance, inches	Simple opening, with or without flange or taper but no plane flanking opening[1]		Simple rectangular opening flanked by plane parallel to axis[2]	
	Exhaust rates, cfm			
	Minor[3]	Major	Minor	Major
2	25	75	25	50
4	50	100	50	75
6	100	200	75	150
8	200	400	150	300
10	300	600	200	400
12	400	800	300	600
15	600	1200	400	800
18	900	1500	600	900

[1] Refers to simple hood corresponding to $Q = V(10X^2+A)$

[2] Refers to hood having a complete flanking plane parallel to the axis which prevents air from flowing from half of the normal air supply zone, corresponding to the contour formula, $Q = V(5X^2+A)$, e.g., rectangular hood resting on bench top.

[3] Minor refers to lower safety factor; major to higher safety factor as explained on Table 6-1.

Table 6-4B
Rates of Exhaust with Single Slots Flanked by Parallel Plane on Which Pulvation Occurs

For X-distance (i.e., width of plane) always less than 1/2 length of slot[1]

Condition	Exhaust
X-distance < 2 ft	125-150 cfm per sq ft plane area
X-distance > 2 ft	75-125 cfm per sq ft plane area

[1] If the X-distance is greater than 1/2 slot length, consider it as a simple opening flanked by a plane (Table 6-4A). With exhaust rate divided between two slots, an additional safety factor is automatically provided.

Practical Design Considerations

In Tables 6-4A and B, all the considerations discussed in the preceding sections have been combined in the suggested exhaust capacities for the common forms of *exterior hoods* for various X-distances.

The bases and limitations of these design figures are as follows:

(1) They refer to an X-distance measured along the hood axis *to the null point.* This is emphasized because it differs from prior published concepts in which X is measured only to the most remote point of *origin.* For non-violent pulvation there is no difference.

(2) Account is taken of the salvage zone by specifying higher control velocities for the smaller X-distances, i.e., lower exhaust rates, and, conversely, lower velocities for the greater X-distances. Thus at an X-distance = 2 inches, a minimum control velocity of 100 fpm was used; at a distance of 48 inches velocities of 30 to 45 fpm are judged adequate. The salvage zone width, at 7,000 cfm (X=4 ft), is about 1 1/2 ft and thus provides a comfortable safety margin.

(3) The values have been deliberately rounded off. It is intended to discourage the use of more than one or two significant figures. Hence, a calculated result of 289 cfm becomes 300 cfm; one of 2,258 cfm becomes 2,300 cfm. The relationships from which the figures are derived do not warrant any closer specification.

(4) The figures apply only to cold pulvation processes, usually multidirectional. They are not applicable to processes where heat imparts buoyancy to the air bearing the contaminant.

(5) "Major" and "Minor" categories which are applied to each limiting figure refer to the same circumstances as in Table 6-1. Minor problems require a lower safety factor and major ones a higher safety factor. The vast majority of industrial problems falling within the cold, multidirectional process category will be classed as minor.

Major problems are those in which considerable quantities of contaminant are generated and also where the permissible atmospheric concentration is low.

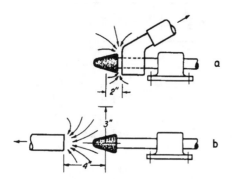

Figure 6-5. Two exhaust hood arrangements for small buffing wheels.

In the next three examples, try to visualize the airflow contours for various types of hoods applied to each process.

Example 4. Two exhaust hood arrangements for small buffing wheels are depicted in Figure 6-5, a and b. How could one go about estimating the appropriate exhaust volume flowrate?

Solution. The exhaust volume flowrate could be estimated by assuming spherical shapes for the area in the formula Q = VA (using Table 6-3 or Table 6-4a).

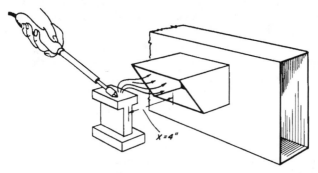

Figure 6-6. Exterior hood for control of flux smoke in soldering requires very small rates of exhaust due to small X-distance and small scale of pulvation (assuming non-drafty locations).

Example 5. Estimate exhaust requirements for control of fumes from a manual soldering operation with the hood arrangement illustrated in Figure 6-6. How would one go about estimating the appropriate exhaust volume flowrate?

Solution. The exhaust volume flowrate could be estimated by assuming spherical shapes for the area in the formula Q = VA (using Table 6-3 or Table 6-4a).

Example 6. Estimate the exhaust rate to be provided for control of nuisance dust originating at the top of barrels with the arrangements shown in Figure 6-7. How would one go about estimating the appropriate exhaust volume flowrate?

Solution. The exhaust volume flowrate could be estimated by assuming a cylindrical shape for the area in the formula Q = VA, use the equations provided in Table 6-3 or Table 6-4a.

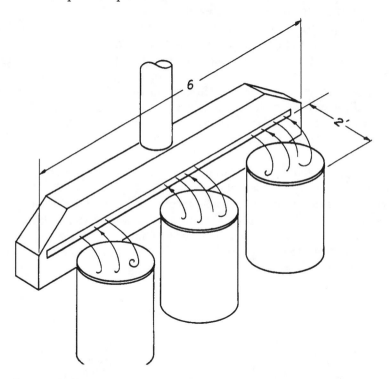

Figure 6-7. Slot arrangement for exhaust of several cylindrical tanks requires excessive exhaust rates compared with exhaust by individual hoods.

Exterior Hood Applications

In addition to the principles describing airflow characteristics of processes and of ventilating systems, the following need attention in ventilation design problems: (1) definition, in exact terms, of the nature of the contaminant (whether dust, fume, gas, sensible heat,

radiant heat, humidity, radiant energy, etc.); (2) facts concerning the
degree of toxicity and amounts being disseminated into the air; (3)
exact location of the individual sources (air analysis data permit a
better evaluation of the problem); (4) possibility of partial or
complete elimination of an offending substance by substituting one
that is not objectionable; (5) changes in equipment which may
eliminate the pulvation source; for instance, the substitution of an
enclosed screw conveyor for an open belt conveyor handling powdered
material; and (6) the use of water, perhaps with wetting agents, to
reduce pulvation to a scale more easily managed by an exhaust
system.

In the following pages, some practical illustrations of the
principles set forth earlier in this chapter are discussed.

The quantities of dust incident to *bench scale processes* like
powder handling are usually small and, therefore, when local
exhaust is required for dust control it is because the dust is highly
toxic, irritating to the skin, eyes, or respiratory tract, or is otherwise
disagreeable; for example, handling of powdered compounds of lead,
like litharge, organic compounds in powder form, such as
unpolymerized resins which may cause dermatitis, irritating or toxic
insecticides, or dye powders.

Powder Handling. In a simple operation like transfer, by scoop or
shovel, of a dusty powder from a box to a pan on scales, there are two
principal sources of dust: one resulting from spillage of excess
material when the scoop or shovel is filled, and one resulting from
discharge into the receptacle. Local exhaust at these two points
might control the dust provided there is no spillage during carriage
between the two points.

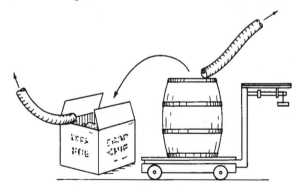

Figure 6-8. Transfer of powder from barrel to box. Adjustable exhaust
hood in the form of flexible hose may withdraw only a portion of the dust
at each point of origin.

In small-scale transfer of material by scoop there is usually insignificant splashing and, therefore, the null point can ordinarily be taken to coincide with the farthest boundary of the container.

If several individual sources of dust exist within a limited area, a single large hood might be considered for the sake of mechanical simplicity. Downdraft exhaust through a perforated bench top may be advantageous in reducing the X-distance.

Weighing, Mixing, Sifting. A typical temporary arrangement for removal of dust or fume originating inside barrels, boxes, tote boxes, weigh pans and other receptacles that are not permanently fixed at one location, is illustrated in Figure 6-8. Generally, it is advantageous to use metal hose which can support some of its own weight and maintain a fixed position without auxiliary support. However, it should be noted that such methods are invariably associated with general untidiness due to materials spillage, which cannot be corrected by an exhaust system. Furthermore, the capture zone of such systems is small and will create fugitive emissions in the absence of careful workpractices.

Such applications should not be used as permanent installations.

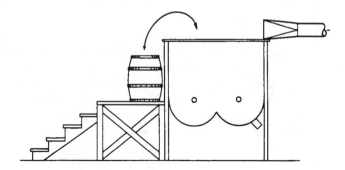

Figure 6-9. Charging powder to a mixer. Slot type exhaust along one edge will not likely provide enough capacity to control dust from the barrel; a separate exhaust would be preferable.

Rigidly positioned hoods can be arranged for *mechanical mixers* by lateral exhaust through a slot at the long edge (Figure 6-9). Exhaust requirements for control during charging, with open top, may be selected on a basis of 75 to 100 cfm per square foot of top area when side baffles are employed. A tightly fitting cover would obviate the need for any exhaust during actual mixing.

Bench Grilles (Downdraft). Moderate exhaust capacities usually suffice for dust control in manual work on small castings of iron, brass,

or bronze. Indeed, air sampling studies may suggest little need for any special ventilation at some operations.

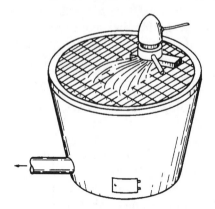

Figure 6-10. Light grinding over a grille-covered table with downdraft exhaust ventilation.

Manual brushing of such objects may produce excessive dust. Power brushing certainly does, and downdraft exhaust benches are commonly employed to control the dust in this work (Figure 6-10). Exhaust rates are usually specified to provide an air velocity through the gross area of perforated bench top of 200 fpm. Experience has indicated this to be satisfactory.

A more general specification is one based on the velocity-distance equation for hoods. It is clear from a little reflection that a downdraft hood of the type illustrated is one for which the equation $Q = V(10X^2 + A)$ applies; also that the velocities of significance in controlling dust are those above the grille surface at the highest level where dust may originate. In most applications this level does not exceed one foot, i.e., the X-distance. For a typical grille of 8 sq ft, the value of Q, at 200 fpm face velocity, would be 1,600 cfm. Use of this value for Q, X = 1, and A = 8, reduces the calculation to the following more rational specification: *exhaust through a downdraft hood at a rate that will create a velocity of 75 to 100 fpm at the null point.* In non-violent operations and relatively small rates of dust production, 75-100 fpm would be acceptable. A vertical baffle to intercept the stream of inertials from the rotating brush is desirable where the downward air velocities induced by the hood are inadequate to divert them into the hood. In heavy duty grinding or scratch brushing such baffles may be imperative.

Barrel Packing. Packing powdered material by gravity flow from overhead storage bins is an extremely dusty operation. If the barrel filling is incidental to the collection of processed material and proceeds *at a low rate*, the dust can be controlled by providing a cover with flexible chute connecting it to the supply chute or bin, with a small exhaust connection to the cover (Figure 6-11). Very low exhaust

rates (50 cfm) are sufficient in this arrangement. However, if it is planned to connect this branch to a larger exhaust system, exhaust rates of 200 to 300 cfm or greater will be necessary to prevent plugging of this line, for the reasons outlined in Chapter 12. In that case there is no basis for economy in exhaust capacity and one of the open arrangements described below might as well be employed.

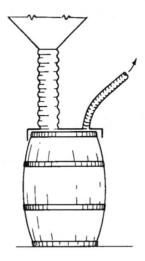

Figure 6-11. Flexible filling chute with small exhaust connection to cover.

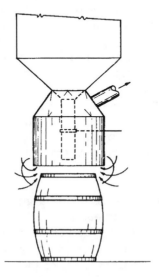

Figure 6-12. Barrel packing by gravity flow and dust control by hood around filling spout. Spillage will cause dust source at floor level, not controlled by this arrangement.

Where packing takes place at
production rates, an overhead
exhaust hood (Figure 6-12) may be
used, with a capacity of 300 to 400
cfm.

In some limited circumstances
the simple arrangement illustrated
in Figure 6-13 may be suitable.

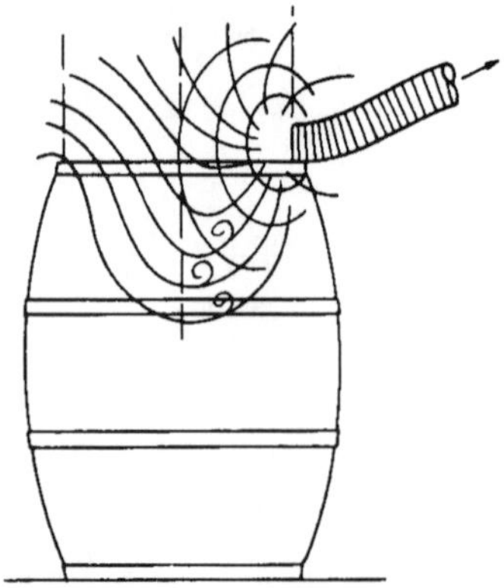

Figure 6-13 Exhaust at the edge of a
receptacle within which dust originates.
To induce velocity, V, downward through
left section, exhaust at a rate $Q = 1.5Va_f$.

An analysis of exhaust requirements can be made, applying some
of the basic principles described in Chapter 5, as follows.

If the end of the hose were to project downward inside the barrel,
air would enter uniformly through the open top, as though the
opening were the open face of a booth, and exhaust requirements
would be estimated on the basis of inducing a suitable face velocity.
Where the end of the hose is disposed at the rim, we may infer from
the airflow principles described in Chapter 5 that the airflow
pattern is somewhat like that illustrated in Figure 6-13. Half the air
flowing toward the hood opening comes from the regions above the
barrel rim, and the other half from the barrel interior, below the
hood. The latter must enter the interior at the more remote portion of
the opening.

Based on the approaches described in this chapter and a control
velocity of 50 fpm, an exhaust capacity would be required equal to 75
fpm through the total open area. On this basis, a barrel 10 inches in
diameter would require about 200 cfm; and on the basis of 75 fpm
control velocity, about 300 cfm.

Bagging Powders. The filling of large bags with pulverized
material may produce large amounts of dust.

Filling of open mouth bags by gravity from an overhead supply
hopper is illustrated in Figure 6-14 where an overhead hood is
provided with action similar to that of the barrel packing hood of
Figure 6-12. Pulvation produces streams of dusty air vertically
upward due to displacement of air from bag interior.   An exhaust rate

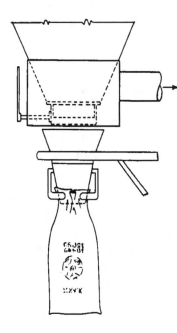

Figure 6-14. Exhaust hood
arrangement for gravity-fill bagging.

of 400 to 500 cfm suffices for control of most of the dust, but, if the dust
has significant toxic properties, the leakage at the bag mouth,
indicated in the figure, might be excessive and an exhaust capacity of
1,000 to 1,500 cfm would be necessary to "reach" the additional
distance from hood opening to this point.

Valve bag filling, while usually producing somewhat less dust,
than open mouth bag filling, still is responsible for excessive amounts.

Studies have shown that exhaust hoods for bagging equipment
with face velocities of 150 to 250 fpm through the front opening of a
filling booth provided effective dust control.

In all forms of filling equipment, spillage of material will create
dust at floor level. Also dust will be seen to originate at the hand
truck or pile onto which the filled bags are thrown. When the dust is
toxic or otherwise objectionable, these sources are important and
require separate attention. Lateral exhaust at both locations is then
desirable. Downdraft exhaust through a grille below the bagging
station is not a good arrangement because the exhaust system and dust
collector are then saddled unnecessarily with large quantities of
spilled material, as distinct from airborne dust.

Floor Scale Processes

Floor Grilles. Pits below floor level covered by exhaust grilles are
commonly provided for floor scale operations like spray painting of

machines or dusty operations on a scale too large to be handled on a bench. They are not essentially different from grilles in the top of work benches except in dimensions; the X-distance is naturally greater and they therefore involve much higher exhaust capacities. The formula $Q = V(10X^2 + A)$ describes their performance, and the quantity A is likely to be significant in relation to the $10X^2$ term in contrast to the smaller exterior hoods discussed in previous sections.

Where large volumes of dusty material are dislodged in an operation over such grilles, the exhaust duct connection below should be shielded to avoid unnecessary withdrawal of material into the duct system (see Chapter 12).

Moveable Hoods. In manipulative, dust or fume-generating processes, like stone-cutting or dressing, and electric welding, the sources of contaminants range over a wide area as the tool roves from one position to another. To conserve exhaust capacity, an arrangement is commonly employed in which each branch of a conventional, central exhaust system terminates in an exhaust hood whose position is adjustable; the worker shifts the position of the hood from time to time as the location of the tool changes, to always keep the hood close to the source of dust or fume.

The efficacy of moveable hoods resides primarily in the quality of supervision, especially during the early period after installation, since this determines the extent to which the worker acquires the necessary awareness of the hood's function and the habit of shifting its position. A well-balanced hood suspension mechanism which is easy to move will help greatly in this development, but superior supervision can ensure successful application of such devices with relatively inferior mechanical arrangements.

A supplementary device to provide greater insurance that the dust-making tool will always operate within the boundaries of the safety zone is an arrangement in which a chain or similar link connects the tool to the edge of the hood, the length of the chain corresponding to the X-distance of the hood and process. The worker is thus reminded, when his tool strays to the permissible distance limits, that the hood must be repositioned. An elastic support for the chain keeps it off the working surface and avoids interference with the work of the tool.

Surfaces of Open Tanks

Tanks contributing to air contamination may include water, water solutions, oil baths, acids, and liquid solvents, e.g., degreasing tanks employing chlorinated hydrocarbon solvents like trichlorethylene.

The discussion of contour velocities presented earlier showed the basic character of the airflow pattern above a tank exhausted by slots disposed along the long edges, and it was explained how exhaust rates in this case are logically expressed in terms of tank area, i.e., cfm per sq ft, or capture velocity at the far edge of the tank, the X-distance from the slot. Both approaches are widely used.

The ANSI Z9 standard [see References and contact AIHA, the ANSI Z9 Secretariat] covering these operations has seemed to this author to be unnecessarily complex and detailed [circa 1955], and it has therefore been condensed into the data shown in Tables 6-5 and 6-6. Principles pertinent to lateral exhaust for boiling aqueous solution, covered by Table 6-5, are treated in Chapter 8. The present summary does not reproduce the recommendations for tanks containing volatile solvents because it is held herein that the design treatment for solvent should be handled according to more exacting techniques, e.g., the dilution principles set forth in Chapter 10.

Table 6-5
Rates of Ventilation by Lateral Exhaust for Boiling Water

	Approximate tank dimensions, length : width			
	>10:1	4:1 to 10:1	2:1	Square
Condition	Recommended cfm per sq foot surface area			
Tank baffled (e.g., a wall)	75-90	100	130	150
Tank not baffled	110-130	140	170	190

Table 6-6
Rates of Ventilation by Lateral Exhaust for Solutions where Violent Gassing Occurs or Solution is near the Boiling Point

	Approximate tank dimensions, length : width			
	>10:1	4:1 to 10:1	2:1	Square
Condition	Recommended cfm per sq foot surface area			
Tank baffled (e.g., a wall)	100-125	140	175	200
Tank not baffled	150-175	190	225	250

[Editor's Note: Recent work by the ANSI Z9 committee should be consulted before using these rules-of-thumb. Contact AIHA for more information. See References and Bibliography.]

Face Velocities at Booths

Design of a partial exhausted enclosure such as the open-front booth provides the best possible solution to most problems. Not only is it more economical in required exhaust capacity due to reduced area of contour (open face area) but, also, the surrounding walls provide stability by preventing impingement of stray air currents on the contamination zone. This fact provides an even greater factor of safety for any given contour velocity (face velocity) over a simple exterior hood employing the same velocity.

Non-violent, cold processes are readily controlled by booth face velocities of about 50 fpm (assuming quiet conditions at the hood face).

Multidirectional, cold pulvation processes surrounded by booths may be analyzed by methods similar to those described for simple exterior hoods. Attention is directed to the magnitude of pulvation in the direction of escape through the booth opening; the null point is noted in relation to the location of the opening, and the exhaust action outside the booth analyzed by the exterior hood method.

Spray Painting Booths

One of the most common industrial applications of open-front booths is for control of the overspray in paint spraying. A violent process, face velocities are determined not by draftiness characteristics of the workroom but by the characteristics, as to violence, of the spraying operation. Face velocities required are dependent not only on the violence of the spray, but also on the magnitude of the total exhaust, for precisely the same reasons this factor is of importance in design of exterior hoods. That is, as the salvage zone is larger and more potent, the larger the exhaust rate, and vice versa.

Some guidance is provided in the ANSI Z9 standard for spray finishing operations. However, values given in the standard are subject to considerable uncertainty where the objects being sprayed are of a shape that results in violent, direct rebound of the overspray.

If, in such a spraying operation, the null point occurs beyond the position held by the worker, he or she may not be adequately protected even though all overspray is captured by the exterior hood action of the booth. In these circumstances a spray gun with an extension arm which positions the worker back beyond the null point should be given consideration.

[Editor's note: Recent work suggests that when painters are positioned perpendicular to the flow of air, they are sometimes better protected than when standing upwind. This is apparently due to the

eddying effect of the air around the painter's body when standing upwind. The use of this side-standing approach should be applied on a case-by-case basis after thorough study of the specific painting operation.]

Finally, it should be noted that most commercially available spray booths are not designed to protect the painter; rather, the booth is intended to control paint overspray. Therefore, painters may need respirators even in the best of booths.

Laboratory Hoods

In laboratory buildings having a large number of hoods, the most cost-effective rate of exhaust is a question of primary importance because of the costs of heating and cooling the make-up air. This matter has received additional attention in recent years with the increasing costs of air conditioning.

The cost of conditioning air is typically manyfold greater than heating the same air volume. There is a variation in ventilation needs between different laboratories at the same time, and the same laboratory at different times. A lab building may, for example, actually require only 10 to 15% of the installed exhaust capacity considering the hood use in the building as a whole at any given hour. The actual need in this hypothetical example is shifting between different laboratories from hour to hour and from day to day and since the hood fans are not shut off during periods of no-use, the entire capacity must be provided. These considerations raise the question as to how much capacity to provide for each hood.

The most common face velocity value employed for specifying hood ventilation rates is 80 to 120 fpm, in accordance with ANSI Z9.5-1992. This is a compromise between the need for containment and the problem of costs (and of eddying around the labworker's body).

It is quite likely that a hood face velocity of 60 fpm could provide adequate fume control if conditions at the face of the hood were perfect (e.g., no cross drafts, uniform distribution of exhaust flow, proper workpractices, and no equipment in the face of the hood). Unfortunately, these conditions are rarely maintained over time, so higher velocities are used to deal with these shortcomings.

Most lab hoods have exhaust slots at the bench level and at the top of the hood. The low exhaust slot has the advantage of providing uniform velocities through the front opening. With no heat release within the hood, and exhaust only from the upper part of the enclosure, velocities through the face opening are greater near the upper rim than near bench level and may be inadequate in situations where the contaminant is released at bench level near the front of the hood.

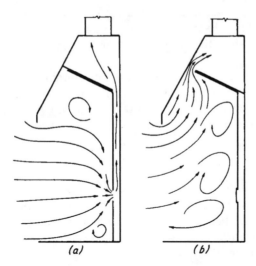

Figure 6-15. Airflow distribution at face of hood with exhaust through a low slot (a) and through the upper part of hood. No heat generation in hood. Relatively lean flow and poor distribution near bench with exhaust at top as in (b).

The majority of operations carried on in hoods produce only small amounts of contaminant, many of which are relatively non-toxic. In many cases the amounts volatilized are so small that mere dilution with the fresh air entering the hood reduces their concentration in the air within the hood to acceptable levels, so that if, due to low hood efficiency, some of this mildly contaminated air escaped from the hood interior instead of passing into the exhaust flue with the main air stream, it would not be noticed.

Thermal Effects

The different situations affecting hood performance in preventing escape of contaminants generated within are related principally to the rate of heat release by burners, hot plates, and the like. Upward moving thermal air currents are generated within the hood at flow rates that increase with increasing use of heat. If the hood fan capacity is less than the upward flow of the hot, contaminated air currents, then interior air will flow outward just under the top edge of the hood face, even while air enters through lower sections. The same thing will occur even with a sufficient rate of hood exhaust if the exhaust connection is below the upper rim of the hood opening. (See Figure 6-16.)

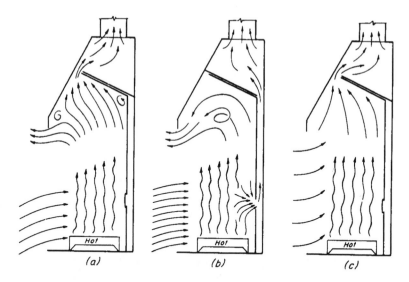

Figure 6-16. Spillage of fumes from hoods in conditions of high heat release on the interior: (a) insufficient exhaust capacity, although proper location of exhaust outlet; (b) sufficient exhaust capacity, but improper exhaust location; (c) same exhaust capacity as (b), properly located at top.

The generation of heat within a hood creates a volume of hot air in the space above the hood face opening which results in the thermostatic pressure effect discussed in Chapter 8. If there are any cracks or holes in this portion of the hood casing, leakage of contaminated air outward may occur in spite of the operation of the exhaust system. This may be significant where the contaminant is either highly malodorous or highly toxic.

Leakage from this cause may be prevented by ensuring that there are no cracks or other openings in the upper portions of the shell casing. Hoods provided by laboratory furniture makers are well constructed in this respect. Homemade hoods frequently leak by this mechanism. The spillage at the upper rim of the hood opening that may occur in conditions of high rate of heat generation is prevented by (a) exhausting from the upper space above the hood opening, and (b) providing a sufficient rate of hood exhaust in relation to the area of the hood opening. A hood with moderate exhaust, properly connected at the top of the hood, in which thermal air spillage occurs may be made to perform satisfactorily by partial closure of the face opening at the top. In hoods with a sash front, this is accomplished by merely lowering the sash.

Chapter 7

Airflow In Materials Handling Systems

[Editor's note: Hemeon was originally assisted by Julio C. Duran when he wrote this chapter.]

Brief discussions of the nature of air motion induced by inertials were provided in Chapters 2 and 3 [and additional materials are found in the Appendix]. The phenomenon is of practical importance in the design of dust exhaust systems where high speed motion of crushed and pulverized material induces airflow of a magnitude which can influence the rate at which air must be exhausted.

A dramatic illustration of this phenomenon is provided in a paper by R. T. Pring ("Dust Control in Ore Concentrating Operations," Technical Print 1225, *Am. Inst. Mining & Met. Engs.*, 1950) which described a portion of an ore crushing, screening and conveying system wherein the fall of crushed material into a surge bin induced tremendous quantities of air. A trial installation of a fan system demonstrated the necessity for exhausting air at a rate of about 35,000 cfm from the interior of the bin to prevent escape of dusty air; this constituted a measure of the rate at which air was being pumped into the bin by the action of falling material.

One sees other illustrations of the same action in many bulk material handling systems. The rate of induced airflow is seldom as large as in the example cited above but a knowledge of its magnitude is important, whether large or small, for a sound design of the exhaust system. An enclosure may have, for example, openings totaling 5 sq ft. In the absence of air induction, an exhaust rate for an inward velocity of 100 fpm might be indicated, resulting in a design exhaust rate of 500 cfm. If, however, air were induced into the enclosure by falling material at 5,000 cfm, an exhaust rate of 500 cfm would accomplish nothing.

A similar problem is represented by the design of a receiving hood for an abrasive cut-off when used in cutting brick and tile. The required exhaust rate is determined by the rate at which dusty air is delivered to the hood by the pump action of the millions of inertials projected in that direction.

One approach for estimating the magnitude of induced airflow in various circumstances or for systematic correlation of practical field data is needed to place these problems on an engineering base.

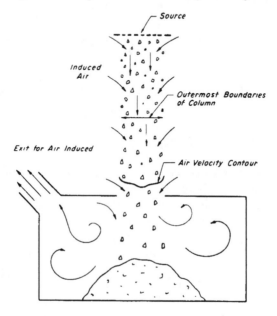

Figure 7-1. Particles falling into an enclosure induce a flow of air which must be exhausted.

[Note: Hemeon's theoretical development has been moved to the Appendix.]

Simplified Approach. To facilitate application of the theory [shown in "Chapter 7" of the Appendix] to practical problems, Table 7-1 has been prepared. The reader is cautioned against the temptation to reduce these data to graphs in order to enable more convenient interpolation, for then he will almost certainly be led to ascribe an accuracy to the estimate that is not intended by [Hemeon]. Until field data have been assembled to check the theoretical estimates of air induction [as developed by Hemeon], the methods provided herein should be employed only to estimate orders of magnitude.

Table 7-1
Induced Airflow in Falling in cfm
Unenclosed Streams of Particles
Solids flow rate, R = 1 lb/sec; specific gravity, z = 1
Multiply cfm by actual (R/z)$^{1/3}$

Falling Distance, s = 2 ft

Stream Area, sq ft	Particle Size, Millimeters	
	1	2.5
1/2	200	100
1	300	200
2	500	300
4	800	500
8	1200	700
15	2000	1000
25	2500	1500
50	4000	2500

Falling Distance, s = 3 ft

Stream Area Sq Feet	Particle Size, Millimeters						
	1	2	5	10	20	50	100
1/2	450	350	220	180	150	80	60
1	750	550	350	300	250	150	100
2	1200	850	550	450	350	200	170
4	1900	1400	850	700	600	350	250
8	3000	2200	1400	1100	900	500	400
15	4600	3300	2000	1700	1400	800	650
25	6400	4600	3000	2500	2000	1000	900
50	9900	7500	4700	4000	3200	1800	1400

Falling Distance, s = 6 ft

Stream Area Sq Feet	Particle Size, Millimeters					
	<2*	5	10	20	50	100
1/2	650	350	300	250	200	150
1	1000	600	500	350	300	200
2	1500	900	700	600	400	350
4	2000	1500	1200	900	700	500
8	4000	2300	1800	1500	1000	900
15	6500	3500	2800	2200	1500	1300
25	9000	5000	4000	3200	2300	1800
50	-	8000	6000	5000	3700	3000

Falling Distance, s = 12 ft

Stream Area Sq Feet	Particle Size, Millimeters					
	<2*	5	10	20	50	100
1/2	800	560	450	350	270	220
1	1300	900	700	600	420	350
2	2000	1500	1100	900	700	550
4	3300	2300	1800	1500	1000	850
8	5100	3600	2900	2300	1700	1300
15	8000	5300	4200	3500	2600	2100
25	-	7800	6000	5000	3600	2900
50	-	-	9900	8000	6000	4700

Falling Distance, s = 20 ft

Stream Area Sq Feet	Particle Size, Millimeters					
	<2*	5	10	20	50	100
1/2	950	800	650	500	370	300
1	1500	1300	1000	800	600	500
2	2400	2000	1600	1300	960	750
4	4000	3300	2600	2000	1500	1200
8	6000	5000	4100	3200	2400	1900
15	9500	8000	6400	5000	3700	2900
25	-	-	9000	7000	5000	4000

Falling Distance, s = 30 ft

Stream Area Sq Feet	Particle Size, Millimeters					
	<2*	5	10	20	50	100
1/2	1100	1000	850	650	500	400
1	1800	1700	1400	1100	800	600
2	2800	2700	2100	1700	1300	1000
4	4500	4400	3500	2700	2000	1600
8	7000	6800	5400	4200	3200	2500
15	-	9900	8200	6600	4800	4000
25	-	-	-	9000	6800	5400

Note: * = multiply cfm by actual $R^{1/3}$; all others, multiply cfm by $(R/z)^{1/3}$

Limitations of the Table [and its theoretical underpinnings]

When evaluating a falling stream, a determination of particle size distribution might be made by screening (unless the information already exists as operating data) and these figures used with the table. However, development of the original power equation was based on the assumption that each particle acts independently of its fellows and this supposes that they are completely separated from each other. Obviously, this can seldom be true. More often than not, a stream of material falls in a column as a semi-compact mass that is only partly "dispersed" or aerated except at its outer boundaries. A notable exception is seen in the material falling below an industrial vibrating screen which, by its very nature, effects a high degree of separation of the individual particles. For this reason (as well as the large area), the circulation of air below a screen is unusually large even though the height of the fall may be small. (Where screens are *completely enclosed*, the induced airflow may result only in internal circulation rather than positive induction through the enclosure.)

Another exception is illustrated by the particles projected at high velocity from a high speed abrasive wheel, where power characteristics are as discussed in a later section. A higher degree of dispersion must occur in this action.

Another departure from theory is evident from a consideration of the widely varying falling velocities of the particles of various sizes in a typical industrial mixture of crushed material. To what extent do

the particles of smaller velocity contribute to the air motion process in relation to that of larger particles hurtling downward through the mass at a velocity several times that of the smaller particles? Evidently very small particles would be transported in the "wake" of the large particles as mentioned in Chapter 2. We conceive, however, that particles of intermediate size do contribute importantly to the air motion even though many larger sizes are moving at greater velocities. Quantitative evaluation of the actual effects in terms of particle size must await systematic experimentation. In the meantime, one can proceed with Table 7-1 [and the equations and figures provided in the Appendix] for estimation of extremes in the knowledge that the calculated results are safely on the conservative side.

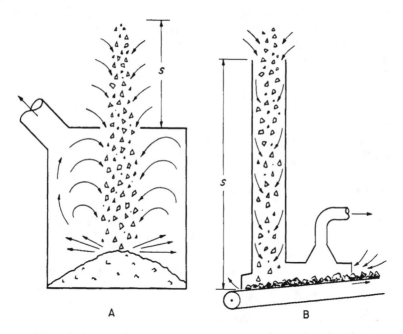

A   B

Figure 7-2. Illustrating effect of enclosure on effective falling distance, s. A. No points of leakage and spacious enclosure facilitating air recirculation. B. Enclosure dimensions suppress free air recirculation, potential points of leakage near bottom.

Establishment of the effective falling distance, s, needs to be given careful consideration because it is not always obvious as will be apparent from the following discussion. Referring to Figure 7-2, diagram A illustrates a situation wherein free circulation of air can occur within the enclosure. Therefore, while air continues to be induced into the column after it has entered the enclosure, air escaping from the bottom of the column on impaction is available by

upward recirculation to supply the needs of that section of the column inside. The net inward airflow induced by falling material is therefore that corresponding to the height of fall from the source to the enclosure entrance, as indicated on the diagram.

Another condition is shown in B where the small cross-sectional dimensions of the enclosure prevent free upward recirculation of air from the bottom. A slight pressure would be required to effect such upward motion counter current to the downward flow of material and induced air, and this would result in leakage of dusty air outward through cracks in the bottom of the enclosure. In order to avoid such leakage it is therefore necessary to exhaust all air that would be induced by free fall through the total height; hence the effective falling distance is as shown. Even though air is exhausted at a rate equal to the air induction rate, dusty air still may escape through bottom crevices due to *splash effect*. It is therefore important in this instance to exhaust an additional amount to prevent it if possible or to recapture air that does escape, and also to provide for an inward flow through the vestibule opening.

The idealized situations discussed above are intended to clarify the significant factors, enabling one to better analyze particular problems. The illustrations provided should not be too readily applied without question to practical problems, since critical observations by many engineers are needed to test the theory provided herein.

Air Induction Laws for Streams of Falling Particles

By a combination of equations [shown in the Appendix], some useful generalization can be *postulated* as follows.

Intermediate Flow Region. Intermediate flow is confined to falling distances not exceeding 4 to 5 feet for particle sizes around 1.5 millimeters (1,500) microns) and/or not exceeding 2-foot falls for particles 2 millimeters (2,000) and larger.

In this region, the rate of air induction is approximately proportional to
  (1) The cube root of rate of solids flow,
  (2) The cube root of specific gravity of solids,
  (3) The square root of particle diameter,
  (4) The 2/3 power of solids stream cross-sectional area,
  (5) The square root of falling distance.

Turbulent Flow Region, Accelerating Velocities. For practical purposes, turbulent flow may be regarded as confined to particles larger than about 2 millimeters and extends in distance, measured in

feet, to s = 2-12 feet. In this region, the rate of air induction is approximately proportional to

(1) The cube root of rate of solids flow,
(2) The cube root of specific gravity of the solids,
(3) The cube root of particle diameter,
(4) The 2/3 power of solids stream cross-sectional area,
(5) The 2/3 power of falling distance.

Constant Velocity Region, All Types of Flow. Constant velocity for particles in all flow regions is taken as starting at about 12 feet.
In this region, the rate of air induction is

(1) Proportional to the cube root of tonnage rate,
(2), (3) Independent of particle size and specific gravity,
(4) Proportional to the 2/3 power of stream cross-sectional area,
(5) Proportional to the cube root of falling distance (corrected for preceding distance in accelerating motion).

Applications

The airflow induced by a grinding wheel or abrasive cutting wheel is due to the high velocity projection of the inertials produced in the grinding process. The equations [shown in the Appendix] could provide a basis for estimating its magnitude. In practice, however, there is little need for such procedures since the required information has already been pretty well developed from practical experience.

The estimates for falling material, on the other hand, could be widely useful, especially for development of knowledge concerning the exhaust requirements of various parts of materials handling systems involving belt conveyors, bucket elevators, classifying screens, crushers and fine grinding mills, and storage bins. The following examples will illustrate some applications.

Example 1. Material having a specific gravity of 2.7 passes at a rate of 60 lb per minute through a 20-mesh (opening = 835 microns) commercial vibrating screen, 4 feet wide and 4 feet long. Assume that the effective particle size is that of the screen opening. The screen is disposed, without enclosure, directly above an opening of about the same size in the top of a storage bin and screened material drops a distance of 3 feet before entering the bin opening. Estimate the maximum rate of air induction due to material falling from screen to bin.

Solution. Note that the solids flow rate is given at 60 lb per minute or R = 1 lb per sec, s = 3 feet, stream area = 16 square feet, and z = 2.7. From Table 7-1 for s = 3 ft and particle size less than 1 millimeter, q ≈ just above 4,600 cfm. Extrapolating, we estimate q ≈ 5,000 cfm for R = 1, z = 1. Therefore

$$q \approx 5,000 \ (R/z)^{1/3} \approx 5,000 \times 0.718 \approx 3600 \text{ cfm}$$

Example 2. Crushed rock having an average diameter of 1 inch (= 25,400 microns) and specific gravity of 3, flowing at a rate of 60 tons per hour, drops through a vertical chute 2 ft square in cross section and 20 ft high. It feeds at the bottom onto a belt conveyor. An enclosure is constructed above the belt and the conveyor discharge point and is to be exhausted. Calculate the maximum air induction rate, assuming that the effective falling height for airflow calculation is the full length of the chute.

Solution. One of two extremes are possible: the flow within the chute acts like plug flow so that the velocity of the material falling also induces air to flow at the same velocity, or the falling material induces air to flow as in unenclosed flow. These two solutions should provide a bound for the "maximum" airflow expected. When looking at the first solution, these large materials would accelerate throughout the chute and the velocity at the chute outlet would be about 20 ft/sec = 1,200 fpm. If the air is also traveling at this velocity, q ≈ VA ≈ 1,200 × 4 ≈ 4,800 cfm.

The second solution is found using Table 7-1. Note that the solids flow rate is given at 60 tons per hour or R = 33 lb per sec, s = 20 feet, stream area = 4 square feet, and z = 3. From Table 7-1 for s = 20 ft and particle size about 25 millimeters, q ≈ 1,900 cfm. Therefore

$$q \approx 1900 \ (R/z)^{1/3} \approx 1900 \times 2.2 \approx 4,200 \text{ cfm}$$

Example 3. Sand (sp gr = 2.7) of uniform particle size, 2 millimeters diameter, drops from the head pulley of a belt conveyor at a rate of 1200 lb per minute into a storage bin. Height of fall from belt to top level of bin opening is 3.5 ft. The cross-sectional area of the stream of sand is 12 inches wide by 3 inches thick, and the bin top opening is 2 ft by 3 ft. Calculate the rate of exhaust from the bin interior required to prevent escape of dust.

Solution. The solution is found using Table 7-1. Note that the solids flow rate is given at 1,200 lbs/min or R = 20 lb per sec, s = 3.5 feet, stream area = 0.25 square feet, and z = 2.7. From Table 7-1 for s = 3.5 ft and particle size about 2 millimeters, q is about 350/2 ≈ 175 cfm. Therefore

$$q \approx 1,900 \; (R/z)^{1/3} \approx 175 \times 1.9 \approx 300 \; \text{cfm}$$

Now the area of opening into the bin is 6 sq ft and in accordance with the considerations discussed in Chapter 4 and implied by the equation

$$Q = VA + q$$

there is required in the present instance an additional exhaust quantity, AV. The area occupied by the stream of sand (0.25 sq ft) can be subtracted. Taking V of 100 fpm, the total exhaust rate is

$$Q \approx 575 + 300 \approx 875 \; \text{cfm}$$

The first and third examples were deliberately designed to dramatize the high importance of the cross-sectional area of the stream of solids. The low rate of induction represented in the last example by the stream of sand is, in fact, insignificant compared with the requirements posed by the size of opening area. Further, it is certain to be an exaggerated figure because of the poor opportunity for the individual particles to fall independently of each other; their fall is rather as a fairly compact mass.

Example 4. If the screening operation of Example 1 were modified to involve a total screen area of 8 sq ft instead of 4 sq ft, with all other conditions the same, what would be the air induction rate?

Solution. The air induction rate is proportional to stream area to the 2/3 power. (See the discussion just before the example problems.) Doubling the stream area, therefore, will increase induced airflow by $(2)^{2/3}$ which is 1.59. The original airflow was determined to be about 3,600 cfm (Example 1). The new airflow resulting from doubling the stream area will therefore be

$$q \approx 1.59 \times 3,600 \approx 5,700 \; \text{cfm}$$

Example 5. In a certain materials handling system, a storage bin is exhausted at a rate of 2,500 cfm which, experience has indicated, is just sufficient to accommodate airflow induced by

fall of material into the bin. If the solids flowrate is doubled without other changes, what rate of exhaust will be required to care for increased air induction?

Solution. Induced airflow is proportional to the cube root of tonnage flow rate. Therefore, the airflow increase will be in the ratio $(2)^{1/3}$ which is 1.26. Therefore, increased air induction (and required exhaust rate) will be

$$q \approx 2{,}500 \times 1.26 \approx 3{,}100 \text{ cfm}$$

As illustrated by the preceding two examples, experience with existing exhaust arrangements often can be extrapolated to new conditions by means of the air induction laws.

Example 6. Crushed material having a specific gravity of z = 3 falls into a bin with an effective drop of s = 12 ft. The solids stream area is 4 sq ft and the flow rate is 30 tons per hour. Detailed particle size data are not readily available but the most common maximum size is 1.5 inches and the fines range down to about 1/4 inch. Fine dust of smaller particle size is also present but is estimated not to exceed 2 or 3% of the total by weight. Estimate the air induction rate.

Solution. In the absence of particle size data, the maximum and minimum sizes can be employed to obtain a range of induced air flows. The fine dust will be arbitrarily ignored. The largest size, 1.5 inches, is equivalent to 38 millimeters; 1/4 inch is equivalent to about 6 millimeters.

Note that the solids flow rate is given at 30 tons/hr or R = 16.7 lb per sec, s = 12 feet, stream area = 4 square feet, and z = 3. From Table 7-1 for s = 12 ft and particle size about 6 millimeters, q is about 2,200 cfm; for a particle size about 38 millimeters, q = 1,250 cfm. Therefore

$$q \approx 2{,}200 \, (R/z)^{1/3} \approx 2{,}200 \times 1.77 \approx 3{,}900 \text{ cfm}$$

and

$$q \approx 1{,}250 \, (R/z)^{1/3} \approx 1{,}250 \times 1.77 \approx 2{,}000 \text{ cfm}$$

An extreme range of air induction rates is thus obtained, based on assumptions that all of the material is of either maximum or minimum particle size. The relatively narrow range of air induction rates, in spite of the extreme assumptions that were made, makes this a useful procedure. One would be justified, in this last example, in

taking the average of the two extremes in the knowledge that a conservative figure results.

Shallow Drop

Where material is permitted to fall only short distances of 2 to 4 feet as in transfer from one belt to another, it is quite unlikely that air induction plays a significant role in determining the rate of air exhaust required. The material drops in a compact stream with little opportunity for mixture with air. On the other hand, *splash* violence at the bottom of its fall will be evident and rates of exhaust, sufficient in amount to *recapture* dusty air at the X-distance corresponding to its *null point* (see Chapter 6), are needed. This is often true even though the dusty material transfer point is largely enclosed, since seldom is it possible to effect tight enclosure of a transfer point to completely eliminate escape orifices.

Additional Applications in the Design of Exhausted Enclosures

In the design of exhausted enclosures, the following aspects require consideration: (1) rate of air induction into the space; (2) location of cracks or other openings in relation to splash effects; (3) design of exhaust which will avoid excess withdrawal of the product; and (4) adequate airflow for dilution of interior concentrations, for visibility or safety from explosions, if a factor.

Induction and Splash

While air induction is often a primary factor in determining the required rate of exhaust from an enclosure, splash may coexist in operations where air induction also prevails, and it may be difficult to distinguish between them. (See Chapter 4.) The distinction, however, is of great practical importance. Splash is that phenomenon which could be corrected by enlarging the enclosure to move the opening beyond the null point of the splash, whereas air induction is independent of the size of the enclosure or location of openings. One might exhaust sufficient air to care for induced air but still fail in preventing escape of dust by splash, and it would not necessarily be prevented by considerable exhaust capacity.

Product Withdrawal

The dust which is to be controlled is frequently a portion of the product itself. It is desirable therefore to design the exhaust system

so that only a minimum quantity of dust will be removed. Indeed, in some applications this is a most critical factor determining success or failure of the system.

The amount of material removed in the exhaust system is, in part, a function of the product of the concentration of dust within the enclosure and the rate of ventilation, or rate of exhaust. Beyond that, a most important factor is to avoid high air velocities in close proximity to material which is not already airborne.

The magnitude of air velocities adjacent to the inlet of an exhaust duct may be deduced by application of the same relations that have been described for exterior flanged exhaust hoods; the opening in this case is flanged by the enclosure wall to which it is attached.

Velocities within the enclosure of less than about 50-75 fpm can, in general, be considered to be without any power to lift material from rest to the airborne state. Velocities of 500-1,000 fpm can raise some dusts from rest into the air stream and withdraw them. In between these values the effects are less certain and variable with the material and its state of agitation.

The rules which follow from these considerations are clear:

(1) Avoid locating an exhaust inlet close to zones where the parent material is in a state of agitation, at chute entrances, or near the point of splash. Note that from the overall standpoint of controlling escape of air from the enclosure, the exhaust connection is equally effective regardless of where it is attached to the enclosure wall.

(2) If the dimensional limitations of the enclosure make it impractical to attach the exhaust at a distance from zones of agitation, an exhaust vestibule or plenum—which is in effect a special low-velocity annex for the main enclosure—should be constructed and the exhaust connected to it. The size of the vestibule in relation to the rate of exhaust should result in air velocities as low as possible, down to 50-75 fpm.

Interior Concentration

The required exhaust rate to prevent escape of contaminant from the enclosure interior is sometimes less than required for desired concentrations of the contaminant within. In some dusty operations, interior visibility may be a factor, as in sandblasting; in others it may be desirable to use the exhaust system to aid in removal of dust by extra flushing action of greater ventilation rates. In rock crusher exhaust systems the removal of fine dust improves the product. In

enclosures where explosive vapors originate, the airflow must be adequate to dilute them to concentrations below the explosive limit.

Bulk Material Handling

In industrial operations involving crushing, transporting, and classifying of bulk materials, one sees the development of very large quantities of dust. In fact within this category some operations occur in which the entire purpose is the manufacture of dust, such as pulverizing plants for reducing quartz rock to pulverized silica. At the other extreme, one notes the crushing of rock for use in concrete, where for economic reasons, it is desired to produce as little fine dust as possible.

One encounters extremes in the toxicity of materials in this category—such as pulverized silica or lead oxide at one end and synthetic abrasive dust and some rock dusts at the other. In some cases high rates of air exhaust which remove quantities of fine dust from the product are welcomed for reasons of product quality, e.g., crushed rock for concrete; in others, any loss of dust represents a loss of the product and may need special arrangements for re-incorporating it in the product stream.

In some handling processes moisture can be used to suppress dust partially or completely; in others no moisture at all can be tolerated. The scale of operations encompasses both the huge and rugged equipment for crushing and handling of some metal ores and the little systems for handling of such products as cosmetic powders or synthetic resin powders. In the design of plants for small-scale handling of materials, it is often possible to provide complete enclosure of all equipment to eliminate the need for any exhaust system or other means.

Coarse Crushers (Jaw, Gyratory, Roll, Etc.). Crushers of this type which break lump material by application of pressure, but which do not include any high speed mechanical elements, are characterized by a mild pulvation action; that is, there are no primary or secondary air currents characteristic of the action itself.

Thus exhaust for a small laboratory jaw crusher or similar equipment would consist of a simple exterior hood below the jaws and one above at the point of feeding, and X-distances would be little greater than that from the face of the hood to the farther boundary of the crusher.

In actual operation of large industrial crushers there are always secondary air currents resulting from the fall of crushed material from the unit to its destination. In usual arrangements the material falls onto a belt or similar conveyor, or else through a chute. Exhaust is

applied to a short length of enclosure above the belt, or to the bin or other equipment enclosure into which material is discharged by chute. The magnitude of the exhaust required depends on the airflow induced by the falling materials. In other words, the exhaust requirements for such crushers depend on how the material is fed or discharged not on the operating characteristics of the crusher itself; the exhaust requirements are really those characteristic of the chute or of the belt conveyor.

Jaw and gyratory crushers reduce the size of such hard materials as rock or synthetic abrasive by the application of pressure. Since there are no high speed mechanical elements, the primary pulvation action is a gentle one. However, as the crushed material drops from the point within the crusher housing where it is formed, to a conveyor below, some air induction will occur which, since it is an integral part of the crusher action, can be regarded as the primary pulvation characteristic of the crusher (although actually it is a secondary stage within the machine itself).

L.P. Hatch ("Silica Dust Control at Hiwassee Dam," *Civil Engineering,* Vol. 10 No. 9, Sept 1940) described the exhaust requirements of a gyratory rock crusher of a 42-inch diameter handling rock at 500 tons per hour and reducing it to a nominal maximum size of 4 $^3/4$ inches (discharge aperture). The exhaust arrangement is illustrated in Figure 7-3. It was found that an airflow of 4,000 and 5,000 cfm was necessary for complete dust control.

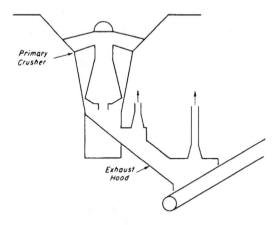

Figure 7-3. Arrangement of exhausting a rock crusher.

Example 7. Calculation of Exhaust Requirements for a Gyratory Rock Crusher. The above experience data (Hatch) provide an opportunity to test the equations developed for this chapter [See Chapter 7 materials in the Appendix]. Assume that the height of

fall, s, in the 42-inch crusher is 6 ft and that the average particle size is 4 inches (= 10 cm = 10^5 microns), and specific gravity, z, is 3.0. Material flow rates were given as 500 tons per hour, so R=280 lb per second.

Particle Size, Microns

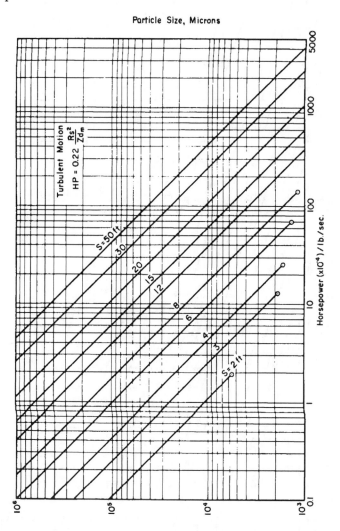

Referring to the graph above ["Figure 7-5" from Appendix], the power for R=1 and z=1 corresponding to particle size of 10 cm is

$$hp = 1 \times 10^{-4}$$

Multiplying by R = 280 and dividing by z = 3 gives

$$hp = 93 \times 10^{-4}$$

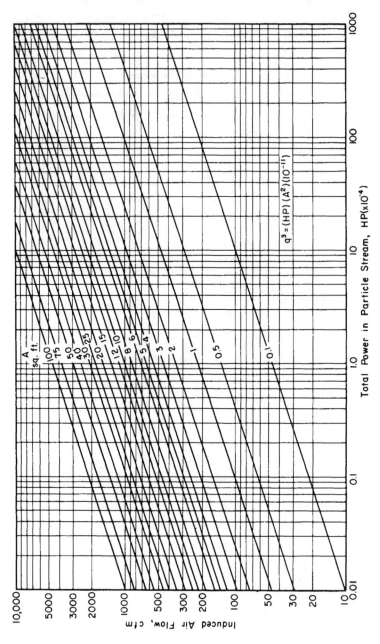

Now take the cross sectional area of the falling material to be that of a 42-inch circle, or about 10 sq ft. Then from the above chart ["Fig.7-7" from the Appendix] the induced airflow would be q = 4,500 cfm.

It is fortuitous that the calculated airflow rate coincides so closely with the exhaust rate Hatch reported necessary in practice for these reasons: (a) The reported exhaust rate must be higher than the rock-induced airflow since an excess is needed to draw air positively inward through apertures in the enclosure. (b) The particle size taken for calculation (4 inches) is the top size; the actual average size may be smaller. (c) The cross-sectional area taken for calculation (10 sq ft) is probably larger than fact and would tend to give a high calculated result. (d) The fall of crushed material is not a free, unimpeded drop; it tumbles downward on slanting interior surfaces of the crusher discharge. This would tend to reduce actual falling velocities, and therefore a calculated result higher than fact is likely.

Perhaps the several factors enumerated above counterbalance each other. In any case, the results give one enough confidence in the method to argue that it is a reasonable basis for correlating actual field practice by experimental confirmation and the eventual development of good handbook data. [Both are yet to be completed.]

Secondary Crushers. Secondary crushing of rock is applied to reduction of material from the primary crusher, after screening out smaller sizes, say $1/2$ inch to 2 inches maximum size. Insofar as dust control requirements are concerned the considerations are identical with those discussed under primary crushers. L.P. Hatch described the operation of a $5^1/2$ foot cone crusher handling 250 tons per hour for reduction of rock to a maximum size of $1^1/2$ inches which was discharged onto a 24-inch belt conveyor. Air was exhausted from a belt enclosure at the point of transfer (Figure 7-11) at a rate of 2,800 cfm. Dust was collected at a rate equivalent to 0.2% of the rock weight. This exhaust rate was found to be occasionally inadequate to confine all dust when rock surged through.

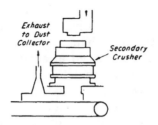

Figure 7-11. Conveyor belt enclosure.

Example 8. Calculations of air induction rates like those performed in the gyratory crusher example can be made with the following assumptions:

Distance = s = 4 feet
Particle size = 1 inch = 2.5 cm = 2.5 x 10^4 microns
Rock flow = 140 lbs per sec   Area = 10 sq ft   SG = 3

Using similar procedures to those previously demonstrated, we estimate

$$Q \sim 4,000 \text{ cfm}$$

Again the result appears to coincide as to order of magnitude with reported experience.

Effect of Water Spraying. Water was routinely sprayed on the rock at points considerably upstream from the secondary crusher; two belt conveyors, a storage pile, and a screening operation intervened. Based on moisture measurements on the fine fraction, the average water content was 0.6% of the total rock weight. (The total water applied was about 9 gallons per ton or 3.7% by weight.)

When the water application was experimentally discontinued, considerable quantities of dust were seen to escape at the transfer enclosure below the secondary crusher. This phenomenon is consistent with some similar observations by the author [Hemeon] and is undoubtedly due to an increased rate of induced airflow produced by that fine particle fraction of the aggregate mixture. In the presence of moisture they are cemented together or to the larger lumps and thus rendered inactive.

Hammer Mills, Disintegrators, etc.

In contrast to the coarse crushers, the high speed mill typified by the hammer mill displays a characteristic integral airflow. The high speed hammers, rotating within the crusher housing, have a capacity like that of a centrifugal fan to induce a flow of air through it along with the pulverized material. Control of dust can only be successfully effected by exhausting from an enclosure into which crushed material is discharged at a rate greater than the induced flow of dusty air. Unfortunately, little data are available on the magnitude of the flow, even on the part of the mill manufacturers. It must range from a few hundred cubic feet per minute for smaller mills up to several thousand cubic feet per minute for the large units. Reasoning from the performance of centrifugal fans, the airflow must be less than that flowrate per revolution generated by the hammer assembly, i.e.,

$$\text{Maximum cu ft per rev} = \pi/4(D^2)W$$

where D is the diameter of the hammer assembly measured from the tip to the shaft, and W is the width of the hammers.

In addition to the integral airflow of the mill itself, there could also be air induction due to the drop of product from the bottom of the mill. The latter, however, would result only in internal air circulation if the crushed material were discharged directly into a bin as is often the case.

In the previously cited description by L.P. Hatch, 42-inch hammer mills (each crushing 110 tons per hour to a fineness of sand, 30% passing 100 mesh) required about 3,500 cfm exhaust per mill from the belt enclosure receiving the crushed product.

Since airflow is inward with the feed, there is no requirement for exhaust at the point that is inherent in the machine; there may however, be an induced flow of air from the preceding element of the equipment; for example, a chute which, if greater than that induced by the mill, would require independent exhaust capacity for the feed enclosure.

Bucket Elevators (Housed)

There are three common types of bucket elevators designated by the adjectives *continuous, centrifugal,* and *perfect discharge.* The centrifugal type operates at such a speed and arrangements of discharge spout at the top that material is thrown out of the buckets as they pass over the top pulley directly into the spout. In accordance with our knowledge of the air induction potential of falling material, it is clear that each bucket of material must induce a certain quantity of dusty air to travel with it through the delivery chute into the next equipment enclosure of a particular system. The magnitude of this airflow would be determined by the various factors that have been cited. It will be noted that the action is not that of falling material starting from rest (since it has an initial velocity imparted to it by the buckets), nor is it that of material projected with initial velocity where gravity effects are of no significance. (Gravity does determine its speed.) However, the magnitude of the airflow induced by this mechanism does not appear to be very great, probably seldom over a few hundred cubic feet per minute, and revisions in the theory presented earlier to fit this particular action do not seem necessary.

This discussion is intended to apply only to that portion of the discharge spout that can be considered as a part of the elevator proper, not including air induction effects due to chutes through which material may subsequently fall.

Air motion within the housing due to the movement of the buckets does not appear to be of significant magnitude. A simple calculation of the cubical contents of the buckets passing a point per unit time

indicates a very low air displacement and, in any case, it is not directed into or out of the elevator housing.

In elevators having separately housed legs, the air motion would, of course, follow the trail of buckets. At the discharge spout there is, however, no inherent tendency for this air to move outward by the motion of the granular material itself as it falls through the spout and its characteristics determine the magnitude of the net airflow through the elevator housing. In summary, then, it is clear that the fan action of rapidly moving buckets within an elevator enclosure causes no appreciable airflow into or out of the enclosure. Such air displacement as does occur is determined by material flow characteristics at elevator discharge spouts or at the loading points at the bottom.

Splash at the Boot. Some leakage of dusty air can sometimes be observed at the boot of an elevator passing outward through the spaces in the elevator housing around the bottom of the shaft. The amount of this leakage is small and would be significant only where the dustiness tolerance is very low—as in the case of a highly toxic or irritating dust. This kind of air leakage is not to be confused with air induction since an enlargement of the elevator boot, which would locate the leakage orifices more remote from the position of the bottom bucket into which this material falls, would eliminate it.

Exhaust from Elevators. It is clear now from the preceding discussion that elevators do not induce large quantities of air except in unusual circumstances and there is therefore little inherent need for exhausting them, providing of course, the housing is dust tight. However, the housing provides a convenient duct (or chimney, if the air is hot within the housing) connecting the elevator boot and the discharge spout, and it is common practice to exhaust moderate quantities of air from them. If the system is one in which it is desirable to minimize the withdrawal of fine dust particles to a dust collector, then it may be acceptable not to attach an exhaust duct to the elevator housing. The entrance of air through the discharge spout, hence through the interior housing toward the exhaust duct, results inevitably in the withdrawal of some particles that would otherwise settle out. In the handling of ordinary material this factor would doubtless be of no consequence, and an exhaust would be warranted to keep the housing under a slight negative pressure.

Unhoused Elevators. Open elevators, commonly used in heavy duty rock and ore crushing plants do not generate dust during ascent and descent of the buckets. Dust sources are at the feed point at the bottom where crushed material is dropped into the buckets and at the

top where the material is discharged to a chute. Dust control requires construction of partial housings at both points for the application of exhaust.

It is difficult to estimate exhaust requirements from such an enclosure because those elevators are employed on operations the scale of which results in violent pulvation where material drops into the buckets and the pulvation distance may be 2 to 3 feet or more. Moreover, there may be considerable air induced by fall of crushed material from the preceding equipment, and the movement of loaded buckets out of the enclosure tends to drag quantities of air which though small are heavily loaded with dust.

It is difficult to construct any enclosure with a minimum of opening and to estimate this area in advance of is construction. A code devised and in force in New York State applying to rock crushing plants specified exhaust rates equivalent to 200 cfm per square foot of opening into the enclosure. [Circa 1950s; 200 fpm is still considered acceptable today where quiet conditions are found at the openings.] Independent estimates should be made based on a) air induction by material fed to the buckets; b) degree of *splash* in relation to walls of the enclosure; c) area of enclosure openings; and d) linear velocity of the stream of buckets leaving the enclosure. Exhaust rates of 1,500 to 2,000 cfm from such enclosures are a common minimum for systems handling about 200 to 300 tons per hour.

At the top of the elevators, considerations apply similar to those at the bottom except that there is little air induction at the discharge. Elevators in this application are generally of the *continuous* type which have a lesser tendency to induce air motion. Spillage of material due to incomplete delivery of material to the receiving chute may result in a bothersome trickle of dust escaping the action of the exhausted enclosure. The main source of dust is from splash of material discharged from the buckets to the chute. The returning buckets drag small quantities of heavily dust laden air out of the enclosure. Rates of exhaust about as great as those which are required at the elevator boot are indicated.

Belt Conveyors

Dust from operation of belt conveyors generally originates mainly at two points: at the tail pulley where it receives material from prior equipment (due to splash and sometimes also due to air induction depending on the manner of material delivery), and at the head pulley where material is discharged to a chute or the equipment succeeding it. There is usually little dust generation along the belt, hence dust control often requires only that the two points of transfer

be housed and exhausted as closely as the equipment arrangement permits.

Exhaust requirements at the head pulley are, in general, small because any air induction tendency is downward and away from this transfer point. Indeed, in some circumstances, it is conceivable that no exhaust at all would be required, for example, where material is discharged directly into a chute, inducing air in the process such that all dust due to transfer is carried into the chute along with the induced air stream. In common circumstances exhaust at a rate of a few hundred cubic feet per minute would be necessary at this location, just sufficient to induce a linear air velocity of around 200 fpm through open areas of the enclosure.

At the tail pulley, exhaust requirements again are determined principally by the amount of air induced by the delivery chute. If transfer to this belt is a mild action due to a shallow drop from the preceding element, then exhaust requirements will be correspondingly less but still greater than at the tail pulley because of the air drag action of the belt and its load leaving the enclosures. As in the case of buckets of an elevator, the quantity of air dragged out is small but since it is heavily laden with dust, important air contamination may result if it is not suppressed.

It is often specified that belt transfer points be enclosed and exhaust at a rate of 400 cfm and 700 cfm per foot of belt width, the lower figure where belt speed, are less than 200 fpm and the higher figure for higher belt speeds. This rule assumes, further, that induced air velocities into the enclosure will not be less than 200 to 300 fpm for adequate control.

In order to minimize withdrawal of useful material by too close proximity of the high velocity air currents at the exhaust duct entrance to the zone of material splash within the enclosure, a low-velocity chamber or vestibule should be provided to separate the splash zone from the high duct velocities.

When finely powdered material is carried on a belt, a residual film tends to remain adherent to the belt surface after passage downward around the head pulley, then it slowly becomes dislodged causing a trickle of dust to fall through the air, much of it becoming airborne or, if a shelf is provided below the belt, this may quickly become filled to the level of the belt. If the material being handled is highly toxic, e.g., pulverized silica or litharge (lead oxide powder), this trickle of dust can be a major source of air contamination and will require a different type of conveyor, such as a screw, or a full, ventilated housing.

Sometimes short belt conveyors are provided with complete enclosures from end to end. In this arrangement material may accumulate under the belt and must be periodically removed by

shovel. There is a major disadvantage to this arrangement. (This is not to be confused with weather covers usually provided to shield outdoor belt conveyors from wind and rain.)

Screw Conveyors

From the standpoint of dust control, the screw conveyor is one of the best elements of a transport system for dusty materials because, when provided with dust-tight covers, it prevents escape of dust without any exhaust. As in the case of other conveying equipment which does not itself induce air flow, this is subject to the manner in which material is delivered to it.

Some pulvation occurs within the churning mass of material, the amount depending on the dustiness characteristics of the powder. If objectionable amounts escape due to leaking covers, the most direct remedy is to provide new covers that are tight.

Gravity Transfer

Chutes, Surge Bins, Feeders. The transfer of dust materials by gravity from one element to another of a bulk materials handling system is probably the greatest single cause of dust. If the drop is a short one, about two to three feet, the quantity of air induced by the fall may be negligible due partly to the small falling distance and partly to the maintenance of a compact unaerated stream of solids. The dust making process would be classified as multidirectional pulvation where only the splash factor needs be given special consideration. A partial enclosure around the transfer point provided with exhaust at a rate of about 1,000 cfm would be expected to take care of most such situations.

When, however, the delivery chute is of considerable length, the flow induced may be great, and will determine the rate at which air should be exhausted from the point of arrest. If discharge is onto a belt conveyor, splash escape of dusty air through openings of the enclosure may be a major problem even when the enclosure is exhausted at a sufficient rate to accommodate the induced air delivered.

Sloping chutes which minimize the tumbling fall of particles through the air would not be expected to have as great a potential for induction of air as vertical chutes, but there is no quantitative basis for evaluating this difference in potential.

Devices that arrest the high velocity tumble of material through chutes are highly advantageous in reducing air induction, splash escape tendencies, and also the quantity of airborne dust. Sometimes, unfortunately, the characteristics of the material require

uninterrupted flow to prevent clogging of the system. Where this is not a factor, a surge bin with an arrangement at its bottom for controlled flow of material is of great aid in the control of dust.

Where opportunity is afforded for application of these principles during the early stages of plant design, much can be accomplished toward simplification of the dust problem, by reducing gravity fall, including surge bins, feeders, and the like. The spiral chute used in coal treatment plants for reducing breakage illustrates a means for gently lowering the level of a stream of dusty material and reducing air induction potential.

Screens. Vibrating and rotary screens may produce tremendous quantities of dust which can be controlled only by complete enclosure. Indeed, it is possible to enclose them so well as to eliminate the need for any exhaust.

The air induction potential of vibrating screens is high because of the large cross-sectional area of the stream of solids. Vibrating screens operate, of course, in a manner that distributes the material passing through the meshes over almost the entire screen area. Reference to the air induction laws reminds one that airflow induced by falling material is proportional to its cross-sectional area raised to the two-thirds power, $A^{2/3}$. More important, the screening action separates particles from each other and tends to cause them to behave more nearly like separate particles, in contrast to the same fine material falling through a chute as a stream of massed clumps. Finally, the particles are often of small size, which, pound for pound, induce much greater air flows than coarse lumps.

Field observations of conditions affecting dust control at materials handling equipment should be carefully recorded.

Chapter 8

Exhaust for Hot Processes

The "canopy hood" in its common usage includes any exhausted opening that is located above a source of contamination, regardless of its shape or height above the source it is intended to service. It may take the form of a large positive-flow roof ventilator directly above a large-scale heat process; a suction grille in a ceiling above a hot, fuming process; or a low-hanging pyramid-shaped hood above a tank.

Canopy Hood as a Partial Enclosure

A low-hanging canopy hood above a *cold* process is essentially the same as a poorly constructed booth or enclosure, having four open faces instead of a single-plane opening (Figure 8-1). Therefore, the design of such a hood is summed up in a description of the amount of open area of the four vertical planes between the lower lip of the hood and the structure below it. That area in square feet, A, and the selected face velocity, V, in feet per minute, specifies the exhaust rate requirement by Q = VA, as for any booth. On the basis of contour characteristics one may see that for many hoods of this type a considerable difference prevails between velocity, near the hood rim, compared with a point near the bottom of the opening.

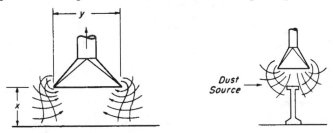

Figure 8-1. (Left) Low hanging canopy hood above a cold process is essentially an enclosure with four open sides, each x by y. (Right) Canopy hood serving a cold process operating as a simple exterior hood.

For this reason, it has been suggested that the area of opening should be figured as 40% greater than that given by the simple area calculation. That particular figure has the limited significance indicated above and is only mentioned at this point because it is often mistakenly assumed to cover the entire problem of canopy hood design.

Canopy Hood as Receiving Hood

In the majority of cases the canopy hood is really intended to function as a receiving hood, and as has been stated before, design of such a hood, beyond the specification of shape, dimensions, and position, consists, principally, in a sound analysis of the rate, q, at which contaminated hot air is delivered to the face. As with air induction by streams of inertials, the receiving hood is disposed at a point along the axis and the exhaust rate is properly specified to be at least equal to the induction rate.

Air velocity control considerations, therefore, have minor impact on the design of receiving hoods (please see Figure 8-2).

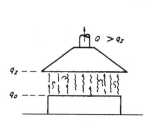
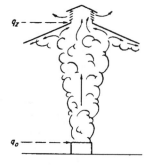

Figure 8-2. Two forms of canopy hood. Positioned above the hot process, they are receiving hoods for air currents set in vertical motion by heat. (Left) Vertical air currents above a hot fuming liquid are characteristic of the hot process, not of the hood. (Right) Opening in roof above a hot process is a form of canopy hood, whether natural draft or motor-driven ventilator.

Canopy hood installations frequently have a common failing. If the contaminant is a visible dust or fume, hot air can be seen curling outward under the rim of the hood to escape into the general room air, exactly analogous to the way water would overflow the rim of a vessel if the supply rate exceeds the rate of withdrawal (Figure 8-3). The cause of such air escape is identical with that in the water tank. Hot air is being delivered to the hood at a rate greater than the rate of exhaust.

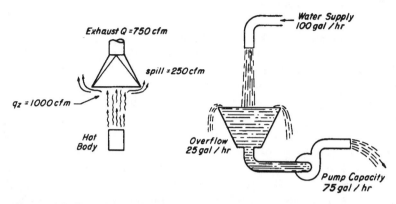

Figure 8-3. Behavior of a canopy receiving hood when the exhaust rate Q is less than the delivery rate q_z. Spillage at the rim (left) is analogous to the water system (right) where water supply rate exceeds pump capacity.

Symbols Used in Equations for This Chapter

A_f = hood face area, or orifice area in an enclosure, sq ft
A_p = cross-sectional area of hot air stream at top of heat source, sq ft
A_s = surface area of hot body, sq ft
B = horizontal width of hot surface, sq ft
D = diameter of hot air column at height, Z, ft
G = rate of steam formation, lb per (sq ft) or (minute)
H' = sensible heat transfer from hot process to air stream, Btu/min
h = buoyant head responsible for air motion in a hot air column, feet of air
h_c = natural convection coefficient, heat transfer, Btu per (hr), (sq ft), or (F)
L = vertical height of heated air in an enclosure, ft
m, m' = height of air column adjacent to a vertical hot surface; width of hot column of air immediately above a horizontal hot surface, ft and inches, respectively
Q = hood exhaust capacity, total cfm
q_o = rate of thermal air motion at top of heat source, cfm
q_z = rate of air motion at hood face after some mixing with room air, cfm
s_h = specific gravity of hot air relative to surrounding air, unitless
T_c, t_c = temperature of room air, °R and °F, respectively
T_h = average temperature of heated air column, °R absolute
T_z, t_z = temperature of hot air stream at top of rising column, °R and °F, respectively
$\Delta T, \Delta t$ = temperature difference between hot body and room air, or between hot air and room air
V = face velocity induced at hood face by exhaust capacity margin over q_z, fpm
v_h, V_h = velocity of hot air stream, ft per second, ft per minute, respectively
V_z = average velocity of rising column of hot gas, fpm
Y = actual distance from hot surface to hood face, ft
Z = distance from a hypothetical point source to hood face, ft
ρ_h = density of heated air, lb per cu ft

[Note: Figures 8-5 and 8-7 have been moved to the Appendix.]

Airflow from Hot Bodies

The mechanisms of convectional heat loss includes molecular conduction through a thin film at the hot surface to the adjacent air which, when warmed, becomes lighter than surrounding air and is forced to rise, its place being taken by cooler air in the vicinity. The resulting rate of airflow is related to the rate of convectional heat loss.

The velocity with which a mass of warm air rises is determined by the force of gravity, and its magnitude is expressible in terms of the falling body formula , $v = \sqrt{2gh}$. Where the velocity of a rigid body moving through air is affected by the frictional air resistance, the motion of a moving mass of air is influenced by the relative turbulence, since that determines the rate of mixing and dilution with ambient air.

Air Streams at Hot Surfaces. Consider the buoyant air currents flanking, for example, a heated vertical panel (Figure 8-4).

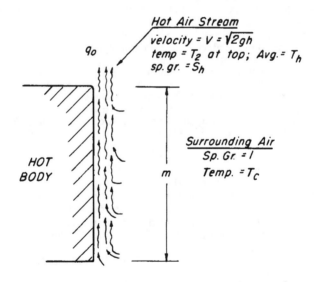

Figure 8-4. Induction of airflow by transfer of sensible heat from a hot vertical surface to surrounding air.

If the specific gravity of the general mass of cooler air surrounding the rising column is called unity, and that of the hot air, s_h, the net buoyant head responsible for velocity is

$$h = (1 - s_h)m$$

where *m* is the height of the column adjacent to the hot source. The value for s_h due to temperature can be expressed in terms of the ratio of *absolute* temperatures

$$s_h = T_c / T_h$$

where subscripts c and h refer to the cooler and hotter air masses. The formula

$$v = \sqrt{2gh}$$

then becomes

$$v_h = \sqrt{2gm \left(1 - \frac{T_c}{T_h}\right)}$$

and, converting v_h (ft per sec) to V_h (ft. per min) and giving g its numerical value 32.2, we have

$$V_h = 480 \sqrt{m \left(1 - \frac{T_c}{T_h}\right)} \tag{1}$$

This can be readily be transformed to an expression for airflow rate, q, by multiplying it by the cross-sectional area of the rising stream. However, the appearance of T_h in the formula raises difficulties in its application to practical problems. Mixing occurs throughout the travel of the column, hence the column temperature varies in proportion to the quantity of cooler air entrained. It is therefore desirable to proceed to elimination of the temperature difference term, if possible.

The ordinary heat capacity formula applies to the same air stream, and may be written

$$t_2 - t_c = T_2 - T_c = \frac{H'}{\rho_h(q_o)(0.24)} \tag{2}$$

where t_2 and T_2 are Fahrenheit and absolute temperatures at top of hot air column; H' is sensible heat transfer from process to air stream in Btu per minute; ρ_h is density of heated air; q_o is cfm of air in motion at top of heat source. Rearrangement of equation (1) gives

$$T_h - T_c = \left(\frac{q_o}{480A_p}\right)^2 \cdot \frac{T_h}{m} \qquad (3)$$

where A_p is the projected area of the object or of the hot air stream.

If equations (2) and (3) could be set equal to each other a useful equation for q_o would result; the validity of this procedure hinges on the difference between T_2 and T_h. The term T_2 represents the final air temperature of the leaving air stream corresponding to the complete absorption of the heat, H', whereas T_h is the average distance (because it is related directly to m) of temperature in the column. The error in assuming equivalence of T_h and T_1 may not be serious for our purposes.

Combining equations (2) and (3) gives

$$q_o{}^3 = \frac{10^6 H' A_p{}^2 m}{T_h \rho_h} \qquad (4)$$

which is further simplified by letting air density = $0.075(T_c/T_h)$ where Tc = 530°R.

The working equation is developed as

$$q_o{}^3 = 2.5 \times 10^4 H' A_p{}^2 m \qquad (5)$$

and taking the cube root

$$q_o = 29\sqrt[3]{H' m A_p{}^2} \qquad (5a)$$

where

q_o = air induction rate at upper limits of hot body, cfm
A_p = cross sectional area of air stream, sq ft
m = height (or width) of column receiving heat, ft (e.g., height of hot body)
H' = convectional heat transfer rate, Btu per minute

Experimental Data

The literature dealing with heat transmission by natural convection includes several studies involving measurement of the temperature gradient adjacent to hot vertical plates and horizontal rods, and some where air velocities were measured in the same zone. Some of these data are shown in Figs. 8-6 and Table 8.1.

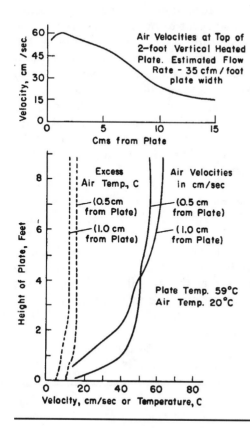

Figure 8-6. Air velocity and air temperature at 0.5 cm and 1.0 cm distant from a heated vertical plate (Griffiths and Davis).

Table 8-1
Rates of Convectional Air Motion at Different Elevations Adjacent to Heated Vertical Plates
(after Griffiths and Davis [G&D])

Panel Conditions	Height from bottom of panel, feet	Calculated airflow q_o, cfm/ft of width
Panel 2.8 ft high, Temperature = 120 F	0.5	1.2
	0.9	1.8
	1.4	2.7
	1.9	3.6
	2.5	3.6
	2.8	5.7
Panel 9 ft high, Temperature = 140 F	3.2	15
	4.5	17
	5.4	22
	6.5	25
	7.5	26
	9.0	35

This available experimental information permits estimation of the rates of thermal air motion adjacent to hot surfaces. Where the study reported only the temperatures in the hot air stream, the probable rate of heat loss was estimated from surface temperatures using well-known formulas, and then the airflow rate was calculated. Where velocity data were also available, calculations could be made on two separate bases. For example, in one case calculation from the temperature data show airflow rate of 2.6 cfm per foot of plate width, and calculations from velocity data, a rate of 1.9 cfm. These are in good agreement with each other considering the inherent impediments to accurate measurements.

The results of these several studies are give in Table 8-2, wherein airflow obtained from the experimental data are compared with the value calculated by equation (5a), employing the appropriate values of m, H′, and A_p given in the table. For our purposes, the agreement between theory and experimental data is entirely satisfactory, especially since the calculated values are higher than the experimental ones, and indicate that the simplifying assumptions employed in the derivation do not seriously detract from it.

Table 8-2
Comparison of Calculated Rates of Vertical Air Motion
with Experimental Data Results

Heat Source	m, ft	H′	A_p^2	Reference	Exper. q_o*	Calculated cfm
Vertical Plate	2.82	2.0	0.01	G & D	5.7	10.2
Vertical Plate	9	7.9	0.25	G & D	35	70
Vertical Plate	0.8	2.5	0.0009	Schmidt	2.6	3.1
Vertical Plate	0.38	0.13	0.00011	Kennard	0.95	1
Vertical Plate	0.38	1.79	0.00011	Kennard	1.4	2.4
Horizontal rod	0.08	0.94	0.00052	Kennard	2.3	2
Steam tank	2.5	970	16	Hatch	>800	900

* Determined from reported temperature or velocity data.

Estimation of Input Data

It is not always obvious how to characterize the required input data. The following paragraphs offer some suggested approaches.

Stream Area. Area of the hot air stream cross section may often be approximated from the dimensions of the hot body. For example, in the case of the horizontal rod of Table 8-2, the width of the stream is very close to the diameter of the rod.

The air stream beside a vertical plane appears to thicken at an angle of 4 to 5 degrees with the plane according to the data of Griffiths and Davis (G&D) and this will serve for estimation of stream areas in this category.

The effective stream area for horizontal planes is uncertain. It is suggested that it be taken to be equal to, or somewhat less, than the area of the plane itself.

Heat Loss—H'. Heat loss, H', is readily estimated from well-established heat transfer formulas. Note the total heat loss includes radiation losses which are not accounted for in this application. Convectional heat loss formulas for shapes are similar in form, and do not vary greatly in absolute values. The h_c coefficients summarized in Table 8-3 are used in the formula

$$H' = \frac{h_c A_s (\Delta t)}{60}$$

where h_c is the convection coefficient obtained from Table 8-3.

Table 8-3
Coefficients of Heat Loss for Use in Estimating Sensible Heat Losses by Natural Convection

Shape or disposition of hot surface	Natural convection heat loss coeff., h_c
Vertical plates, over 2 feet high	$0.3\Delta t^{0.25}$
Vertical plates, less than 2 feet high (X = height in feet)	$0.28(\Delta t/X)^{0.25}$
Horizontal plates, facing upwards	$0.38\Delta t^{0.25}$
Horizontal plates, facing downwards	$0.2\Delta t^{0.25}$
Single horizontal cylinders (X' = diameter in inches)	$0.42(\Delta t/X')^{0.25}$
Vertical cylinders, over 2 feet high (X' = diameter in inches)	$0.4(\Delta t/X')^{0.25}$

In the last formula, A_s is the surface area actually *emitting heat*, which is not to be confused with area of stream cross section, A_p; Δt is temperature difference between hot skin temperature and room air temperature; H', as before, is Btu per min. The prime exponent has been used throughout this chapter to remind the reader that it is the

minute rate, whereas heat losses are conventionally calculated on an hourly basis.

[Note: Figure 8-7 has been moved to the Appendix. It is a graphical representation of Table 8-3.]

Values for m. In the case of vertical planes, m is obviously the height of the plane itself. For horizontal cylinders it is the diameter of the cylinder. No data for horizontal planes is included in Table 8-2, and it is not immediately obvious what the dimension of m should be.

Comparison of the coefficients for large vertical and horizontal plates shows that heat transfer is greater by only about 25% for horizontal planes which, together with relation between V and m shown in equation (1), suggests that the dimension for m may be the same as though it were disposed vertically, i.e., the horizontal diameter. The uncertainty regarding the proper value for stream area, referred to above, is considerably greater, and indeed of more importance because the airflow rate, q, is proportional to the two-thirds power of area and only to the cube root of m. There is a need for experimental investigation on these points.

The preceding discussion is concerned with the magnitude of air flow, q_o, at the upper boundary of a hot body. The flowrate, q_z, at the hood level, after mixing and dilution has occurred, is discussed in a later section of this chapter.

Low Canopy Hoods

One category of canopy hood includes those that are mounted above horizontal hot surfaces with vertical distances not much greater than three feet. Some simplified relationships can be derived on the basis of the assumptions indicated in the following. Considering the rising air column to be about the area of the horizontal heated surface below, it can reasonably be inferred that little mixing with room air occurs during its short vertical travel and that the total airflow rate, q_z, at the level of the hood is only a little greater than the initial induction rate, q_o, hence an equation for q_o above horizontal heated surfaces would be a useful indication of the exhaust requirements for such low hoods.

Where no heat from steam is involved, the value of convectional heat from a horizontal plane is given by

$$H' = \frac{0.38}{60} A_s (\Delta t)^{5/4}$$

where the value of h_c has been inserted in the last equation from Table 8-3, A_s represents not only the horizontal hot surface from which the convection currents originate but also the face area of the canopy hood. When this expression is substituted in equation (5), noting that in this instance $A_s = A_p$, one obtains

$$\frac{q}{A_s} = 5.4 \sqrt[3]{m(\Delta t)^{5/4}} \qquad (6)$$

Steaming Tanks

A similar simplification can be produced for the case where the heat is provided by steam from a tank of hot water. Letting G = pounds steam evaporated per minute per square foot, and taking the latent heat to be 1,000 Btu per pound, we have $H' = 1,000 G A_s$. Substituting this value in equation (5), one obtains

$$\frac{q}{A_s} = 290 \sqrt[3]{Gm} \qquad (7)$$

It is emphasized again that these relationships require substantiation by experimental investigations. At this writing [1963], such confirmation of them is lacking. It is of interest to compare them with a rule-of-thumb basis that has long been used by practical exhaust system contractors, the old "blow-duct men." Their rule was that the exhaust rate from a canopy hood should be enough to induce a hood face velocity (the horizontal, plane area) of 200 fpm. Like most such rules, it often failed in actual practice, but it is reasonable to assume that its originators had some experience supporting the figure. The theoretical relationships indicate that it would frequently be of the correct order of magnitude.

Booths and Partial Enclosures Above Hot Surfaces. Booth-type hoods which enclose the space above hot surfaces, such as pots of molten metal are canopy receiving hoods identical in their characteristics with ordinary canopy hoods which lack sidewalls. This may not at first be obvious but is easily demonstrated. Consider a canopy hood without side shields, above a rectangular tank of molten metal, (Figure 8-8). Clearly, a vertical baffle could be inserted, extending from the hood face down to the hot surface, without modifying the airflow pattern in any way. The baffle coincides with a plane of airflow symmetry. The baffle described, however, has resulted in a series of booths over the tank. It follows, therefore, that specifying face velocity at a booth, or similar partial enclosure, over

a hot surface is an incorrect basis. Like any canopy hood, the ventilation rate should be specified in relation to the flowrate of thermally induced air flow.

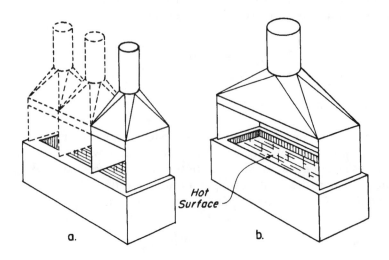

Figure 8-8. Three booth type hoods (left) operate like a single hood, which is like half of a canopy hood (right).

Small Area Heat Sources Under Large, Low Hoods

Previous discussion has dealt with heated horizontal surfaces with areas about the same as the face area, in plan, of the hood. Additional factors need consideration where the induced column of hot air is small in area relative to the hood, the area of the latter being determined by other considerations such as the following:

(a) The source of heated air is in motion, or is intermittently sifted, within certain boundaries. For example, in pouring lead or high-lead bronze by ladle into molds disposed in a long row, a long narrow canopy hood might be employed to receive the fume-laden air regardless of the location of the ladle source. Similarly, pouring may take place in molds moving on a conveyor, but the required pouring rate does not permit a fixed position for the ladle.

(b) The source may be fixed, but adjacent operations or contaminant sources of different character dictate the need for a hood larger than necessary for the hot air column alone. A hot plate in a laboratory hood illustrates this case. It has already been shown that a booth enclosing a hot body requires the same kind of analysis as canopy receiving hoods.

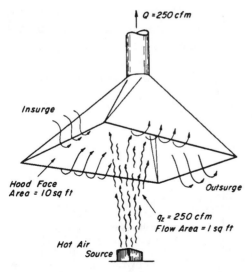

Figure 8-9. Airflow pattern in a large area hood receiving a hot air stream of small dimensions where the hood exhaust rate is just equal to the thermal induction rate. There is no *net* flow in other areas since $Q = q_z$. However, surges will occur and these carry contaminated air out of the hood space into the room.

The key to performance of large hoods receiving hot air of small cross section relative to the hood face area is illustrated by the following example (Figure 8-9). Suppose a column of contaminated hot air of one square foot area, flows at a rate of 250 cfm into a hood having a face area of ten square feet which is exhausted at a rate, also, of 250 cfm. There is one important factor in this situation operating to prevent complete control of the contaminated air stream. Since inflow to and outflow from the hood interior are equal, and inflow takes place entirely through one-tenth of the hood face area, there is no *net* air motion inward nor outward through the remaining nine-tenths of the hood face.

Actually, turbulence universally prevalent in any air space will cause an undulating spillage of interior air, hence there will be both inward and outward flow, *averaging zero*, through the entire remaining area of the hood face. The greater the area of the hood face, the greater the chance for extraneous air currents to create a kind of "tidal wave" motion in the air under the hood, and result in spillage near the edges. Since the air within the hood interior will always be contaminated air, the spillage will carry some contaminant with it.

In a hood having appreciable vertical distance between its bottom face and the high velocity throat, the events pictured in Figure 8-10 can occur. If the contaminant is totally contained in the column of flowrate q_z as it enters at the bottom face of the hood, and if the exhaust rate $Q = q_z$, then it will be seen that mixing continues inside the hood space if the walls of the hood do not prevent it; that the mixing occurs only with interior air, and results therefore only in a kind of internal recirculation. Suppose that by the time the column

reaches the throat of the hood its flowrate has become $2q_z$. Since the exhaust capacity is only q_z, half the column is rejected and passes down to mix with later sections of the column lower down. Thus the whole interior space of such a hood is filled with highly contaminated air. If it could be prevented from escaping from within, the condition would be satisfactory.

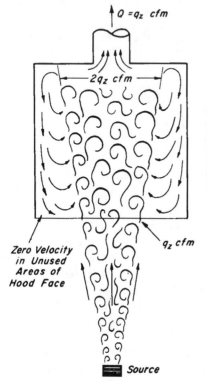

Figure 8-10. Canopy type hood, demonstrating internal circulation which results in filling of the hood space with contaminated air. The exhaust rate, Q, is just equal to the induction flowrate, q_z. Tidal spillage may occur through unused areas of the hood opening.

If there is appreciable space between the outer edges of the hood and the rising air column, and $Q = q_z$ exactly, then there will be no inward (i.e., upward) velocity into the hood, through those sections of the plane of its face not actively passing the entering column. Therefore, there is no force to overcome escape tendencies of the downward moving recirculation air.

It is clear from this analysis that dimensions of the hood face should not be greater than those of the column cross section, unless the exhaust rate, Q, is made greater than q_z, by an amount sufficient to create a brisk velocity through the inactive marginal areas. This means exhaust capacities equal to the hot air supply rate plus an amount sufficient to create controlling face velocities.

$$Q = q_z + VA_f$$

where Q is the required exhaust rate, q_z is the hot air induction rate, V is the desired face velocity and A_f is the area of the hood face or, to be more precise, the area not occupied by the entering hot air currents.

Values of V required are somewhat uncertain within limits as with controlling velocities in other situations—exterior hoods and cold process enclosures. Velocities of 100 to 150 fpm may be adequate in most cases where room air turbulence due to drafts is not excessive. Higher values would be indicated where very large quantities of a contaminant like lead fumes, having a low permissible concentration, are being generated.

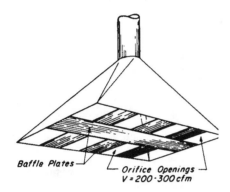

Figure 8-11. Canopy receiving hood for a hot column of relatively small dimensions, as in Figure 8-9. Baffle plates reducing the net open area facilitate economical use of higher velocities into the hood interior which prevent tidal surge and spillage of interior air.

The device illustrated in Figure 8-11 may occasionally be of value to eliminate tidal surges from the hood interior while minimizing the rate of exhaust required. It applies to the case where for mechanical reasons, the gross hood area must be large in relation to that of the rising air column. Baffles across the face of the hood are installed to the extent that velocities through the net area of openings are 200 to 300 fpm while the gross face velocity (cfm per square foot of gross face area) is 75 to 100 fpm.

High Canopy Hoods

The heated air stream originating from the surfaces of a hot body or from condensation of steam mixes turbulently with the surrounding air as it moves upward, becoming more and more diluted with increasing vertical distance.

The "Low Hood" classification is intended to cover those situations in which the relative dimensions of hot body, air column, and receiving hood minimize the importance of such mixing.

The "High Hood" group includes those where the vertical distance between source of heated air and the hood is great, relative

to the diameter of the air column at its base. Then the vertical flow
of contaminated air at the hood level is a function of both original
flow at the hot surface and the quantity of air with which it has
become mixed. For example, the contaminant may be entirely
contained in an air stream of 1,000 cfm at the hot body; after some
vertical travel it may have set in motion, and become mixed with, an
additional flow of 10,000 cfm of surrounding air, all of which must be
accommodated by the hood. (See Figure 8-12).

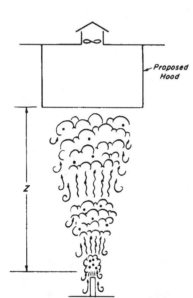

Figure 8-12. High canopy hood
above hot process wherein the
same quantity of contaminant is
carried in a constantly growing
stream of air. Concentrations are
reduced as air rises and grows in
volume flowrate. The hood must
be able to exhaust the entire plume
of air if the contaminant is to be
controlled.

Sutton (*J. Meteor.*, Vol. 7, No. 5, 307-312, Oct. 1950) has described
the mathematics of this problem having particularly in mind the
rate of velocity decay in a hot gas column rising from a chimney top in
the open atmosphere, where there is no wind to cause deflection from
the vertical. Constants for the equation are assigned from
experimental data available from eddy diffusion studies in
horizontal wind, i.e., the same kind of turbulence is assumed. His
expression is based on the conception of a point source of heat so that
an *effective distance* term must be derived in the case of a source
having finite dimensions by extrapolation backwards to a
hypothetical point source.

The expanding gases are assumed in this development to take the
form of a circular cone, the outer boundary of which is arbitrarily
defined as the locus of points where the concentration has decreased
to 1/10 of the concentration at the axis. The boundary is further
specified as the "instantaneous" one as distinguished from a more

distant "time" boundary which allows for oscillation of the cone to and fro in relation to an ideal axis. Sutton analyzed the experimental data of Schmidt on air motion above a heated coil and found agreement with theory.

Transformation of Sutton's equation into US units, and after condensation of terms, gives the following equation for average velocity, V_z, of the rising column in feet per minute.

$$V_z \approx \frac{37}{Z^{0.29}} \sqrt[3]{H'} \qquad (8)$$

where Z is the *effective height* from the hypothetical point source to the location of interest, in feet; and H' is the rate of sensible heat transfer to the air column, in Btu per minute (see Figure 8-13).

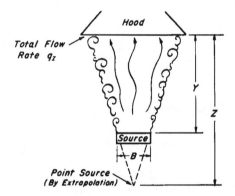

Figure 8-13. Dimensions pertaining to equations describing total air motion at varying distances above a heat source (Equations 10-11).

The diameter (instantaneous) of the column, in feet, is

$$D \approx \frac{Z^{0.88}}{2} \qquad (9)$$

A combination of these two equations by $q_z = VA$ yields an expression for airflow rate q_z at effective height, Z,

$$q_z \approx 7.4Z^{3/2} \cdot \sqrt[3]{H'} \qquad (10)$$

The extrapolated distance from the actual heat source to the hypothetical point source may be taken as, approximately,

$$Z \approx Y + 2B \qquad (11)$$

where Y is the distance from heat source to hood face, and B is source width.

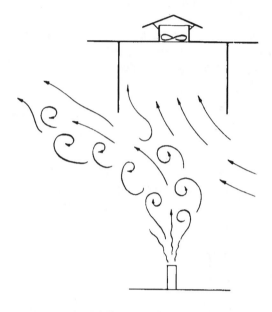

Figure 8-14. A common failing of high canopy hoods occurs when cross drafts divert the rising column of warm air from the vertical and the hood is missed.

Diversion from Vertical Flow by Cross Currents

Major hazards to successful performance of canopy hoods are horizontal currents of air which may tip the column away from the vertical and cause it to miss the hood (Figure 8-14). Large rising streams of air at high temperature with their greater vertical momentum are less easily diverted than smaller and cooler ones; also, obviously, the action is likely to be more serious, the greater the vertical distance of travel. If the diversion is likely to be only slight, larger hoods may take care of it, subject to the face velocity limitations discussed under "Low Hoods" previously.

The same hazard may be encountered in low hoods, but there is more frequent opportunity for use of partial side walls as baffles to intercept room air currents.

Example 1. A round furnace 6 ft in diameter discharges fumes into the atmosphere which rise rapidly in a column of hot air. It is proposed to erect a canopy hood for ventilation but it cannot be placed closer than 15 ft above the furnace. The furnace releases *sensible* heat to the air stream at a rate of 1,000 Btu per minute (radiant heat is not considered). Calculate the rate of exhaust required and dimensions of the hood.

Solution. The distance Z is estimated by

$$Z \approx Y + 2B = 15' + 2(6') \approx 27' \quad \text{and} \quad H' = 1{,}000 \, \text{btu/min (given)}$$

$q_z \approx 10{,}000$ cfm after inserting these values into Equation 10.

The diameter of the column is estimated by $D = Z^{0.88}/2$.

This gives $D \approx 9$ feet.

According to definitions written in the derivation, this pertains to the instantaneous position of the column, allowing for no "waver." The designer might select a hood diameter about 3 feet wider on each side, or $D_h = 15$ feet.

The exhaust rate must be sufficient to accommodate 10,000 cfm arriving, plus an additional amount sufficient to induce moderate inward velocities through the "dead" areas of the hood opening, in accordance with the expression

$$Q = q_z + VA_f$$

where A_f is the difference between a 9 ft plume and the 15 ft hood, the hood size selected. Let $V = 150$ fpm for the extra area. Then

$$Q = 10{,}000 + VA_f = 10{,}000 + (176 \, \text{sq ft} - 64 \, \text{sq ft})150 = 27{,}000 \, \text{cfm}$$

These figures, speaking for themselves, illustrate one reason why high canopy hoods are such a poor control method. In this example, about three times as much exhaust capacity is devoted to taking care of increased hood size due to the hazard of column deflection from cross drafts.

Double Canopy Hoods. The spillage of hot air at the rim of the canopy receiving hood as illustrated in Figure 8-3 suggested to somebody long ago the remedy of the so-called double hood in which the central inner space of the main hood is blocked off by an inner hood of slightly smaller dimensions. Thereby, the airflow is concentrated at the rim of the hood structure, with only a portion permitted to pass to the inner space through a small orifice in the apex of the inner hood. It was reasoned that since hot air escaped at the rim of ordinary hoods, this would be prevented by disposing the major part of the exhaust capacity at this point. This form of hood, and the reasoning behind it has had rather wide acceptance, and it is

therefore appropriate to examine it carefully at this point to separate fact from fancy.

It should be clear to the reader from the preceding exposition in this chapter that a canopy receiving hood can function properly only when the buoyant flow of hot, contaminated air at hood level, q_z, is matched by an exhaust capacity, Q, at least as great in magnitude. It immediately follows that these facts cannot be circumvented by any specially designed shape of hood. If, therefore, spillage occurs at the rim of a plain hood, this will not be prevented by inserting an inner piece to form a double hood.

It next becomes necessary to raise the question whether the double hood has advantages over a plain hood when both are designed to exhaust at proper rates, sufficient to accommodate the induced air flow, q_z. We have seen that a large hood is subject to the hazard of spillage from its interior by tidal surges, the remedy for which is in the exhaust of additional air, and, if possible, by reduction of the net open area by including baffles in the hood face.

The inner piece of a double hood could be said to function as a hood face baffle but this was clearly not intended. Moreover the baffle can function effectively, probably more so, by assuming the simpler shape of Figure 8-11. The areas of orifices in a double hood are designed for air velocities of 1,000 to 2,000 fpm; for resisting tidal spillage as discussed herein, such velocities need not be greater than 200 to 300 fpm. Clearly, then, the double hood cannot be justified on the basis of its baffling function.

The velocity contour characteristics of a double hood are readily derived by recognizing that, fundamentally, it consists in an endless slot, the Q-V-X relation for which Q per foot = 6.3 VX. From this we can calculate the dimensions of significant contours, say the distance downward from the rim of the 75 fpm contour as a definition of the distance below which there is no local exhaust effect. As commonly designed, double hoods are arranged to exhaust 200 to 400 cfm per linear foot of rim slot. On these bases we calculate the 75 fpm contour to extend only 5 to 10 inches below the rim. This is a small distance, indeed, in comparison with the usual 6- to 7-foot distance between floor level and hood.

We are forced to conclude from the foregoing reasoning that double canopy hoods not only do not function as usually pictured, but do not have any advantages over plain hoods.

Lateral Exhaust for Hot Processes

The horizontal deflection of heated air into an exterior hood involves an interaction of forces that is quite different from the case where heat is absent. It is now insufficient to specify a velocity at

some X-distance which will overcome the dispersing effect of room air currents. Rather, it is necessary to induce a stream of air toward the hood which will overcome the tendency of the hot contaminated air to rise buoyantly away from the source and the hood.

Deflection may occur due to the viscous drag of the exhaust air streaming toward the hood or it may result from turbulent mixing of the hood-induced stream with the hot air. In some cases one mechanism predominates over the other; sometimes a combination of the two operates. The relationships have not been quantitatively investigated and it is not, therefore, possible to offer equations for application to local exhaust problems. A qualitative consideration of the nature of significant forces makes it apparent that the required exhaust rate of an exterior hood must be a function of the X-distance, the area of the hot air column, and the temperature difference between hot air and room air.

Furthermore, Q must be greater than q_o, the initial air induction rate, which is to say that, as with a canopy hood, the contaminant is only exhausted by withdrawal of all the air with which it is mixed. The lateral exhaust hood must, in addition, induce the necessary deflecting force.

Example 2. A hot cylinder, 2 ft in diameter and 3 ft high, with surface temperature of 1650 F stands vertically on the floor. Toxic fumes are volatilized from its surface and are carried in the vertically moving hot air currents set in motion by the hot cylinder. In connection with plans for a canopy hood, estimate the rate of flow of convectional air currents above the cylinder. A_s = areas of side and top = 22 sq ft; diameter of cylinder = X' = 24"; area of the air column, A_p, may be taken to be about equal to the cylinder, 3.1 sq ft; the value, m, is about the height of the cylinder, or 3 feet.

Solution. Equation 5a applies.

$$q_o = 29 \sqrt[3]{H'mA_p^{\,2}}$$

The convectional heat loss, H', is obtained from the equation

$$H' = \frac{h_c A_s (\Delta t)}{60}$$

For a cylinder, h_c is determined as (from Table 8-3)

$$h_c = 0.4(\frac{\Delta t}{X'})^{1/4} = 0.4(\frac{1650-70}{24})^{1/4} = 1.14$$

therefore

$$H' = \frac{h_c A_s(\Delta t)}{60} = \frac{1.14 \times 22 \times 1580}{60} = 660 \text{ Btu per min}$$

and

$$q_o = 29 \sqrt[3]{H' m A_p^2} = 29 \sqrt[3]{660 \times 3 \times 3.1^2} = 770 \text{ cfm}$$

If the hood were to be suspended within a couple of feet or so above the top of the cylinder, we can assume no significant additional mixing with surrounding air and would thus specify an exhaust rate into the hood to exceed the delivery rate, 770 cfm, by a suitable margin, say 1,000 cfm.

If the hood face dimensions are made a little larger than the hot air column, for overhang, say a square hood 3 ft by 3 ft, then an additional exhaust is required as indicated by the formula

$$Q = q_0 + AV$$

If V = 150 fpm,

$$Q = 770 + (9 - 3.1)150 = 1700 \text{ cfm as a rounded number.}$$

Example 3. A bath of molten lead maintained at 1,000°F is to be provided with a *low canopy hood*. The pot is rectangular, 4 ft long by 3 ft wide. Estimate the minimum required rate of exhaust through the hood to remove lead fumes.

Equation (6) can be used. Note that m ≈ 3.5 feet, Δt = 930°F, $A_s = A_p$ = 12 sq feet.

$$\frac{q_o}{A_s} = 5.4 \sqrt[3]{m(\Delta t)^{5/4}} = 5.4 \sqrt[3]{3.5(930)^{5/4}} = 140 \text{ Btu/min} \bullet \text{sq ft}$$

$$q_0 = 140 \times 12 = 1700 \text{ cfm}$$

The total exhaust rate should be specified as some value in excess, say 2,500 cfm.

Example 4. A tank 6 ft by 3 ft contains vigorously boiling water in which steam evaporates at a rate of 10 lb per sq ft • hr. A low canopy hood is to be installed for steam removal. Estimate the minimum required rate of exhaust through the hood.

Solution. For a low hood, one may assume $q_o = q_z$ and use equation (7). Note that $m \approx 4.5$ feet, $G = 1/6$ lb/sq ft • min, $A_s = A_p = 18$ sq feet.

$$\frac{q_o}{A_s} = 290\sqrt[3]{Gm} = 290\sqrt[3]{(1/6)(4.5)} = 260$$

and

$$q_o = 260 \times 18 = 4,700 \text{ cfm.}$$

Enclosures for Hot Processes

Exhausted enclosures, in which the air is hotter than the outside air, may permit the escape of air, with contaminant, near the top in some circumstances. This phenomenon may sometimes be observed in drying ovens having small openings or cracks at the top, or which are open at the ends; in canopy hoods with access openings near the top; and even from the top portion of the doorway to a room in which large amounts of heat are generated and from which air is being exhausted. It may also be observed at the top of the face of an exhausted booth which contains a heated process.

It is of some importance in the design of ventilation for hoods and enclosures to understand the thermal forces responsible for such escape.

Liquid Analog

Consider the pressure-flow characteristics of the combination of vessels and liquids illustrated in Figure 8-15.

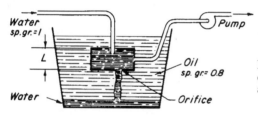

Figure 8-15. Liquid analog demonstrating leakage through an orifice.

Water is continuously supplied to the inner vessel, from which a pump withdraws it. An orifice at the bottom permits leakage of a continuous stream downward through the surrounding oil, the density of which is less than that of the water.

If there were no oil present the net head on the orifice would be L feet if water. The oil, however, exerts a buoyant effect, so that the net head is

$$h_w = L(s_w - s_k) \tag{12}$$

where h_w = net head, feet of water
        L = height of water above orifice, feet
        s_w, s_k = specific gravities of water and oil

The velocity through the orifice will, therefore, be according to the elementary hydraulic formula

$$v = C\sqrt{2gh}$$

where C is the orifice coefficient.

It will be noted that the rate of supply of water, and the rate at which it is pumped away, has no relation to the rate of leakage through the orifice at the bottom, except as it may affect the magnitude of L.

Heated Air

If the diagram for liquids was inverted, it could represent the conditions in an exhausted enclosure or canopy hood which is filled with hot air or gases from the process (Figure 8-16). The oil corresponds to the cooler air of the room being drawn into the hood by the exhaust fan and the water to the interior air which has absorbed process heat.

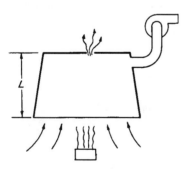

Figure 8-16. Leakage promoted by thermal head.

The pressure-flow relations for this situation are analogous to that of the liquid combination

$$h_{th} = L(1 - s_h) \tag{13}$$

where h_{th} is the thermostatic head, feet of air, s_h is the specific gravity of the interior hot air relative to that of the cooler room air, and L is the height in feet, of the mass of air in the enclosure.

But

$$s_h = s_c \times \frac{T_c}{T_h} = 1 \times \frac{T_c}{T_h}$$

where s_c = specific gravity of the cooler air of the room, and is equal to unity, since it is made the point of reference

T_c = absolute temperature of cooler air
T_h = absolute temperature of hot, interior air.

The original equation then becomes

$$h_{th} = L(1 - \frac{T_c}{T_h}) = L(\frac{T_h - T_c}{T_h}) = \frac{L \Delta T}{T_h} \tag{14}$$

The value of h_{th} so obtained is in feet of air and may therefore be used in the equation

$$v = C \sqrt{2gh}$$

to ascertain velocity through the orifice. Therefore

$$V = C \sqrt{2 \times 32.2 \times \frac{L \Delta t}{T_h}} \times \frac{60 \, min}{hr} = 480C \sqrt{\frac{L \Delta t}{T_h}} \tag{15}$$

Note that V = velocity in feet per minute.

Values of L, change in t, T_h, C

The value L cannot be greater than the vertical depth of the hood or enclosure although it might be less. Ordinarily it will be taken as equal to the height of the enclosure, or the distance from the bottom of the opening where room air enters, to the location of the orifice.

The magnitude of Δt is fixed by the rate of generation of heat, H', in Btu per minute, and the total rate of exhaust, Q, cfm. Generally, T_h = 530 + Δt. C is usually assumed to be 0.6, the coefficient of entry for a sharp edged orifice.

Such simplifications lead to an estimate for the velocity of air through an opening in a canopy hood in terms of the heat transfer H':

$$V \approx 20(\frac{LH'}{A_f})^{1/3} \tag{16}$$

for temperatures up to 200 degrees F where L = height of hot air column, feet; H' = sensible heat transfer to air stream, Btu per minute; A_f = area of sharp-edged openings, sq ft.

Preventing Leakage in Enclosing Hoods

The most important application of the relationships developed in the preceding section is in estimating the possibilities of spill in a hood or enclosure being designed, and in designing preventive measures.

It is clear, in the first place, that the enclosure should be so constructed that there is no possibility of cracks or other orifices in its upper section. This is especially important in design of ovens.

Where orifices cannot be avoided for operational reasons, leakage can be prevented by adequate exhaust as described below.

The force responsible for spill of air through the orifice is, as has been demonstrated, the thermostatic head. If the exhaust arrangement is such that a negative static pressure is created inside the enclosure that is equal to, or somewhat greater than the thermostatic head, then leakage will be prevented. The velocity of escape through orifices described in Equation (15) is therefore that which must be induced by an exhaust system to overcome the air escape tendencies. In other words, Equation (15) is the required face velocity for heated enclosures where leakage through openings at the upper part of the enclosure is to be prevented.

Example 5. A rectangular laboratory type hood is 6 ft high from bench top to the flat top of the enclosure. The face opening is 4 ft wide and 3 ft high from the bench top. It encloses a hot plate which burns natural gas at a rate of 0.1 cfm. Air in the top of the hood is 79° F, nine degrees above ambient. Air is exhausted from the back of the hood at a rate to induce a face velocity of 80 fpm. (See Figure 8-17.) Cracks at the top of the hood amount to 0.05 sq

ft. Calculate the approximate velocity and volume of escape of interior air through these cracks. Assume $\Delta SP = 0$.

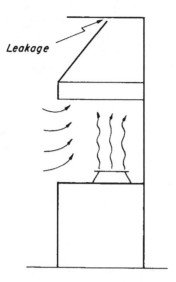

Leakage

Figure 8-17. Example.

Solution. Use Equation (15) which assumes the cracks behave as sharp edge orifices where C = 0.6. The height of hot air column, L, may be taken to be that from the top edge of the opening to the top of the enclosure (6' - 3' = 3 ft). The exhaust rate, Q = 100(4 x 3) = 1,200 cfm. The absolute temperature of interior air, T_h, is 530 + 9 = 539 degrees. Then

$$V \approx 480C \sqrt{\frac{L \Delta t}{T_h}} \approx 480 \times 0.6 \times \sqrt{\frac{3(9)}{539}} \approx 60 \text{ fpm}$$

Rate of air escape through cracks is, then, 0.05 • 60 = 3 cfm.

The estimation can also be made using Equation (16) which yields the same result within approximately 10%. While this is a very small air flow, it may be significant where the interior air is highly malodorous, or where radio-active contaminants are being handled.

Applications of the Theory Developed in This Chapter

The principles for high canopy hoods can be applied to a consideration of a ventilation arrangement for removal of dust through roof openings resulting from the shake out of a hot steel casting.

If the casting is sufficiently large and hot, the convection currents set up around it may carry the light dust particles aloft to roof level where exhaust fans can be employed to transfer the dusty air to the outside or to air cleaners. Since the dusty air stream becomes diluted by turbulent mixing with surrounding air during its upward ascent, enormous ventilation rates through the roof are required to accommodate the total flow of air. These facts are illustrated in the following example.

Example 6. A hot flat casting is considered to have an idealized shape and dimensions of a horizontal plane, 8 ft on a side. Calculate the magnitude of vertical airflow set in motion at roof level 25 ft above the casting. Assume the surface temperature to be 1,000°F. The surface area, A_s, is 64 ft^2.

Solution. Referring to Table 8-3, we may estimate H′ by using equation

$$H' = \frac{0.38}{60} A_s (\Delta t)^{5/4} = \frac{0.38}{60} 64(930)^{5/4} = 2,000 \text{ btu/min}$$

We may apply equation (10), to estimate the airflow at roof level. The value of Z will be Y + 2B = 25 + 2 (8), or about 40 ft. Using these values

$$q_z \approx 7.4Z^{3/2} \cdot \sqrt[3]{H'} \approx 7.4(40)^{3/2} \cdot \sqrt[3]{2,000} \approx 24,000 \text{ cfm}$$

From equation (8), the velocity at roof level, where Z = 40, can be estimated, using, as before, H′ = 2,000 Btu/min.

$$V_z = \frac{37}{Z^{0.29}} \sqrt[3]{H'} \cdot \frac{37}{40^{0.29}} \sqrt[3]{2,000} = 160 \text{ fpm}$$

Because of the assumptions on which the derivation of these formulas were based, we note that these figures represent the airflow at instantaneous locations. Thus, from the equation, $D = Z^{0.88}/2$, we estimate the diameter of the hot air column at roof level to be about 13 ft, but, in fact, a lateral wavering of the column would cause the effective diameter, or "time-diameter," to be some multiple of this value.

In any case these calculations provide a basis for illustrating the nature of these ventilation problems which can be summarized as follows. If ventilation in the roof were to be installed to withdraw hot dusty air streams from shakeout of castings of this size regardless

of location, the installed exhaust capacity would need to be in the vicinity of 160 fpm per square foot of roof area, or of floor area, devoted to this work. Such a rate would be prohibitive for large areas.

On the other hand a central or fixed shakeout location results in reduced floor area and might be practical if the arrangement includes curtain walls to effect partial isolation of the upper zone thus protecting it from cross drafts. Exhaust rates to meet the preceding requirements may easily amount to 400,000 to 500,000 cfm for the shakeout area of a steel foundry making large castings (e.g., 200,000 to 300,000 pounds or larger).

It should be noted that the preceding example dealt with a very large and hot casting. As one passes downward on the size scale, the buoyancy and volume of convection currents is reduced and before long a point is reached where direct removal through the roof would be quite impractical.

Lateral Exhaust (Hot)

Systematic studies on effective rate of lateral exhaust for hot processes are practically non-existent. In a study of exhaust requirements by exterior hood for control of welding fumes, Tebbens and Drinker observed the following data, relating necessary control velocity at the arc, to the power input:

Rod size, inches	Power input, watts	Control velocity, fpm
5/32	4,500	75
1/4	9,900	100
5/16	19,000	125

Noting that the rate of heat generation, H, is proportional to the power input, it is clear that the latter is a critical factor. While these data are only fragmentary, they suggest that the required control velocity at electric welding is given by the following:

$$V = 75 \sqrt[3]{\frac{watts}{4500}}$$

The theoretical air induction rate, q_o, has been shown to be proportional to the cube root of the rate of heat transfer to the air

stream and it is, therefore, interesting to note that in the above expression for required control velocity, the heat input (watts) also appears as the cube root.

In the absence of completely developed relationships for lateral exhaust of hot processes, one may approach a given problem somewhat speculatively as is done in the following example.

Example 7. Lateral exhaust is to be designed for a lead pot in which some refining operations take place. The pot is 10.5 ft inside diameter, having a circular area of 83 sq ft and the metal is normally at a temperature of 900F.

Solution. The natural convection rate of airflow above the molten surface may be estimated by $H' = h_c A_s \Delta t$ (see also Table 8-3), then $H' = 0.38(800)^{1.25} \times 83 / 60 = 2300$ Btu per min; $A_p^2 = 6800$; and m = 10.5; and using equation (5a):

$$q_o = 16,000 \text{ cfm}$$

Such a calculation indicates the order of magnitude required of a *canopy hood* exhaust, physically the most favorable. When one turns to consideration of a lateral exhaust arrangement, it seems entirely reasonable to suppose that the rate of exhaust must be at least equal to this value, if not a great deal more.

It is, however, apparent that the hot contaminated air originating in the center of the pot would be particularly resistant to the action of peripheral exhaust even at high rates and special mechanical arrangements would be needed to counter that situation.

Exhaust for Heated Enclosures

The principles elicited from the analysis of thermostatic pressures inside enclosures can be applied to an understanding of the behavior of industrial ovens and to a derivation of exhaust requirements from the interior.

Mathematical speculation is facilitated on new and untried exhaust arrangements as is illustrated in the following example.

Example 8. Consideration is being given to a scheme for exhausting the interior of a 20-ton electric furnace having an open space around the three electrodes of 1.8 sq ft. Assume cracks around the doorway add 2.2 for a total area of 4 sq ft. During melting down, 2200 kilowatts (which is equivalent to about 125,000 Btu per min) is supplied. The proportion of this heat that is given up to the air inside the furnace is not known. A major

proportion of this heat is devoted to heating the furnace and furnace contents. During subsequent refining, heat input is about 30,000 Btu per min. Assume 20,000 Btu per min available to heating of furnace atmosphere. Calculate thermal escape velocity of interior gases, i.e., required inward velocity to prevent escape of gases, taking L = 10 ft.

Solution. Assume interior air temperature is 800°F.

Then

$$V = 14 \, (LH'/A_f)^{0.33} = 14 \, (10 \times 20,000)/4)^{0.33} = 510 \text{ fpm}$$

This assumes a temperature of 800F. Checking this for V = 510 fpm, or

$$Q = 510 \times 4 = 2,040 \text{ cfm}$$

$$t = 58 \, (H'/Q) = 58 \times 20,000 \, / \, 2,000 = 580°F$$

which would indicate a coefficient of 16 in the formula above instead of 14, an insignificant refinement in the calculation. It is therefore, concluded that an exhaust rate of 2,000 to 2,500 cfm would prevent the escape of fume-laden air from this furnace interior.

Chapter 9

Characteristics of Free Air Jets, Push-Pull Exhaust Systems, and Air Curtains

Terms and Units for Chapter 9

A_o = cross-sectional area of discharge nozzle, sq ft

A_z = cross-sectional area of total jet stream at X, sq ft

D = diameter of a circular nozzle, ft

F = force of total jet stream, lb

L_s = Length of (push) slot or device

M/t = mass flowrate of jet stream, lbs per second

N = distance, no. of slot diameters or slot widths ($N = X/D$ or X/W)

q_o = airflow discharged from jet nozzle, cfm

q_z = total airflow in a jet stream at distance X, cfm

Q_{exh} = total air exhausted from exhaust (pull) hood

V_{max} = maximum (center line) air velocity of a jet at some specified section, fpm

V_o = average air velocity at discharge nozzle, fpm

V_z, V_x = average air velocity across entire section of a jet at a specified section, fpm

W = slot width, ft

X = distance measured along axis of an air jet, from nozzle or slot, ft

ρ = air density, lb per cu ft

θ = one half the included angle of a jet stream

Free air jets are those streams of air in free space that issue from a supply source such as the end of a duct, slot, or hole in a vessel or duct under air pressure. A knowledge of the characteristics of such air

streams with respect to velocity-distance relationships, change in volume flowrate, etc., is of value in several types of problems.

Expansion Mechanism

A jet of air mixes at its periphery with the surrounding air of the space so that with increasing distance from its source the rate of airflow constantly increases and the velocity decreases. The mixing action is one of simple turbulence rather than an effect of negative static pressure. Air from the stationary surrounding mass moves gently inward to the periphery of the expanding stream for its full length to replace that kicked into motion by mixing. Continuance of the motion is due solely to the momentum of the primary air. Velocities at the axis are maximum as in the flow of water in ducts.

Jet Shape

Nozzles of circular or square cross section result in a stream that is circular and conical, and the angle of expansion is known from the theory of momentum conservation and confirmed by experimental observation.

Air streams from nozzles that are rectangular, as in the case of slots, do not retain that shape. If the length-width ratio is not excessive, relative to the travel distance, the jet becomes circular by more rapid mixing and divergence at its longer boundaries.

There are three characteristic stages measured along the axis of the jet: a short preliminary stage during which the core of primary air is dissolved by mixing with room air, a second stage exhibited only by streams from slot-shaped nozzles during which the stream becomes circular in cross-section, and a third stage common to all streams—whatever the shape of the nozzle—in which mixing and expansion occurs as a cone-shaped stream. These are illustrated in Figs. 9-1 and 9-2.

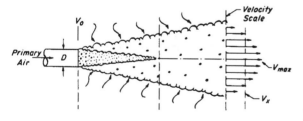

Figure 9-1. A jet of air issuing from a round or square nozzle displays first and third mixing phases, the boundary defined by the point of disappearance of the core of primary air. An intermediate or second phase occurs when the jet from a slot assumes a round shape, as in Figure 9-2.

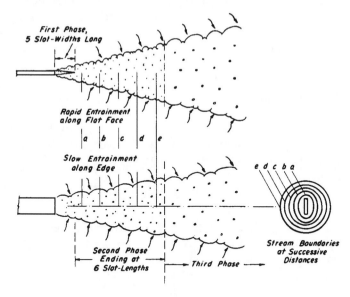

Figure 9-2. A second mixing phase occurs where the jet is slot-shaped in which the air stream gradually assumes a circular cross-section.

Force or Momentum of Jet Streams

The force exerted by an air jet is the momentum per unit time and is described by

$$F = \frac{Mv}{t}$$

where F is the total force of the stream in pounds; M is the mass in lbs; v is the velocity in feet per second; and t the time in seconds.

The mass per unit time, M/t, is equivalent to $VA\rho/g$, where A is the cross-sectional area of the stream, in square feet, ρ is the density of air in pounds per cubic foot, and g, the gravitational constant. The conversion of v fps to V fpm, and using density = 0.075 lb per cu ft gives

$$F = (\frac{V}{60})^2 A \frac{\rho}{g}$$

or, consolidating

$$F = 0.65 \times 10^{-6} V^2 A \quad \text{at STP.} \tag{1}$$

It has been established from theoretical considerations and confirmed experimentally that the momentum of the total jet stream is the same at all sections at reasonable distances from the nozzle. (This refers to the total stream area, so the force per unit area is not constant.)

Air Entrainment Ratio

The total rate of airflow in the stream in relation to the primary airflow issuing from the nozzle can be deduced from the principle of the *conservation of momentum* cited above, that is,

$$V_o{}^2 A_o = V_x{}^2 A_x$$

where the zero subscripts refer to conditions at the nozzle and subscript x to any point measured along the axis. Since air flow, q, is equal to the product AV one may write

$$V_o q_o = V_x q_x$$

and rearranging

$$q_x/q_o = V_o/V_x \qquad (2)$$

Third Phase Expansion

The third phase of jet expansion illustrated in Figure 9-1 and 9-2 occurs throughout almost the entire length of a circular or square nozzle excepting only an insignificant first stage about 5 diameters in length. In the case of slots this phase occurs only *after* the jet has become circular in cross section which requires a distance of about 6 slot lengths.

From the studies of McElroy, and Tuve and Priester (McElroy, G.E., "Airflow at Discharge of Fan-Duct Lines in Mines," *USBOM Rept. Invest.*, 3730, Nov., 1943; Tuve, G.L. And G.B. Priester, "Control of Air Streams in Large Spaces," *ASHVE Journal*, Jan., 1944), the average stream velocity, V_x, can be derived as follows:

$$V_{max} \approx \frac{KV_o\sqrt{A_o}}{X} \quad \text{and} \quad \frac{V_{max}}{V_x} \approx 2.7 \quad \text{(from above papers)}$$

Therefore

$$V_x \approx \frac{KV_o\sqrt{A_o}}{2.7X}$$

Substituting this value in Equation (2) leads to an expression for entrainment ratio:

$$\frac{q_x}{q_0} \approx \frac{2.7X}{K\sqrt{A_o}}$$

The value of K varies between 4 and 7. The practical range of jet application herein embraces V_x values from 100 fpm and greater for the most part and for this range K varies between 5 and 7. An average value of $K = 6$ can be used practically and the equation above becomes

$$\frac{q_x}{q_0} \approx \frac{0.45X}{\sqrt{A_o}} \tag{3}$$

Considering $\sqrt{A_o}$ to be the nozzle diameter (which is actually so for square nozzles and approximately so for circular nozzles), the entrainment ratio is proportional directly to X/D, the number of nozzles diameters, N.

$$\frac{q_x}{q_0} \approx \frac{N}{2} \qquad \text{(approximately)}$$

Two-Sided Second Phase Expansion for Slots

The second phase expansion is illustrated in Figure 9-2 and exists up to about 6 slot lengths. It is the phase of greatest interest in blow and exhaust ventilation applications discussed later. The first phase extends for a distance of only about 5 slot widths and is therefore of little practical significance. McElroy suggests for this expansion phase, the following:

$$V_{max} \approx 2.45\, V_o \left(\frac{W}{X}\right)^{0.5} \quad \text{and} \quad \frac{V_{max}}{V_x} \approx 2.5$$

from which

$$\frac{V_x}{V_o} = \left(\frac{W}{X}\right)^{0.5} = \left(\frac{1}{N}\right)^{0.5}$$

Substitution in Equation (2) gives

$$\frac{q_x}{q_0} \approx \sqrt{\frac{X}{W}} \approx \sqrt{N} \tag{4}$$

where W is slot width in feet, N is the number of sloth widths, and other terms as before.

One-Sided Second Phase Expansion for Slots

The velocity relationships in a jet issuing from a slot-shaped nozzle expanding against a parallel plane flanking its path are based again on McElroy's data. As in two-sided expansion they are expected to apply up to a distance of about 6 slot lengths.

$$V_{max} \approx 2.26 \left(\frac{W}{X}\right)^{0.36} \quad \text{and} \quad \frac{V_{max}}{V_x} \approx 1.88$$

which gives

$$\frac{V_x}{V_o} \approx 1.2 \left(\frac{W}{X}\right)^{0.36}$$

Substituting in Equation (2) gives an expression for entrainment ratio:

$$\frac{q_x}{q_0} \approx 0.83 \left(\frac{X}{W}\right)^{0.36} \approx 0.83N^{0.36} \tag{5}$$

Entrainment equations are provided in Table 9-1 and Figure 9-3.

Table 9-1.
Velocity Characteristics of Free Air Jets

Type of nozzle	Distance of Applicability	Ratio, q_z/q_o
Circular or square	>5 diameters from nozzle	$\dfrac{N}{2} \approx \dfrac{0.45X}{\sqrt{A_o}}$
Slot, expanding on both sides	Less than 6 slot lengths	$\sqrt{\dfrac{X}{W}} \approx \sqrt{N}$
Slot, expanding on one side only	Less than 6 slot lengths	$0.83\left(\dfrac{X}{W}\right)^{0.36} \approx \dfrac{5}{6}N^{0.36}$

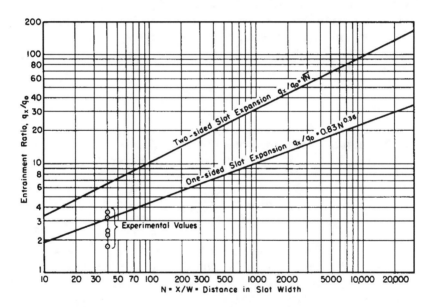

Figure 9-3. Air entrainment in jets from slot-shaped nozzles (applying up to six slot lengths).

Angle of Jet Boundaries

The diverging lines that bound a section of a jet stream form characteristic angles in various circumstances. That information is needed in later applications in order to determine the over-all cross-sectional dimensions of the jet stream structure and the size of the receiving hood. Forthman (reported by McElroy) found expansion at an angle of 18 degrees. Forthman, as again cited by McElroy, on expansion from a slot flanked by a plane observed an angle of nine

degrees. [Unpublished observations of push-pull systems by the editor suggests angles of 10-20 degrees for 2-sided jets from slots and 5-10 degrees on one-sided expansions within six slot lengths. Note also (as explained by Hemeon in a later section) that the jet tends to hug the physical surface it parallels as it also expands in size and volume.]

For applications where two-sided expansion occurs, and up to 5 or 6 slot lengths, it would seem advisable to assume a design angle of 15 degrees to be on the safe side.

Jet Combined with Hood

The contrast between velocity characteristics of jets and those of exhaust hoods is marked. The air circulating fan provides a common demonstration of the long distance effect of air being blown. The rate of airflow from a jet, q_o cfm, to create an average velocity, V_x at a distance X is

$$q_o \approx \frac{V_x X \sqrt{A_o}}{2}$$

The ratio of this expression to that for a plain exhaust hood is a quantitative comparison of their velocity characteristics

$$\text{Ratio} \frac{Q(\text{exhaust})}{q_o(\text{blowing})} \approx \frac{20X}{\sqrt{A_o}}$$

Thus, for a 3-inch circular duct nozzle, the ratio is 90X, which means that it would require 90 times as great a rate of exhaust to create a given velocity, at a distance of 1 foot, as it would by the jet; 180 times at a distance of 2 feet, etc.

These comparisons lead naturally to the idea of combining a blowing system with an exhausting system, wherein a jet of air is caused to blow across a zone of contamination, forcing the contaminant to a point where an exhaust hood can "reach" it.

With the background of the preceding section on jet characteristics and those of exhaust hoods, it is possible to examine systematically the various pertinent design factors.

Factors that may be considered are as follows (see Figure 9-4):

(1) The increase in airflow rate of the jet as a result of entrainment (by the primary air from the nozzle) of secondary air from the surrounding space requires that the exhaust rate at least be equal to the total of primary and secondary air.

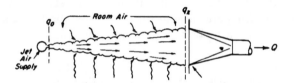

Figure 9-4. Some design considerations in a blow-and-exhaust system. Capacity of exhaust hood Q must exceed air in motion, q_x. The hood structure must be large enough to intercept the jet air stream completely.

(2) The jet must be *completely engulfed* at its point of arrival at the hood, either by the face of the hood or by a flange, because residual velocities will normally be greater than the intercepting capacity of hood velocity contours (due only to excess exhaust capacity, if provided).

(3) The jet stream will be composed largely of contaminated air, in the proportion that secondary air bears to primary air. Therefore any objects passing through the jet result in splash and dissemination of contaminated air into the surrounding space.

(4) If dusty objects pass through the jet, splash of dust from the surface of the object during passage will occur, perhaps destroying any advantage otherwise accruing.

(5) The shape of a slot-shaped nozzle is not retained by the jet. It becomes circular after about 6 slot lengths, because of more rapid divergence in a central plane at right angles to the slot.

6) Not least important, the advantages of a jet-assisted system over a simple exhaust system may sometimes be economically unsound when the added cost of blowing equipment is included in the comparison.

Blow and Exhaust Above Open Tanks

Application of this system to open top tanks has been the most common, where it has been referred to as the push-pull system. The several design factors are considered in the following paragraphs:

(a) A basis is needed for design of the blowing jet, i.e., the flowrate, q_o, and also the nozzle velocity, V_o. It will be noted that these two terms define both nozzle area, A_o, and the force or momentum of the jet, $F = 0.65 \times 10^{-6} V^2 A$.

(b) The average velocity of the jet at its destination, V_x, independent of any induced by the exhaust hood, may be significant.

(c) A basis for estimating the magnitude of total jet air in motion, q_x, at jet destination is needed, for this determines the rate of exhaust, Q, required.

(d) The rate of increase of jet dimensions determines its cross section at its terminus and this in turn indicates the face dimensions required of the receiving hood and any flanking baffles.

Jet Force in Design. The magnitude of the jet force, F, I believe, determines the degree to which deflection of that stream may occur or be resisted. The force tending to deflect it are those possessed by stray air currents, and, in the case of jets applied above tanks of hot liquid, the buoyant forces due to the increase in jet stream temperature as it gains heat in passage across the tank surface. It seems quite likely, therefore, that a specification of the minimum value of F to be employed in the design of a blow-and-exhaust system is an important element. Unfortunately there is no theoretical development to guide one to suitable values. [Hemeon goes on to justify a minimum F = 0.25 lbs, certainly a reasonable number for engineering applications in industrial ventilation.]

Inspection of the formulas relating jet force, F, to nozzle air flow, q_o, from Equation (1),

$$q_o = 1.54 \times 10^6 \frac{F}{V_o} \tag{6}$$

indicates clearly that there are an infinite number of combinations of q_o and V_o with any fixed F value. A high nozzle velocity requires a low primary airflow and visa versa. The choice is narrowed by the consideration of jet velocities developed in the following paragraphs.

Ultimate Jet Velocity, V_z. Although the total force of the jet stream is theoretically constant at all distances, the force per unit area is constantly decreasing as the total area increases. This intensity, F/A_z, is proportional to V_x^2, and must be—in my opinion— a significant design factor. As the jet enlarges with distance, and as velocity decreases, it must eventually become so low relative to velocities of room air it is directionally unreliable. The average jet velocity might be described as reflecting the cohesiveness of the tail end of the stream, the directional energy reserve remaining in the stream at that point.

Reasoning by analogy to the lowest control velocities useful in local exhaust hood design, it seems doubtful that V_x values lower than 100-200 fpm should ever be employed. [250 fpm is suggested.]

Primary Air Flow, q_o. If minimum values of F and V_x are specified, they can be combined in Equation (6) as follows. Pending

more certain information, an arbitrary selection for minimum values of $F = 0.25$ and $V_x = 250$ is suggested. Then using the V_z/V_0 relationships for one-sided and two-sided expansion, respectively, we get

$$\text{Minimum } q_0 \approx \frac{2000}{N^{0.36}}$$

based on the one-sided expansion velocity relationships, and

$$\text{Minimum } q_0 \approx \frac{1540}{\sqrt{N}} \text{ or about } \approx 1500\sqrt{\frac{W}{X}} \qquad (7)$$

based on the two-sided relationships. The latter expression is more convenient and because the approximate nature of the equation is as valid as the former one. Either serves simply as a convenient way of summarizing the particular assumptions as to F and V_x in the expression for q_0.

Air Entrainment, q_x. A consideration of the relationship of a slot-shaped jet to the tank and liquid barrier below might persuade one that the rate of air entrainment is well described by one-sided expansion, Equation (5). If a rectangular tank were filled *to the brim* with liquid, a slot-nozzle disposed on one long edge would certainly represent the conditions on which Equation (5) is based. The industrial tank has its liquid surface several inches below the rim and in such a case one-sided expansion might still prevail. There would certainly be a circulation of air in the space between the jet and the liquid surface and, one would expect, a lessened opportunity for a net gain of air into that space. (See Figure 9-5).

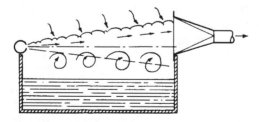

Figure 9-5. Blow and exhaust at the top of a liquid tank illustrating ideal conception of a "one-sided" jet expansion. The space between jet and liquid surface is highly turbulent but normally sealed off, See also Figure 9-6.

Ege and Silverman (Ege, J.F. and Silverman, L.S., "Design of Push-Pull Exhaust Systems," *Heating & Ventilating*, Oct. 1950, p. 73) have described modeling studies in which the negative pressure

developed in the space below the jet caused a marked downward deflection of the jet toward and along the liquid surface (Figure 9-6).

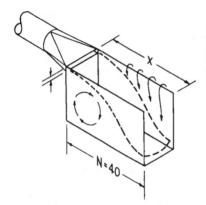

Figure 9-6. Downward deflection of jet stream in a model of an endless slot above a tank. Dotted lines represent the axis of the expanding jet.
Evidently the space below, containing the vortex, is under negative pressure. (After Ege and Silverman.)

They also found increasing entrainment ratios as the space was deepened. This suggests that the development of the negative pressure between the jet and the surface tended to cause "leakage" of outside air into that space, air which would feed the bottom side of the jet thus tending somewhat toward two-sided expansion. If this were true one would expect the higher entrainment values to occur at a point intermediate between the curves for one-sided and two-sided expansion. Their highest and lowest values plotted on Fig 9-3 at N = 40 do suggest this interpretation.

Jet Expansion Angle. The final element of jet performance to be considered is the size of the jet at the location of the receiving hood for this determines the size of the hood with baffles. As has been observed previously, the jet stream must be completely intercepted by the structure of the receiving hood, which may include flanges around the hood opening. In a previous section it was suggested than an angle of 15° should be sufficient for a two-sided expansion. From trigonometry, using this angle, the height of the hood structure, Y, above the plane of the nozzle could therefore be, in round numbers,

$$Y = X/4 \qquad\qquad (8)$$

[Editor's note: This analysis by Hemeon has not been confirmed experimentally and can lead to inconsistencies with his other formulas. Equation (8) should not be used for design without reference to the more recent literature or to in-house engineering studies.]

The following examples illustrate the use of the approximations developed in Chapter 9.

Example 1. A push-pull hood was applied to a tank 40 inches wide (3.33'). The blowing arrangement consists of a small duct along one top edge with $1/16$ -inch holes drilled on $1^1/2$ -inch centers discharging air horizontally over the tank surface at a rate of 1.3 cfm per linear foot toward the opposite edge where a long exhaust hood is located. Evaluate the hood using Hemeon's approach.

Solution. The area of holes is 1.7×10^{-4} sq ft per foot of tank length, and the initial velocity, V_o, is therefore 7,650 fpm. The number of slot widths across the tank is

$$N = X/W = 3.33' / (1.7 \times 10^{-4}) \approx 20,000$$

Note that in this case W is the width of a continuous slot equivalent in area to that of the holes. Assuming one-sided expansion, read from Figure 9-3 that

$$q_x/q_o = 30$$

Since $q_o = 1.3$ cfm per foot,

$$q_x = 1.3 \times 30 = 40 \text{ cfm per foot}$$

The average velocity, V_x, at the opposite edge of the tank is estimated by

$$\frac{V_x}{V_o} \approx 1.2 \left(\frac{W}{X}\right)^{0.36}$$

Note that this ratio is the reciprocal of q_x/q_o given in Figure 9-3 already established at about 30. Therefore,

$$V_x \approx 7,650/30 \approx 250 \text{ fpm}$$

The force of the jet

$$F \approx 0.65 \times 10^{-6} A V^2$$

Since this value is constant at all distances, it is convenient to calculate at the point of jet origin where (from $q_o = 1.3$ cfm and $V_o = 7,650$ fpm)

$$F \approx 0.65 \times 10^{-6} A_o V_o{}^2 \approx 0.65 \times 10^{-6} q_o V_o \approx 0.0065 \text{ lbs}$$

Example 2. It was desired to apply a push-pull exhaust system to control a mist of soluble oil and steam arising from an aluminum blooming mill operation producing flat sheets. Steam from water-cooling of stock and oil mist emits at the rolls which are partially enclosed at sides and top by the rolling mill housing.

A push-pull system was installed with a slot-type jet running $L_s = 8.5$ ft arranged to discharge a jet of air across the top opening of the mill, a horizontal distance of 11 ft. A receiving hood on the opposite side, also flanking the mill housing has a face opening 8-1/2 ft wide by 3 ft high. Side baffles are arranged along the throw distance to channel the air jet. The total jet air supplied for 8.5 linear feet was 400 cfm at a slot velocity of 3,000 fpm, and total exhaust rate through the receiving hood was 15,000 cfm. Evaluate the system using Hemeon's approach.

Solution. In terms of the jet factors previously discussed

$Q_0 = 400/8.5 = 47$ cfm/ft $X = 11$ feet

$A_0/ft = Q_0$ per foot $/ V_0 = W = 47 / 3,000 = 0.0157$ feet

$N = X/W \approx 700$

Unlike jets above tanks the jet expansion here is two-sided. Hence referring to Figure 9-3, opposite an N value of 700, read

Entrainment Ratio $q_x/q_0 = 27$

and since $q_0 = 47$ cfm/ft for a length of 8.5 ft, the total jet flow

$q_x \approx 27 \times 47/ft \times L_s \approx 1270$ cfm/ft $\times 8.5$ feet $\approx 11,000$ cfm

(which is close to the actual exhaust capacity $Q_{exh} = 15,000$ cfm).

Example 3. A push-pull hood will control hot oil mist and smoke rising from a machine. The push-pull system consists of a horizontal air curtain Ls = 21 ft long at an elevation of 13 ft above floor level, blowing over the machine, a distance of X = 14 ft to a lateral exhaust receiving hood. Side baffles are used to contain the jet to a horizontal width of 21 feet. The jet entrainment design data are:

2-sided expansion	$q_0 = 33$ cfm/ft	$V_0 = 6,000$ fpm
X distance = 14 ft	Ls = 21	

Determine the airflow rate needed at the receiving hood using Hemeon's methods.

Solution. This is a two-sided expansion.

$A_o/ft = Q_o$ per foot $/ V_o = W = 33 / 6,000 = 0.0055$ ft

$N = X/W \approx 14/0.0055 \approx 2,550$

Referring to Figure 9-3, opposite an N value of 2,550, read

Entrainment Ratio $q_x/q_o \approx 50$

and since $q_o = 33$ cfm/ft for a length of 21 ft, the total jet flow

$q_x \approx 50$ x 33 cfm/ft x $L_s \approx 1,650$ cfm/ft x 14 feet $\approx 23,000$ cfm

The installed exhaust capacity Q_{exh} must be something greater, say 30,000 cfm.

The jet velocity, V_x, at the hood face eleven feet along the jet is

$V_x \approx V_o (1/N)^{0.5} \approx 6,000(1/50)^{0.5} \approx 850$ fpm

(which is higher than the recommended minimum $V_o > 250$ fpm).

Air Curtains

The notion that a stream of air flowing as a sheet can act like a solid barrier is clearly false from a consideration of jet stream characteristics previously described. The so-called air curtain can, however, serve a useful function in certain circumstances. Its nature, limitations and possible applications are as follows.

Since a jet stream entrains surrounding air, if the latter is contaminated then the jet stream itself will consist of partly contaminated air, the extent of which can be calculated by reference to previously established jet relationships. Clearly, therefore, a jet stream cannot be used to separate a zone of contaminated air from clean air, as one would employ a solid barrier. This is illustrated by the analysis of such a scheme in the following example.

Example 4. A vertical curtain of air issuing from a slot in the floor, 4 ft long and 1/2 inch wide, is arranged as a barrier in a 4-ft doorway between two rooms in one of which dust is produced

making an average concentration in that room of 50 million particles per cubic foot of air. The velocity of air leaving the slot is V_o = 6,000 fpm. An exhaust opening is arranged overhead at a height of 7 ft (Figure 9-7).

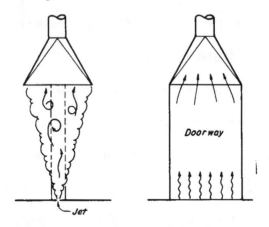

Figure 9-7. Illustration of Example 4.

Evaluate the situation using Hemeon's approaches.

Solution.   N = 84"/0.5" = 168   From Figure 9-3 q_x/q_o = 13

Since q_o = 1,000 cfm, there will be in excess of 10,000 cfm of total air in motion at the level of the hood.

Dusty air will be entrained from one side of the jet. The total volume of entrained air is 10,000$^+$ cfm less 1,000 from the push jet ≈ 9,000$^+$ cfm, one-half from each side. There will, therefore, be about 4,500 cfm of dusty air, mixed with 10,000$^+$ cfm total air. If any of this air is accidentally spilled into the other room, contamination will result.

[Editor's note: Several other push-pull approaches have been developed and are found in recent references such as ACGIH's *Ventilation Manual*, 23rd Edition. Mostly empirical in nature, they give reasonably similar results to Hemeon's methods.]

Chapter 10

General Exhaust Ventilation

The term *general exhaust ventilation* (GEV) has acquired a specific meaning in comparison to *local exhaust ventilation* (LEV), and refers to the ventilation process in which the general space is constantly flushed with fresh, outdoor air. *Natural ventilation* is effected by natural air movement—wind and temperature differences —through wall and roof openings. *Mechanical ventilation* is induced by motor driven fans.

The basis for the engineering design of general ventilation is related to *dilution* and, if the requisite data are available, simple calculations can be made. If a stream of water receives a trickle of salt at a rate of one gram per minute and it is desired to establish a brine solution concentration of one gram per thousand gallons of water, then water flow of 1,000 gallons per minute is necessary.

Similarly, if the space to be ventilated is continuously contaminated by a substance at a rate of one milligram per minute and if it is known that concentrations in air lower than one milligram per cubic meter (35.3 cu ft) are acceptable, one can calculate that the space needs to be ventilated at a rate of one cubic meter per minute. The formula is $Q = q/C_a$ where q is the emission rate, C_a is the acceptable concentration, and Q is the dilution rate, same units as q.

This approach provides a basis for judging whether general ventilation is a practical method in a given circumstance, even despite the fact that necessary input data may be inexact.

Analysis of Ventilation Needs

Attainment of a good understanding of the character of a given ventilation problem is required before a proper basis for ventilation can be laid. It is necessary to know exactly why the ventilation is required, that is, (1) what is the contaminant, (2) what is its objectionable property, (3) to what concentration or intensity should it be reduced, (4) at what points in the space does it originate, and (5) by what mechanism does it contaminate the atmosphere?

Nature of Contaminants. The classes of contaminants which may be effected by dilution ventilation are discussed in Chapter 1. They are (a) molecular dispersions of solvent vapors and chemical gases, (b) air dispersions of particulate matter, i.e., dusts, fumes, and smoke, (c) sensible heat, (d) radiant heat, and (e) water vapor.

In many instances it is necessary to make searching inquiry into the exact nature of the condition for which ventilation is sought, especially where thermal factors are involved and where the information sources are untrained individuals. The factors in problems of thermal comfort are air temperature, humidity, air velocity, and radiant heat intensity. Any of these may also be accompanied by some objectionable odorous physical contaminant.

These observations are illustrated by a consideration of the term *stuffy*, commonly applied to a poorly ventilated space, and of the term *drafty*.

Stuffy may, in practice, be interpreted to mean any of the following conditions, or combinations of them:

(1) The air temperature may be above the level of comfort, with low air velocities (e.g., T > 76° F; V < 25 fpm).

(2) Radiant heat, perhaps of low intensity, impinges on parts of the body, especially the face, together with a low degree of air motion.

(3) An uncomfortably elevated relative humidity may prevail (e.g., greater than 60 to 70% RH).

(4) A concentration of some airborne substance (e.g., body odor, tobacco smoke, chemical gas or vapor) causes unpleasant or irritating odors to pervade the atmosphere.

The first three of these may be alleviated by an increase in air velocity by circulating fans or by an increase in ventilation air, provided the intensity of the thermal factor is not too great. Probably because of this fact, an increase in air motion—air velocities—is often confused in the lay mind with improved ventilation, whereas it is clear that a disagreeable atmosphere caused by air contaminants is not usually improved by a mere increase in air velocities.

The remedy for a stuffy atmosphere depends in fact on which of the elements or combination of them is responsible for the condition. It may simply require an increased ventilation rate, or shields to intercept radiant heat, or local exhaust for removal of hot air or steam at the points of origin, or lowering the temperature, or combinations of these with, perhaps, circulating fans.

A drafty space may mean excessive air velocities in combination with low air temperatures; or simply low air temperatures below the zone of comfort; or it could be a situation in which the body heat is dissipated by excessive radiation to a nearby cold surface, e.g., a window pane in cold weather. As in the problem of analyzing a stuffy

condition, so with drafts, it is necessary to understand the exact causes by going beyond mere acceptance of the term itself before sound corrective measures can be designed.

Origin of Contaminant. An evaluation of the *sources* of the contaminant and the mechanism of emission or contamination are essential elements in the study of many ventilation problems. The gathering of such information will often lead to a more cost-effective solution to the problem than general ventilation alone—local exhaust, enclosure of a process element to suppress escape of the contaminant, shielding of a source of radiant heat, or even a slight modification of an element of equipment, for example. Such information will also lead to a quantitative or semi-quantitative estimate of rates of evaporation or pulvation and thus to a rough calculation, at least, of the general ventilation rates that would be required for dilution.

The generalities of the preceding paragraphs are illustrated in the following experiences of the author.

Example 1. A complaint occurred in a small production machine shop where "additional ventilation" was desired because "the air is cold." The space was already provided with exhaust fans and one supply fan arranged to deliver fresh air from outdoors through a short duct. Also, pedestal circulating fans were provided for creating cooling air currents where desired. There seemed no reason at all for the judgment that additional ventilation was needed.

Investigation disclosed that cutting oil mists frequently were generated from heavy-cut lathe work, and that nearby workmen attempted to alleviate the unpleasant characteristics by devising directional air currents to their own supposed advantage with the pedestal fans. The effect of these fans was to cause annoyance of cool air drafts to other workers in cold weather.

The problem was reduced to one of removing the oil mists, the original cause of the difficulties, by local exhaust.

Example 2. In a large assembly room, some 200 female workers were employed, the majority equipped with tools for soldering wire connections. Complaints of poor ventilation were widespread in the central portion of the room. In peripheral areas of the room, draftiness in cool weather was the complaint. Generous ventilation was provided in the form of exhaust fans in roof monitors.

It appeared from the subsequent analysis that flux fumes from soldering, coupled with the radiant heat from soldering irons, were the disagreeable factors affecting the majority of workers in the central areas of the room, and these elements, localized close to each worker, were unaffected by the general ventilation supplied to the room. Complaints of draftiness in the areas nearest the outside walls were readily explained by the fact that no equipment had been provided for supplying warmed air to the space to replace that removed by exhaust fans. Air leaked in through door and window cracks and was warmed by radiators only after passage into the central areas of the room.

Local exhaust at each soldering station for removal of flux fumes and a warm air supply system were the remedies indicated.

Significant Air Currents

The reader is reminded again that the entire philosophy underlying the subject matter treated in this volume is that there is no inherent need for guesswork in ventilation design; that the various pertinent factors can be welded into concrete principles that can be applied in an engineering way to the solution of these problems. This statement is made here because so many concepts have crept into the practice of ventilation design that do not meet these criteria and we shall deal with some of them in this section.

The term, *significant air currents,* applied to problems in general ventilation, means—those air currents having a constant position and direction, and of such magnitude that they are measurable, and either (a) can be exploited in the engineering design for useful purposes, or (b) are a factor deleterious to the desired air conditions and must, therefore, be counteracted.

By this definition we exclude those random air currents which are constantly shifting and therefore completely unpredictable. In brief we exclude those air currents that can be diagrammatically represented as existing but which are best disposed of in the picturesque expression, "trained arrows."

Classification of Air Currents

The several types of significant air currents are those discussed in the following paragraphs. Their minimum velocity depends on the volume rate of flow and on the magnitude of other air currents encountered by them. For example, air velocities lower than 25 to 50 fpm are rarely significant in industry; although, in mine ventilation, general air velocities in the range of 15 to 20 fpm in air ways are considered of significance. Air streams having velocities of 50 to 75

fpm may not be significant in a space where interfering air streams have much greater values.

Local Exhaust Envelopes. The zone surrounding an exhaust hood or wall fan providing general ventilation can be considered as defining one area of significant air velocities because it may meet all the requirements of the definition presented; the exception is in those cases where the air currents are not exploited for useful purposes in the system design. In dimension they are usually insignificant compared with those of rooms being ventilated. The dimension can be approximately fixed by calculation for some velocity contour, such as 50 fpm, in accordance with relations outlined in Chapter 5.

Supply Air Streams. Streams of make-up air entering a ventilated space can be dealt with quantitatively since they are basically air jets, the laws of which were discussed in Chapter 9. Such streams may originate by the action of supply air grilles in walls or ceilings, of simply duct outlets of an air supply system, or of a fan in an outer wall arranged to blow outside air into the space; they may also occur in a haphazard manner, entering through door or window openings, or even by simple leakage through cracks.

Natural Convection Above Hot Surfaces. The hot air currents arising from hot bodies such as furnaces are classifiable as significant even though their velocity may not be accurately calculated. In a very still atmosphere, the warm air currents arising from a human body might be of significant magnitude although it is difficult to conceive of any practical consequences. The warm air currents arising from insulated low temperature ovens might be significant except in space where other air currents are of overpowering magnitude. (See Figure 10-1.)

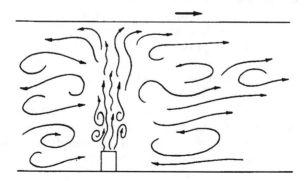

Figure 10-1. Air drift; air currents from point to point are chaotic in direction but the net component of motion is from left to right.

The discussion in Chapter 8 has demonstrated the importance of including in one's estimate of the behavior of rising hot air currents the fact of turbulent mixing with surrounding air. This makes it possible to recognize that a rising hot air stream is not a coherent system that passes directly outward through openings in a roof monitor or through exhaust fans.

In addition to the definable air currents already discussed, there are random air currents which cannot be managed by engineering calculations because of their indefinable and unstable characteristics and low velocity. They must therefore be regarded as of minor significance in ventilation design. Some consideration of a qualitative nature may, however, be accorded to them in connection with the problem of distribution of ventilation air. Thus air supplied at one end of a large room will tend to "drift" toward the exhaust at the other end of the room (Figure 10-1). But open windows or air currents of thermal origin may upset even this prediction. The influence of differing density of large masses of hot and cold air on the distribution of ventilating air is another example of air drift.

Air Distribution

It has been shown, in the discussion of vapor densities in Chapter 2, that the specific gravity of solvent vapors relative to air is of little significance in most cases in determining the flow of air streams contaminated by such vapor. It is incorrect, therefore, to base the design of ventilation system on the assumption that such vapors will flow downward by the influence of gravity and collect in higher concentrations at floor level. (It can flow downward for a few inches before dilution takes control.) That this does not occur in practice is amply illustrated by some air analyses made a number of years ago in several industrial establishments by a former colleague. Two simultaneous air samples were taken over a period of 30 to 45 minutes at floor level and at breathing level, respectively. Comparative results showed that the difference is slight.

Furthermore, it is important to note that in none of these cases was there any mechanical ventilation in the immediate vicinity. Existence of exhaust at floor level would wipe out even these slight differentials. In one instance where a system of floor level exhaust was in operation, simultaneous air samples were taken, one at breathing level and the other from within the exhaust duct. The latter would be a good representation of concentrations at floor level. The difference between the two was insignificant:

In exhaust duct: 240 ppm.   At breathing level: 220 ppm

It should be clear from this discussion that installation of duct work providing for exhaust near the floor level is poor design. An airborne vapor simply does not settle to the floor at ppm concentrations.

The principal reason for the observations cited above is that such relatively small concentrations of vapor are tolerable. Thus the atmosphere of work places is almost always 99.99% air and 0.01% or less solvent vapor. In the exceptional case of an accident in which large quantities of vapor are released, the statements made would be invalid, but this is not a problem with which the designer of general or process ventilation usually is concerned.

Warm and Cold Air. In contrast with the preceding considerations the relative density of warmer and colder air is often of significance in the design of ventilation. An analogy can be drawn between temperature differentials and solvent vapor concentrations which helps to illustrate these points.

Example 3. Calculate the concentration of benzene vapor in air that would have the same density relative to pure air as a mass of air at 70°F relative to air at 80°F.

Solution. It will be noted, first, that the density of 70°F air relative to 80°F air is described by the ratio of the equivalent absolute temperature:

$$\frac{460 + 80}{460 + 70} = 1.02$$

Hence the problem reduces to judging the concentration, x, of benzene air mixture having this density. Taking the apparent molecular weight of air at 28.9, and that of benzol at 78, we find

$$\frac{78x + 28.9(1-x)}{28.9} = 1.02$$

Thus, $x = 0.0122$ or 1.22% by volume or 12,200 ppm

In some poorly ventilated establishments, concentrations of vapors may occasionally be found as high as 1,000 ppm but this is highly unusual. The more common maximum in conditions of poor ventilation is about 100 ppm, and even this is not an "average" situation.

On the other hand, temperature differentials of 5 to 10 degrees F between working level and ceiling of a work room are not unusual. It is

thus further demonstrated how insignificant are the density characteristics of vapors ordinarily evolved in industrial operations; on the other hand common experience demonstrates the ease with which masses of air at different temperatures become stratified, with cooler air near the floor and warm air above, when hot air is being continuously generated, and in the absence of a strong air circulating mechanism.

Mechanical Air Circulation. Where temperature stratification prevails, a device that brings about continuous mixing of all air layers would be advantageous from the standpoint of heating in cold weather. Circulating fans disposed to blow air near the floor upward, toward the ceiling, would bring this about. Indeed, it is surprising that this arrangement or its equivalent is not in common industrial use.

An exhaust system with hoods at working level tends to minimize stratification by direct withdrawal of the cooler air at lower levels if the air supply is permitted to enter and mix with the warm air at the upper levels. In this sense the air motion at lower levels may be classified as "significant," although its characteristics make it an exception to the definition of significant air currents previously offered, i.e., beyond the significant velocity contours of the exhaust hoods, the air currents are not of measurable magnitude and do not have constant position and direction.

Supply Air Currents. The characteristics of air jets make it possible to consider quantitatively the behavior of streams of air supplied to a space, replacing air removed by exhaust system. The air jet relations apply whether the air enters the space from a duct system equipped with a blower, or haphazardly through openings in the wall, because of negative pressure created in the room by operation of the exhaust system.

Consider, for example, the three similar situations illustrated in Figure 10-2.

"A" represents a room with a fan exhausting 1,000 cfm, with an opening in one wall through which replacement air enters by the influence of negative pressure induced by the exhaust fan.

If the room has no other opening for the entrance of air, the effect is identical with that illustrated in "B" in which a blowing fan supplies air to the room at a rate of 1,000 cfm, the air escaping through the outlet opening shown, under influence of the positive pressure created by the supply fan.

In another room, "C," other openings in the walls exist and therefore two fans are illustrated to produce the identical air

currents shown in "A" and "B." One fan supplies 1,000 cfm, and the other exhausts at an equal rate.

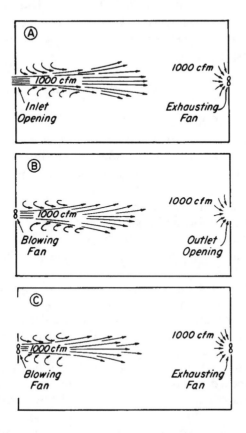

Figure 10-2. Illustrating the relative importance of air currents in blowing and exhausting systems. The situation respecting air currents is practically identical in cases A, B, and C, although the function of a fan is different in each case. Rooms A and B are air tight except at the inlet and outlet.

Analysis may be made of the significant air currents of the supply air stream by use of the air jet formulas, and if desired those of limited dimensions around the exhaust opening or exhaust fan. Between them are the random air currents of indefinable direction and unmeasurable, low velocity. For practical purposes, therefore, they would receive no consideration in the engineering analysis of such a situation.

The air jet formulas are also useful to a consideration of the influence of various arrangements for air supply on the problem of controlling discomforting drafts. From the jet formula (see Chapter 9)

$$V_c = KV_o(A_o)^{0.5}/X$$

we may adopt some maximum air velocity, V_o, such as 50 fpm, judged to represent the limit of safety as to comfort, rearrange the terms, and get an expression for the distance, X_{50}, at which a maximum velocity of 50 fpm would be found:

$$X_{50} = KV_o(A_o)^{0.5}/50 \text{ fpm}$$

This shows that for a given jet discharge velocity, V_o, the effective distance is proportional to the square root of orifice dimensions, $A_o^{0.5}$, and indicates how to minimize this critical distance; that is, how to introduce supply air with a minimum of air disturbance.

Where no arrangement is made for air supply and it leaks into the space through window and door cracks, the latter may be thought of as air jets. The fact of their small dimensions, A_o, explains how such haphazard arrangements so frequently get by without serious difficulty.

The advantage of a small orifice area for the supply air is well illustrated in small rooms (like small lab rooms), in which the equipment requires very large rates of exhaust relative to the size of the room. If there is a dead space of only limited dimensions into which the supply air is to be discharged satisfactory results can be obtained by discharge through a large number of holes of small area (e.g., 1 inch), disposed in a panel of size sufficient to accommodate all the necessary holes. The individual jets spend their velocity within a short distance.

A warm air supply is often distributed to various parts of a large mill building by means of a duct system with discharge branches at intervals. Sometimes this arrangement is unnecessary. The function of such a system is partly to deliver portions of the air supply to various areas, and partly to effect jet mixing while avoiding difficulties of high velocity drafts of air that might result if it were all discharged from a simple orifice of large dimensions, A_o. If, however, there is adequate overhead dead space in which the jet can completely spend its velocity energy without impinging on workers, then one consideration is eliminated.

If, further, the space is one from which air is removed at working level by exhaust systems at a high rate, adequate distribution may be effected, depending on the location of exhaust points in the room.

Short Circuiting. The possibility of inefficiency in distribution of supply air due to short-circuiting of that air stream from its point of

entrance directly to the exhaust is implied in much of the preceding discussion.

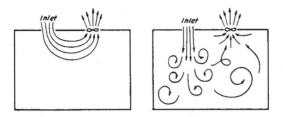

Figure 10-3. Arrangement in which a cool-air inlet is in close proximity to the point of exhaust, illustrating a common misconception concerning short-circuiting. (A warm air supply may, however, short circuit if inlet velocities are low and inlet temperatures are high compared to the room).

It may be helpful to dispose of some prevalent misconceptions which often exaggerate the point. In Figure 10-3 (right), the correct direction of entrance air streams is depicted, whereas sometimes it is mistakenly represented as being bent toward the exhaust outlet. The entering air stream is motivated solely by the negative pressure within the room (or the momentum imparted to it in the case of a stream from a blower system) and its temperature. It has no "knowledge" of the existence of an outlet path, e.g., the fan. The fan can exert no force beyond limits of the velocity contours, 25 to 50 fpm for example, which as we have seen is a very short distance. Therefore, it follows that the supply air will stream into the space in a straight line until its momentum has been spent, or it strikes the opposite wall, or until its temperature has reached the level of the space.

Room Volume and Ventilation Rate

The volume of a room to be ventilated and the ventilation rate are frequently related to the use of the ratio cfm ventilation per cu ft of room volume which is the "number of air changes per minute" so commonly referred to in discussions of ventilation requirements. Unfortunately, though in widespread use over many years, it is more often employed on an incorrect base than properly. It becomes necessary at this point, therefore, to examine in some detail the relations between ventilation rate and room volume.

Concentration Build-Up Rate. Consider a ventilated space into which a substance is being continuously injected either with the ventilation air or by any other means within the space. The substance may be, for example, a solvent evaporating from sources within the

space; or it might be such a gas as carbon monoxide from operation of automotive equipment; or it could be heat energy. The process we are considering could also be represented by a tank through which fresh water is flowing while some soluble salt is continuously being added.

The relation between concentration, c, and time, t, can be developed from elementary calculus in which the assumption is made that the substance is instantaneously mixed with the total air in the space. The differential increase in the total quantity of substance in the room in time interval, dt, is equal to the rate of generation of addition, minus the rate of removal in the air leaving the space (the rate of which is the same as the ventilation rate).

$$p \, dc = G \, dt - Qc/dt$$

where c = concentration of substance expressed as a fraction, at any time, t
  p = volume of the space
  G = rate of generation or addition of the substance
  Q = ventilation rate.

When rearranged

$$\int_{c_o}^{c} \frac{dc}{G - Qc} = \frac{1}{p} \int_{o}^{t} dt$$

and after integration, we get

$$\log_{10} \frac{G - Qc}{G - Qc_o} = -\frac{1}{2.303} \frac{Q}{p} t \tag{1}$$

and for the case where the initial concentration, c_o, is zero,

$$\log_{10} \frac{G - Qc}{G} = -\frac{1}{2.303} \frac{Q}{p} t \tag{2}$$

This formula discloses that the rate of air change, Q/p, in terms of volume of the room is significant in determining the rate at which the concentration increases with time, where at the start, concentration is zero or at a low level.

In the examples that follow, the use of these relations are demonstrated. The results may be in error to the extent that the air leaving the space (due to the problem of incomplete mixing) is of greater or less concentration than the overall average.

Rate of Purging. A similar treatment may be applied to the case where the air of the space is initially charged with contaminant or heat units and it is desired to estimate the rate of decrease of concentration over a period, by some fixed rate of ventilation.

In this case, the basic expression states that the differential change in total quantity of substance in the space, d (pc) or pdc, in time, dt, is equal to the differential quantity removed in air leaving the space.

$$p \, dc = Qc \, dt$$

Similar treatment of this equation results in:

$$\int_{c_o}^{c} \frac{dc}{c} = \frac{Q}{p} \int_{0}^{t} dt$$

and

$$\log_{10} \frac{c}{c_o} = \frac{1}{2.303} \frac{Q}{p} t \qquad (3)$$

Again, the rate of air change, Q/p, is significant in determining how rapidly concentration, c, changes with time. And again, a final mixing factor will be required to achieve reality and safety.

Example 4. In the conditions given below calculate the time required for vapor concentrations in a room to attain a value of $c = 100$ ppm. Assume pure air at the start.

Solution. Room volume, $p = 150,000$ cu ft; ventilation rate, $Q = 8,000$ cfm; rate of vapor generation into the room, $G = 1.35$ cfm.

Substitution of these data in Equation (1)

$$\log_{10} \frac{G - Qc}{G - Qc_o} = - \frac{1}{2.303} \frac{Q}{p} t$$

and solving for t gives $t = 17$ minutes.

Example 5. In the room of the last example, assuming the air contaminating process is terminated but ventilation continues at the same rate, calculate the time required for the concentration to decrease from 100 ppm to 50 ppm.

Solution. Use Equation (3); insert $c_o = 100$ ppm, $c = 50$ ppm, $Q = 8,000$ cfm, $p = 150,000$ cu feet, and solve: $t = 13$ minutes.

Summary of General Ventilation Principles

In the following paragraphs, summarizing the principles that have been discussed, a number of negative statements are necessary because of false notions that have been frequently encountered in practice. Some of them have already been illustrated in detail; others are best demonstrated by actual examples which appear in the following sections.

(1) General ventilation is essentially a process of dilution, and if one could obtain data on the evolution rate of the contaminant into the air space, one could readily compute the rate of ventilation necessary to effect any specified concentration. In practice, this information is always readily obtainable except in the case of industrial solvents evaporation. Nevertheless it is necessary that any estimate of required ventilation be confined to the dilution principle, which means an emission estimate should be made, even if only a "ballpark" quantity.

(2) Ventilation estimates based simply on room volume, i.e., on a specification of rate of air change, have little real validity. Calculations of rate of concentration increase or decrease can be made; they involve not only air change rates but, also, rate of evolution of contaminant or heat units. Such calculations have infrequent practical application in the design of industrial ventilation.

(3) The terms *significant air currents* and *air drift* have a specific meaning in this discussion which cannot be summarized in a few words. The significance of that portion of the discussion is that caution is necessary in ascribing direction and velocity to air currents in a space to avoid descriptions that are purely imaginary. We have borrowed the term "trained arrows" to illustrate the hazard of qualitative treatment. Significant air currents can be predicted as to position, direction, velocity and quantity. Air drift characteristics of a space may be highly uncertain.

(4) For all practical purposes, solvent-air mixtures of vapors are not heavier than air, as commonly believed, and ventilation by exhaust at floor level for removal of such vapors is based on fallacious reasoning.

(5) Differences in density between warm air and cold air are real, and deserve primary consideration in the planning of heating for cold weather.

Industrial Applications

The principal applications of general ventilation are in the control of vapors from industrial solvents, preventing excessive temperature increase, and in controlling some fume or dust-producing

operations where the quantity of contaminant is moderately small and so wide-spread as to render local exhaust difficult and costly. Sometimes the result is satisfactory; often it falls short of really good conditions, even though it might get by.

Ventilation for Industrial Solvents

Many of the principles previously discussed in this chapter are very well illustrated by the problem of ventilation for control of solvent vapors; other principles not heretofore taken up will be outlined.

Solvents appear in modern synthetic varnishes and lacquer, in liquid compositions for coating of fabrics and the like, in industrial adhesives and cements, and innumerable other industrial processes and products. A prime function of solvents at some stage in the application process is to evaporate into the atmosphere leaving behind a physically transformed material formerly associated with it. It is nearly always possible to estimate the rate of use of total liquid composition and, from knowledge of its solvent composition, to determine the rate of evaporation. This makes it especially simple, therefore, to handle solvent vapor ventilation problems on a quantitative basis.

From basic chemical principles we know that a weight of any liquid equal to its molecular weight produces, on evaporation, the same volume of vapor as all other liquids. If the weight is taken in grams, the unit quantity is called the gram-molecular weight, and the volume of vapor is 22.4 liters at 0°C and 760 mm pressure.

If the weight is taken in pounds, as will be most convenient for ventilation purposes, the volume of vapor formed by one pound-molecular weight is about 359 cu. ft. at 32°F and normal pressure. At a room temperature of 70°F, the volume is about 387 cu. ft. (BP = 760 mm Hg).

Example 6. What volume of vapor would be generated by the evaporation of 1 pound of the solvent benzol at room temperature?

Solution. The formula for benzol is C_6H_6, and its molecular weight is, therefore, 78. One pound is 1/78 of a pound molecular weight. The volume of vapor, is then,

1/78 x 387 = 5.0 cu feet (rounded for significant figures, and at sea level pressure, STP)

The calculation of ventilation rate required to produce a specified concentration is illustrated in the next example.

Example 7. Calculate the ventilation rate required to attain an acceptable concentration of 5 parts per million in an exhaust duct where benzol is evaporating at a rate of 1 pound per hour in a hood connected to the duct. (Room temp and sea level pressure.)

Solution. From the previous example it was calculated that 1 pound of benzol forms 5.0 cu ft of vapor. We, therefore, require such an airflow that 5.0 cu ft of vapor per hour will be diluted to 5 ppm.

Necessary cubic feet of air per hour = $Q = q/C$ or

$Q = 5.0 / (5 \times 10^{-6}) = 1{,}000{,}000$ cu ft per hour = 17,000 cfm

Summary Statement at Steady State Conditions

The complete calculation for required ventilation rate to attain a given atmospheric concentration is given below. The concentration selected is called the *Acceptable Concentration*, C_a. (Assumes standard conditions—70° F, 760 mm Hg, dry air—and perfect mixing.)

Q (cfm required) = (lbs evaporated/minute) x $(387/MW)$ x $(10^6/C_a)$

Convectional Dilution at Source

The relation between evaporation and ventilation rates describes average conditions in the room that result when complete mixing of vapor and ventilation air has been effected. It is certainly the concentration that would be found at the point where ventilation air leaves the space, through fans for example, for, at such a location, complete mixing will have become effective.

However, we are concerned with concentrations that prevail in the breathing zone of workers. If they are engaged at tasks that involve evaporation only a short distance from their faces, local concentrations of vapor may be high at the breathing zone while the general average concentration in the room is low. The task of ventilation design, therefore, is not complete with a calculation of general ventilation requirements. It is necessary to explore the question of local concentrations caused by work requiring the worker to maintain a position close to a source of solvent vapor. In tasks where the worker's position is constantly changing, the average exposure will certainly be represented by the average vapor concentration in the room. When he is employed at a bench at work involving evaporation from the articles he handles, the exposure may be materially higher than the room average.

The factors which determine the concentration in the breathing zone when close to the source are, in principle, the identical ones which determine the over-all concentration. The space between the source of vapor and the breathing zone is typically ventilated by the action of natural air currents of low air velocity and random direction. Mixing with vapor from the nearby source, concentrations are lowered in direct proportion to the volume of natural ventilation and inversely with the volume rate of vapor evolution.

Analysis of air samples taken in the breathing zone during normal working operations would determine whether concentrations were excessive or not, but in the planning of new operations it would be helpful to have a method for estimating the need for ventilation measures supplementary to the general ventilation. In the following description, a method is outlined permitting a *tentative* estimate of this nature.

Convection Ventilation Rate

Consider the evaporation of a solvent and the relative concentration of its vapors at varying distances from a flat surface, e.g., a floor or bench top, and the obvious fact that concentrations of vapors continually decrease with distance from the source. The cause of the decrease is in the action of convection currents of air that are always active and have minimum velocities of about 20 fpm in what we consider still atmospheres. Molecular diffusion is by comparison so slow as to be completely negligible.

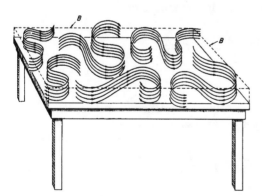

Figure 10-4. Idealized representation of chaotic convection currents above a surface of a table. Only horizontal components of velocity are shown. The rectangular "enclosure" B is the boundary convection area, a combination of five imaginary planes constituting a frame of reference for evaluating the magnitude of the "convection ventilation rate."

In Figure 10-4 convection currents are represented above the surface of a plane where solvent evaporates, idealized to the extent that only horizontal components of actual currents are shown. At any given instant of time the inward flow air must exactly equal the outward flow from any volume of space we may wish to consider.

Assume the space in Figure 10-4 to be enclosed by an imaginary surface that permits free movement of air currents inward and outward—the boundaries located so that concentrations of vapor are approximately equal at all points of the surface. Assume that air currents in the vicinity are chaotic and cause approximately equal dispersion in all directions horizontally and vertically, and are active within the space, not only to mix incoming surrounding air with vapors but also to carry them both upward and laterally.

The quality of any air that passes outward through the imaginary boundary is related to the rate of evaporation and to the rate of ventilation of the space. If v is allowed to represent the average velocity of the convection currents in this vicinity and a, the total area of the surface, it follows that the average convection ventilation rate, Q_c, for the space within the boundary is

$$Q_c = \frac{v\,a}{2}$$

The factor 2 in the denominator is warranted because only half the area is available for inward flow at any instant. The other half permits equal flow outward, assuming uniform velocities. Relative to the space we are considering, and assuming the surrounding air to be also somewhat contaminated by evaporation from the present or from other sources in the room, the following is also true:

cfm vapor from ambient air entering the space + cfm vapor originating within the space = cfm vapor leaving the space, or

$$C_a \frac{v\,a}{2} + q = C_x \frac{v\,a}{2}$$

In the third term, va/2 is actually augmented by the vapor volume, q, but that quantity is often insignificant and is, therefore, neglected in order to simplify the expression,

$$(C_x - C_a) = \frac{2q}{v\,a} \tag{4}$$

where

C_a = average concentration of vapor in ambient air entering space (volume fraction)
C_x = average concentration in air leaving the space at boundary
v = average velocity of convection currents, fpm
q = cfm vapor formed by evaporation within the space
a = area of bounding surface, i.e., Boundary Convection Area, sq ft.

Natural convection currents in occupied spaces are known to equal or exceed 10 to 20 fpm at all times and it will be conservative, therefore, to let v = 15 fpm *as a minimum* under most conditions.

If we can estimate the area of the surface that constitutes the locus of all points of equal concentration, where the height of the surface is selected to coincide with the worker's breathing level, then we can estimate the concentration that would enter the breathing zone.

Boundary Convection Area (BCA)

The true locus of all points of equal concentration would be a curved surface, the geometry of which depends on mathematics too complex to be useful. The simplifying assumptions we propose below are all selected for a conservative result and consideration of them will make it apparent that greater accuracy in estimation of the true boundary convection area is unwarranted.

Point Source

The evaporation of solvent from a point source on an infinite plane would result in a BCA that is hemispherical in shape, provided we assume equal dispersion of vapor in all directions from the source. The assumption is incorrect for there is, in fact, a tendency for flattening of the surface that is the locus, close to the source, due to the slightly greater density of the air-vapor mixture. (The cooling effect of the evaporating solvent, and the presence of higher solvent vapor content of the mixture account for this condition.)

The simplification, however, leads to a mathematical treatment that is simple and therefore practical for application to design problems. The nature of the inaccuracies is satisfactory because they lead to conservative conclusions.

The boundary convection area, then, for the case where the source of vapor is small in area relative to the distance, x, between it and the boundary surface which coincides with the breathing zone, approaches that of a hemisphere.

$$BCA = 2 \pi x^2$$

We shall call this case, "small area evaporation."
Area Source

"Large area evaporation" is a suitable term describing the more common case where the distance from the source of vapor (bench or floor) is small compared with the dimensions of the area of

evaporation. As the distance, x, approaches zero the boundary convection area approaches, as a limit, the area of the evaporating surface itself. Conversely, as the distance increases, the sides of the BCA begin to assume importance. The surface will, obviously, be curved but its geometry is complex. It appears that a rectangular surface represents well enough the shape of the boundary convection area, permitting a rough estimation of its area. The rectangle is one whose base encloses the source of evaporation, and whose top is a plane of the same area and at a height "x" that coincides with the breathing level of the worker.

The boundary convection area for this case, therefore, where the area of evaporation is large relative to the distance, x, from it to the breathing zone is well represented as follows:

$$BCA = LW + 2x(L + W)$$

where
L = Length of evaporation area
W = Width of evaporation area

Strip Source

There is one other form occasionally encountered wherein the solvent evaporates from a long narrow strip, as in the application of cement to long seams in certain rubber fabricating operations. We may simplify the picture of this case to the extent of considering the source as a line (without area) as we considered a point source above. The boundary convection area would then be

$$BCA = \pi L$$

with the symbols having the same significance as above.

Operations are frequently encountered that do not appear to fit any of these forms. In such cases one can arbitrarily select a shape resembling it most closely. For example, in many of the operations incident to the fabrication of fuel cells for aircraft, the cementing results in solvent evaporation from a cubical shape. An imaginary plane can be passed through the principal cementing areas, perpendicular to a line from the worker's breathing zone to the source of vapors, and the plane considered as a solid barrier, as though it were a bench or floor. This will result in a calculated result that is conservative and has the virtue of simplicity.

Convection Velocity

For estimating the convection ventilation rate, use a convection velocity of at least 15 fpm in the absence of a determined value. This is a reasonable figure for a space having a low rate of ventilation relative to its dimensions and errs on the side of safety.

In many cases a higher convection velocity will be seen.

Working Distance

The working distance between the worker's breathing zone and the source of vapor is a factor of importance since it determines in part the BCA.

Example 8. Cleaning is to be done on a bench with solvent evaporating from a small metal part. One worker's breathing zone is often about two feet above the source above the bench (which approximates a large plane). There is no dedicated ventilation, but general and transient mixing is assumed to be about v = 20 fpm. The solvent is emitted at a rate of q = 0.015 cfm. The PEL = 100 ppm. The question is whether these conditions are adequate to prevent unacceptable concentrations at the worker's breathing level.

Solution. We can roughly estimate the concentration at worker's breathing level, by the relationships given above. The input data and assumptions are as follows:

From the descriptions given above, the operation is classed as "close work," and a "point source." C_a is near zero, v = 20 fpm, and the area of convection approximates a hemisphere, $a \approx 2\pi x^2$.

$$(C_x - C_a) \approx \frac{2q}{v\,a} \approx \frac{2 \times 0.015}{20 \times 2\pi 2^2} \approx 0.0000597 \approx 60\,\text{ppm}$$

This would be unacceptable.

Example 9. Suppose for the last example that general ventilation is provided to further dilute the vapor (dilution air now flows past the zone of emission at v = 50 fpm) and, further, the worker is moved three feet from the source. Under worst case conditions (the air moves directly from the source towards the worker's breathing zone), what exposure might be expected?

Solution. Again, we can roughly estimate the concentration at the worker's breathing level by the relationships given above. The input data and assumptions are as follows:

From the descriptions given above, the operation is classed as "close work," and a "point source." C_a is near zero, $v = 50$ fpm, and the area of convection approximates a hemisphere, $a \approx 2\pi x^2$.

$$(C_x - C_a) \approx \frac{2q}{v\,a} \approx \frac{2 \times 0.015}{50 \times 2\pi 3^2} \approx 0.0000106 \approx 10\,ppm$$

This may be acceptable, given the "worst case" assumption. (Simply ask the worker not to stand "downwind" of the source.)

It is clear from the analysis presented that the magnitude of convection velocities is of great importance in determining concentrations of vapor in the breathing zone. Take 15 fpm as a very conservative value for a still atmosphere, although values double this figure are probably common.

The dispersion of vapors may be increased three to four times by the simple expedient of mechanically inducing air currents of higher velocity, as by blowing, in the space between the breathing level and the source.

A small propeller fan driven by a pneumatic motor might be a practical device for this purpose. (The presence of flammable vapors may rule out an electrically driven fan.)

An air jet at a low rate of airflow can also serve the same end. Located at a suitable distance from the critical zone it will entrain from surrounding air several times its original volume while its average velocity decreases to a reasonable level.

The comfort of the worker in winter time is a limiting factor in any scheme to increase velocities by fan or jet and, since human sensitivity to drafts is related to air temperature, such devices cannot be installed haphazardly. If proper analysis indicates a maximum permissible velocity of 50 fpm, for example, then the estimated concentration must be recalculated on the new basis to ascertain whether the necessary conditions will be attained.

Local exhaust without duct work designed to draw all vapors from their sources and discharge them back into the room has been historically used for these purposes. It is the exact opposite of the blowing systems discussed above. In ordinary factory problems an arrangement of this type would only be acceptable in special circumstances where there were great difficulties in running exhaust ducts to the exterior, or where only one or two benches were involved.

Space, also, must be available for discharge of exhaust air in such a manner as to avoid discomforting drafts.

Safety Factors

In special cases it may be desired to produce a room atmosphere with a very high degree of air purity, and it might be reasoned, falsely, that these dilution principles should then be ignored in favor of simple local exhaust calculations. It must be emphasized that the principles discussed in these sections apply even in such cases. If higher standards of air purity are desired in some circumstances, the designer can consider modification of the standard dilution rate for his special requirements but should not ignore the calculations procedure because it may still indicate the way to important savings.

It should be apparent from the forgoing exposition that the design of ventilation for the control of solvent vapors is susceptible to more exact·procedures than in almost any other subject in the field of industrial ventilation.

Typical Solvent Use Rates

Where information is lacking as to the rate at which solvent is evaporated in a process, reference may be made to the data given in Table 10-1 which provides a rough scale of typical use rates. Manual application of solvent cements and the like seldom exceed the 10-fold range, 0.02 to 0.2 pints per minute per worker.

Table 10-1

Solvent Application Rates in Certain Typical Industrial Operations

Operation	Pints/minute/worker
Manual, small brushing, cementing, "fussy" work	0.02 - 0.03
Manual, large-brush operations ("daubing")	0.2
Manual, gross applications, maximum use rate by hand	0.75 - 1.5
Mechanical coating operation	0.33 - 2
Spray finishing machinery	0.25 - 0.5

Natural Ventilation

The movement of air through a building, in the absence of mechanical ventilation, occurs by the action of either of two forces— the play of wind through windows or other openings and the chimney action resulting from a difference in temperature between outdoors and within the building.

In the very nature of things, reliance on natural ventilation is pretty much restricted to the alleviation of excessive heat because it is inoperative as a natural system in cold weather due to the necessity of heating the supply air. An exception applies to the case where there is a large release of process heat to the general atmosphere. In other words, the conditions that are controlling are the most unfavorable ones, and winter time heating capacity required is the determining factor in this case.

Magnitude of Natural Ventilation. It is useful to visualize the magnitude of natural ventilation to assist in better comprehension of the magnitude of a winter time ventilation problem.

Wind Effect. Consider a building with windows having an area 6 ft by 6 ft, subject to the natural ventilation resulting from a wind of 15 miles per hour (i.e., 1,320 fpm), using also 36 sq ft, 1,320 fpm, and a restriction coefficient of 1/2 to 2/3 (the orifice coefficient of the window opening) gives a rate of air motion equal to 24,000 to 32,000 cfm.

A calculation of the cost of equivalent mechanical ventilation in cold weather would include the initial cost of fan equipment, cost of fan power, initial cost of air heating equipment and cost of fuel for heating the air, and lead to a more complete appreciation of the magnitude of warm weather ventilation universally available without any cost. More important, comparisons of this kind provide a useful perspective to the problem of cold weather ventilation and help one avoid errors of judgment from drawing false conclusions about a problem based on warm weather observations.

Ventilation in Air Conditioned Spaces

Where a plant is air conditioned, consideration of the ventilation requirements for dust and fume control is of much greater importance than otherwise. In an ordinary heated space, a design error resulting in too little ventilation may be corrected on a relatively simple basis by an increase in heating capacity. But the ventilating capacity of air conditioning equipment is rigidly fixed, and can only be increased at great expense. In fact, the exhaust requirements for dust and fume control may well determine feasibility of air conditioning in the first place, and if errors are made in this estimate, the result may be disastrously expensive.

Chapter 11

Estimating Losses in Local Exhaust Duct Systems

Terms and Units used in Chapter 11

h'_v = velocity head, feet of air

h_v = velocity head, inches of water [VP]

h'_L = head loss, feet of air

h_L = head loss, inches of water [SP_{loss}]

h_t = total pressure, inches of water [TP]

h_s = static pressure, inches of water [SP]

h_{sf} = fan static pressure, [FSP]

h_{tf} = fan total pressure, [FTP]

v = velocity, ft/second

V = velocity, ft/min

N = number of duct diameters represented by duct length, L

D = duct diameter, ft

ρ = air density, lbs/ft^3

A = duct cross-sectional area, ft^2

K, K_1, K_2 = constants of proportionality, factors, coefficients

Q = air flow, ft^3/min

STP = standard conditions, i.e., 70° F, 29.92" Hg, dry air

[Editor's note: A number of static pressure loss factors or coefficients were developed or reported by Hemeon and are retained in the text. To his credit, almost all are still used (or are usable), but many have been expanded or refined by industrial hygienists and engineers since 1963. Check the more recent references for latest information.]

After the required rates of ventilation for various locations in an area have been established, as well as the types of hoods—shapes, size, volume flowrates and hood static pressures—it is then necessary to design the ductwork, estimate system losses and static pressure needs, and select the fan. [Calculations in this chapter assume STP.]

The design procedures are based on the need to estimate the required total static pressure which—together with the required airflow capacity Q—defines the duty required of a fan. The approach is illustrated in the following design problem for a simple exhaust system.

Example 1. A hood is to exhaust Q = 1,200 cfm. Duct velocities should be targeted for V = 2,000 fpm. The galvanized duct will run 25 ft to a centrifugal fan near the wall which will discharge through a short stack to the out-of-doors (Figure 11-1). What are anticipated flowrate and static pressure requirements at the fan? (All problems at standard conditions, STP.)

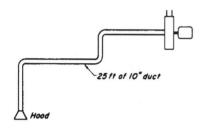

Solution. After selection of the path of the duct work, the next step is to fix the diameter of the duct. This can be done by reference to the duct air velocity. The required cross-sectional area is then

Figure 11-1. Ductwork.   A = Q/V = 1,200/2,000 = 0.60 sq ft

Select a duct diameter of a whole number, in inches, that is closest to this area. A 10-inch duct has an area of 0.5454 sq ft, an 11-inch duct, 0.660 sq ft. In this case we shall specify a 10-inch duct which would make the air velocity 2,200 fpm instead of the target figure of 2,000 fpm. (V = Q/A = 1,200/0.5454 = 2,200 fpm)

The flow conditions are now completely described and we would proceed to estimate the static pressure required by reference to the tables and charts presented and discussed later.

The static pressure required for 1,200 cfm flowing through 25 feet of 10 inch duct with 3 elbows might be estimated as follows: (Actual calculations will be shown later in the chapter.)

Acceleration to duct velocity =	0.30 inches w.g.
Entrance loss =	0.08
Friction loss =	0.18
Elbow losses =	0.81
Loss in short stack =	0.05
Total	1.42 inches w.g.

The required fan duty is, then, to exhaust Q = 1,200 cfm against a static pressure at the fan of 1.42 inches water gage pressure.

The performance of all fans is thus described in units of airflow and total static pressure and therefore the figures derived constitute a nearly complete specification for the selection of a fan.

The motor capacity required to operate the fan at this duty is directly proportional to the product: flow x pressure. In Example 1 it would be slightly less than 1/2 hp if the fan had an efficiency around 60%. Had we selected a duct size of 8 inches, with an air velocity near 3,400 fpm, the total resistance would have been closer to two inches of water. The fan duty specification would then have read "to exhaust 1,200 cfm against 2 inches w.g." The motor capacity would need to be about 3/4 hp in this case.

The methods whereby these calculations can be applied to practical design problems, and the basis of derivation are given in the following sections.

Wall Friction Loss of Ducts

We learn from elementary hydraulics that the friction loss due to the flow of any fluid in a duct is given by the Fanning-D'Arcy equation

$$h'_L = f \frac{L}{D} \frac{v^2}{2g} \tag{1}$$

where h'_L is the loss in feet of flowing fluid; f is a dimensionless friction loss coefficient; L is the length of duct, in feet; D is duct diameter in feet; v is velocity, in feet per second; g is 32.2 ft per sec^2.

Velocity pressure, or velocity head, is the pressure equivalent to that exerted by any air velocity, divorced from static pressure. From the elementary relation

$$v = \sqrt{2g \, h'_v}$$

it is

$$h'_v = \frac{v^2}{2g}$$

In equation (1) note that $L/D = N$ is the number of duct diameters represented by the length, L. It is now convenient to derive the ratio h'_L/h'_v which expresses a friction loss in the units "number of velocity heads," or numbers of velocity pressures. That ratio is obtained by dividing the first equation by the velocity head equation, which gives

$$\frac{h'_L}{h'_v} = \text{friction loss (in VP)} = fN \qquad (2)$$

The value of the coefficient, f, varies over considerable range and depends on the roughness of the wall surface as well as the Reynolds number. The range applicable to industrial exhaust systems extends from f = 0.01 to 0.04 or $^1/100$ to $^1/25$. When considered in conjunction with Equation (1), we note that these f values state, in effect, that "friction losses expressed in velocity heads, range from N/100 to N/25; that is, there is a friction loss of one velocity head in from 25 to 100 duct diameters, depending on the surface roughness and Reynolds number.

It is impractical to define surface roughness except in terms of the duct material and its construction.

Velocity Head or Pressure Units. The expression "$v^2/2g$" describes velocity pressure or head in feet of air, h'_v. In ventilating work, the velocity pressure, h_v or VP, is expressed in inches of water and velocity, V, in feet per minute (fpm). Transformation is effected by use of the relative densities of water (62.4) and air (0.075), 12 inches per foot, and v = V/60.

$$h'_v \text{ (in feet of fluid flowing)} = \frac{v^2}{2g}$$

$$h_v \text{ (in inches of water pressure)} = \left(\frac{V}{60}\right)^2 \times \frac{0.075}{62.4} \times \frac{12}{2 \times 32.2}$$

$$= \left(\frac{V}{4005}\right)^2 \text{ at standard conditions}$$

For the case where air density ρ is different from 0.075 lb per cu ft (STP), the expression becomes

$$h_v = \left(\frac{V}{1096}\right)^2 \rho$$

Friction Loss Chart

The friction loss chart (Figure 11-2, derived from Wright's Approximation) is convenient, and gives the friction loss directly in inches of water per 100 feet of duct length.

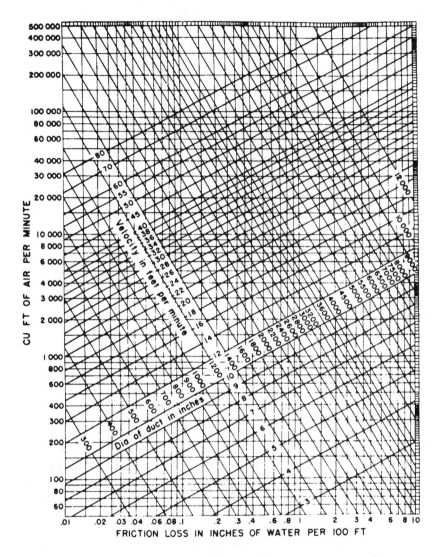

Figure 11-2. Air friction chart for round straight galvanized duct for average construction (1945) with about 40 joints per 100 feet of duct based also on dry air at BP = 29.92" Hg and 70 degrees F. [Note: variations of Wright's Approximation have been published numerous times over the years and can still be seen in recent references and texts.]

Examples 2 to 4. Find the following values from the chart:

Conditions	Friction Loss
(a) Q = 500 cfm in a D = 6" duct:	1.8" w.g. per 100 feet
(b) Q = 2,000 cfm in a D = 10" duct:	1.9" w.g. per 100 feet
(c) Q = 1,200 cfm with V = 3,000 fpm	1.5" w.g. per 100 feet

Friction Loss as Number Of Diameters for 1 VP Loss

Table 11-1 shows friction loss for galvanized duct at STP in terms of "number of duct diameters to create approximately one VP of static pressure loss."

Table 11-1
Friction Loss in Average Galvanized Iron Duct

Duct size, inches	No. of Diameters	Duct size, inches	No. of Diameters
3	39	12	54
4	42	14	56
5	44	16	58
6	46	18	60
7	48	20	61
8	50	25	64
9	52	30	66
10	52	35	68
11	53	40	71

Correction for Wall Roughness

From the data of Madison and Elliott (ASHVE Journal Section, *Heat. Pip. Air Conditioning.*, Dec. 1947) the correction factors shown in Table 11-2 for duct of various degrees of roughness were derived for the range of velocities commonly encountered in industrial exhaust systems. They are to be used as multipliers for the values found from the friction loss chart. [Ed's note: Additional correction factors (but not more accurate) are found in more recent references.]

Table 11-2

Multipliers for Friction Loss Obtained from Chart, Figure 11-2, Accounting for Varying Roughness

Type of Ductwork	Approximate Correction Factor, R
Medium smooth, e.g., steel duct without joints	0.9
Medium rough, e.g., average concrete surface	1.5
Very rough, e.g., riveted steel	2
Extremely rough, e.g., flex duct	3

It will be clear from scrutiny of these figures that one is not warranted in employing a value of friction loss containing more than two significant figures.

Effect of Dust or Fume on Friction Loss

The interior of ducts exhausting mists from electroplating tanks or the like accumulate a crust of dry salts which can markedly modify the surface smoothness of the ducts. Systems handling some fine dry dusts are similarly affected. The extent to which the friction factor is increased by these accumulations has never been reported to our knowledge, but it may be considerable. It seems not unlikely, by analogy to the conditions described in Table 11-2 that friction could be increased by a factor of 1.5 or even more. [Ed's Note: Such data are still not generally available.]

Shock and Turbulence Losses in Duct Fittings

In addition to the energy loss due to friction at the wall of the duct there are losses in various fittings like elbows, enlargements, and contractions in the duct area and points of entrance (hoods) from the atmosphere into the duct system. These are commonly termed *shock losses* in contradistinction to wall friction loss. The term connotes loss due to turbulence in fittings which destroys the forward motion of some elements of air stream and hence requires energy to reestablish its forward motion.

In nearly all common fittings, shock losses are caused by an expansion, change in direction, or enlargement of the area of the forward moving air stream. An understanding of the nature of this loss is fundamental to so many different shapes of fittings we shall discuss it first.

Loss in Abrupt Expansion

Where a sudden enlargement of cross-sectional area occurs in a duct, there is a loss in energy due to the shock depicted in Figure 11-3.

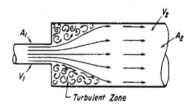

Figure 11-3. Air currents and energy loss in an abrupt enlargement.

In common parlance, the stream is said to *expand*. A better term than *expansion* would be *enlargement of the zone of unidirectional forward air movement* because it avoids any false connotation that the expansion is accompanied by a change in air volume due to reduced pressure. The enlargement of the area of the air stream occurs by entrainment of air from the space between its boundaries and the wall of the duct. This space, therefore, is in a state of turbulence with no net forward motion, due to its contribution of air to the unidirectional jet stream and, simultaneously, its acceptance, from the same source, of air equal in amount to that contributed. Chaotic circulation in this space results.

The energy loss which, of course, is transformed into heat and is readily measured by the difference in readings of two total pressure measurements disposed upstream and downstream, is derived from a theoretical base, well known in hydraulics. The derivation is based on the laws of behavior of fluid jets. The final equation, commonly known as the Carnot-Borda loss, is

$$h_L = \left(\frac{V_1 - V_2}{4005} \right)^2 \qquad \text{at STP}$$

where h_L is the loss in total head in inches of water, and V_1 and V_2 are velocities in feet per minute.

It will be noted that when V_2 is very low or approaching zero, h_L approaches $(V_1/4{,}005)^2$ which simply states that the total initial velocity pressure is lost when the air stream discharges into a large space where the forward velocity is zero, such as when air exits a stack.

Alternate forms of the equation describe the loss in terms of duct dimensions.

$$h_L = \left(1 - \frac{A_1}{A_2} \right) \left(\frac{V_1}{4005} \right)^2$$

or

$$h_L = \left(\frac{1 - D_1^2}{D_2^2} \right) \left(\frac{V_1}{4005} \right)^2$$

where

A is duct area, and D the diameter of the duct in sections 1 and 2.

Loss in a Sharp-Edged Orifice

When air flows through a sharp-edged orifice as illustrated in Figure 11-4 there is little energy loss due to shock or wall friction up

to the point of maximum contraction of stream diameter (the vena contracta), for there is a negligible area of solid surface in contact with the stream. There is, in other words, a nearly perfect transformation of energy in the process of acceleration from the low velocity to higher velocity at the vena contracta.

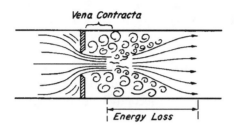

Figure 11-4. Air currents and energy loss for a thin plate orifice where the vena contracta is small relative to the duct diameter (or downstream space). As the orifice approaches the diameter of the duct, abrupt enlargement from the vena contracta is less extreme and energy loss is correspondingly less. Note that $h_L = 2.8h_{v(orifice)}$.

The forward-moving air stream is then permitted to enlarge its diameter and a turbulent action ensues that is practically identical with that in the abrupt expansion fitting previously discussed. If the dimensions of the stream and its expansion limits are known, it is possible to calculate the energy loss by the formulas for abrupt expansion.

The diameter of the vena contracta is about 77.5% of the orifice diameter and the narrowest section of the vena contracta is located about one half to one times the orifice diameter downstream from it, depending on the rate of flow.

In ordinary calculations, the loss through a sharp-edged orifice would be expressed in terms of the orifice dimensions, more specifically as a number of velocity heads based on orifice velocity. It is found to be about 2.8 velocity heads (2.8 x VP) when the orifice plate is thin, and sharp in arrangements where the orifice is small relative to the duct (orifice diameter less than about 1/4 the duct diameter). As such, velocities before and after the orifice may have velocity pressures close to zero. If the plate has appreciable thickness the loss may be closer to 2.5 velocity heads.

Example 5. Compute the loss caused by a sharp-edged orifice where D = 3 inches in diameter and where the orifice velocity is $V_1 = 4,000$ fpm and $VP_1 = 1.0''$ w.g.

Solution. The diameter of the vena contracta is 0.775 x 3 inches and since the orifice D = 3 inches, the ratio of areas is simply $(0.775)^2$, i.e., 0.601. This also represents the ratio of velocities at orifice and at vena contracta. In other words, the velocity at vena contracta is 1/0.601 times that at the orifice. Specifically

$V_2 = 1/.601 \times 4{,}000 = 6{,}655$ fpm. Note that $VP_2 = 2.8''$ w.g.

Where air passes from one chamber to another through a small orifice, the directional energy in the vena contracta jet is completely dissipated (lost) within the second chamber in random eddy motion. The formula for energy or head loss is

$$h_L = \left(\frac{V_1}{4005}\right)^2$$

where V_1, refers to the velocity of the vena contracta. Therefore, in the last example

$$h_L = 2.8\, h_v = 2.8 \times 1.0'' = 2.8''\text{ w.g.}$$

In this connection, note that the head loss, at discharge from the end of a duct into a room, is one velocity head $VP = (V/4{,}005)^2$, based on the velocity in the duct at discharge (STP).

Loss at Duct Entrance from the Hood

The conditions at an entrance to a duct are entirely similar to the flow through a sharp-edged orifice in that there is little or no loss in energy up to the location of the vena contracta a fraction of a diameter inward from the plane of the entrance (Figure 11-5). Beyond that point, the turbulence of abrupt expansion occurs and this accounts for the over-all energy loss for such an entrance.

The loss at such an entrance, measured downstream two or three diameters from the plane of the entrance, is found experimentally to be about 0.93VP (0.93 velocity heads) for round ducts and about 1.07VP for square ducts. One will infer the degree of contraction at the vena contracta by assuming, as before, that all loss occurs after that section, by the process of turbulent abrupt expansion, with no loss in that portion of the passage between the still atmosphere of the outside space and the vena contracta a short distance within the duct.

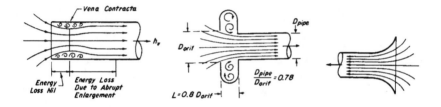

Figure 11-5 (left). Air currents at a duct entrance, depicting the energy loss due to an abrupt expansion from a vena contracta.

Figure 11-6 (center). A fancified shape at a duct entrance illustrating conservation of velocity where a duct is arranged in position and dimensions to coincide with the vena contracta of the orifice and thus prevent a subsequent abrupt enlargement loss. (Actual construction is not warranted.)

Figure 11-7 (right). The bell-mouthed opening illustrates the mechanism of energy losses by showing how they are avoided in a shape that prevents an abrupt enlargement from a vena contracta.

Conserving Velocity of the Vena Contracta

In the present discussion, we are depicting the consequences in energy loss of airflow arrangements in which the high-velocity, contracted stream is allowed to waste its energy turbulently, either totally as for the orifice between two large chambers, or partially as for the orifice in a duct. In the next example, illustrated in Figure 11-6, a mythical duct entrance fitting is shown designed to minimize or eliminate energy loss. Its shape and dimensions are based on an attempt to draw air into a duct without forming a vena contracta in the duct which is the event causing energy loss in an ordinary arrangement.

There is insignificant loss in passage through the initial sharp-edged orifice up to the section where maximum contraction occurs. This section is made to coincide with the entrance to the duct proper, thus fully utilizing the velocity energy at the vena contracta. Frictional losses of the type due to contact between a stream and a solid surface cannot occur because there is no surface to develop friction; shock losses due to abrupt expansion from a vena contracta are eliminated due to the particular dimensions selected.

In an actual model it is unlikely that one could avoid interchange of air from the contracting stream and the turbulent reservoir between orifice plate and the duct end. Also it would be difficult to adjust the various dimensions to effect the ideal conditions depicted. The principle of conservation of velocity energy is, however, illustrated and that is the sole purpose in considering such a strange shape.

Bell Mouth Entrance

If now, at the boundary of the contracting stream, we construct a curved surface to coincide with the curvature of the stream extending from the original orifice to the vena contracta, the stream characteristics are hardly altered (Figure 11-7). The solid surface washed by the air stream results in some wall friction loss but it is very small because of the small wall area. The energy loss then for a shape like this is, like the simple orifice, close to zero. It should be noted, however, that the bell mouth entrance is designed to conserve the energy of the maximum velocity, whereas sharp-edged orifices as usually employed are permitted to expend it in turbulent expansion.

The duct diameter after the bell mouth entrance corresponds to the diameter of the vena contracta of the orifice. If overlooked it may lead to confusion concerning flow coefficients.

The reader could logically expect at this point a continuation in discussion of losses in other entrance shapes, i.e., hoods with tapering sides. However, for the sake of maintaining continuity in development of the logic of shock losses and to lay proper ground for consideration, that discussion will be postponed until after we have taken up airflow behavior in contractions.

Losses in Tapering Contractions

In the fitting shown in Figure 11-8, the area of a duct is reduced to one of smaller diameter by means of tapering connection between the two. There is no energy loss in this section other than the wall friction losses that would be negligible in pieces of ordinary dimensions. The absence of shock loss during acceleration of the stream velocity is consistent with the general observation in fluid flow, and is typified, it will be recalled, in a similar absence of loss in the sharp-edged orifice up to the vena contracta and in the bell mouth entrance piece.

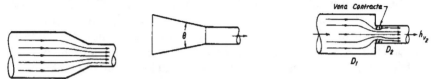

Figure 11-8 (left). Tapering contraction in a duct.

Figure 11-9 (center). A long taper hood has the two potential losses shown. They are negligible for all angles up to about 120 degrees and the maximum loss is about 0.5VP.

Figure 11-10 (right). Abrupt contraction in a duct. Low loss is in contrast with that in abrupt enlargements.

After passage through the contraction to the smaller duct one might expect some loss resulting from abrupt enlargement to the walls of the duct. In fact, however, experimental evidence indicates it to be low. For tapers which include angles up to 45° or so, the loss is very low (1/20 of the velocity head in the smaller duct). Even for tapers with included angles up to 135° evidence indicates losses not exceeding about 1/4 velocity head (smaller duct).

Losses in Abrupt Contraction

The abrupt contraction can be considered as an extension of the previous subject matter, the tapering contraction, in which the included angle is 180°. Such a fitting (Figure 11-10) permits formation of a vena contracta definitely of smaller area than the lesser duct, since an energy loss of up to 1/2 velocity head (smaller duct velocity) has been observed where the difference in duct size is appreciable, i.e., where the larger duct is 3 times the diameter of the smaller.

As the two duct sections approach each other in size one could reasonably expect a lessening of the tendency to contraction of the stream within the smaller duct, and, therefore, less energy loss due to expansion from vena contracta dimensions to those of the duct wall. The available data support this expectation. Losses in such *abrupt contractions*, in relation to relative duct dimensions, are given in Fan Engineering (Fan Engineering, 5th Edition, p. 124, Buffalo Forge Co., 1948) as:

Ratio of Duct Diameters:	2 and 3 : 1	1.5 and 1.75 : 1	1.25 : 1
Loss of VP in smaller duct	0.5	0.33	0.25

[Editor's note: These are acceptable, but more detailed loss factors are found in the more recent references.]

Losses of Hood Entrance Fittings

We are now ready to resume consideration of losses in other types of entrances, the bell mouth entrance and the plain unflanged entrance having been covered in earlier paragraphs.

A hood consisting of a small angle taper preceding the duct proper and of liberal length (Figure 11-9) only differs mechanically slightly from a plain end. It is clear that energy losses up to the usual point of reference, one or two diameters beyond the junction, consist of (a) wall friction in the tapering piece and (b) contraction loss at the junction or

neck. These losses should be considered separately in a basic treatment and added together to derive the overall energy loss.

For illustration consider a taper wherein the face diameter is three times the duct diameter. Wall friction can be considered negligible, in fact, is always negligible in hoods of ordinary shapes. Of the loss remaining, that due to the contraction at the neck will be recognized as very low due to the small angle of the preceding taper. This was discussed under tapering contractions where it was noted that such a loss amounts to no more than about 5% of a velocity head.

There remains one (1.0) velocity head at the duct entrance based on the acceleration of air to the velocity through that section.

Large Angle, Long Taper. As the hood angle becomes greater, formation of a vena contracta occurs beyond the neck. Using data obtained in part from analysis of Brandt's data (Brandt, A. L., "Energy Losses at Suction Hoods, *HPAC*, Sept., 1946) on entry coefficients for hoods having included angles above 90° and relatively long tapering sections, losses at the junction of taper and straight duct have been obtained as follows:

Included angle, degrees	Convergence loss, % of VP in duct
Up to 45	5
90	10
115	13
136	26
180	50

[Editor's note: These are acceptable, but more detailed loss factors are found in the more recent references.]

Tapering Enlargements

Unlike the tapering contraction in duct size where energy losses are negligible, losses are difficult to avoid in the tapering enlargement. This is due to inertia of the air and concomitant absence of any natural tendencies, except turbulent mixing, for the air stream to fill the enlarged area. It is exactly analogous to the situation in the abrupt enlargement.

Losses may be minimized by employing tapers of very small angles. However, the smaller the angle of the taper, the larger the taper required for a given enlargement. Wall friction becomes an appreciable fraction of the total loss in the case of tapers having angles around 10° to 12° because of their increasing length, and when the angle is 2° to 5° and less, wall friction accounts for nearly all of the loss.

The losses are best expressed as a fraction of the loss that would have occurred in an abrupt enlargement from the smaller to larger area. Figure 11-11 presents the experimental results of Gibson on water flow, with which studies by Kratz and Fellows on air are in agreement.

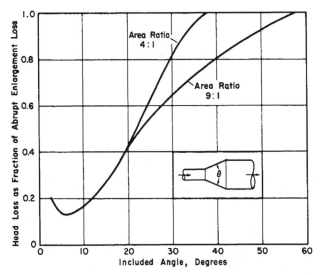

Figure 11-11. Losses in a tapering enlargement expressed as a fraction of $\left(\dfrac{V_1 - V_2}{4005}\right)^2$.

It will be noted that we adhere to our practice of describing pressure changes in terms of velocity head loss (a loss factor x VP) rather than in terms of static pressure regain, as is sometimes done. In our experience this method is much more likely to avoid confusion.

Elbow Losses

The flow of air through elbows cannot be analyzed readily on a theoretical base, hence our knowledge of the magnitude of energy losses is entirely empirical although adequate. [Editor's note: See the latest loss factor data in more recent references.]

It has been shown that a double helical motion of the air stream occurs downstream from the elbow (Figure 11-12) after the vena contracta formed in the elbow. It is not difficult to visualize this motion carrying the air from the far wall against which it has initially impinged.

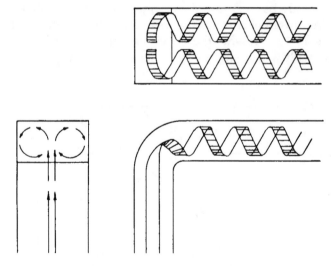

Figure 11-12. Helical flow following an elbow (after Barry).

These processes involve an increase in velocity through a vena contracta and in the bending stream, at the expense of static pressure, but much of the latter is regained a few diameters downstream from the elbow, as the velocity reassumes a normal pattern. The practical implication of this is important, because if an air stream is permitted to discharge to the atmosphere immediately after an elbow, the resulting elbow loss will be 2 to 3 times as much as in an arrangement where a few diameters of duct are supplied after the elbow in which static pressure regain can take place.

Elbows of square cross section have somewhat different losses than are found in round duct elbows, but this difference is small. In square section miter elbows, the loss is about 20% greater than for round section miter elbows; but in curved elbows of center-line radius greater than 1.0, the loss in square section elbows is but 3/4 to 1/2 of that in the round.

The loss in elbows bending through angles different from 90° is about proportional to the angle of bend up to turns of 120° to 150°. A turn of 180° has a loss of 1.65 times that of a 90° elbow, however, instead of 2.0 times, as might be expected.

Losses at Branch Junctions

Alden drew attention to the nature of shock losses at that section of an exhaust system where a branch duct connects to the main, contributing additional air flow. The branch is most commonly

connected to the main at an angle of 45°. Sometimes the angle is made as little as 30° and in large piping where duct fabrication problems exist, 90° connections are sometimes [mis]used.

A loss (or gain) can occur in both the main and the branch ducts, the magnitudes of which are related to the relative volume of air being discharged from the branch into the main, their velocities, and to the angle of discharge to the main. [Editor's note: Figure 11-13 shows an older approach, now not recommended. Today most use the "ACGIH tee" which has the branch entering at the center of a 15-degree taper. See recent references for details and loss calculations.]

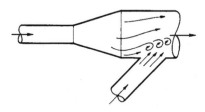

Figure 11-13. Head or pressure loss in a main duct due to entrance from a branch duct.

Summary. The various shock losses discussed in preceding sections are summarized in Figure 11-14. [See references for more recent versions of these loss factors.]

Airflow in Branches

From the pressure loss data given in the previous sections, one can readily calculate the airflow that can circulate through any section of a duct system, if the pressure drop is known together with the dimensions and shapes of this section. This kind of calculation is the reverse, it will be noted, of the usual problem in which resistance is calculated from a specified airflow. The examples following will serve as introduction to the main discussion to follow. (All examples at STP.)

Example 6. What airflow is expected in a 50-ft. section of straight 10-inch round duct, when the total pressure at the upstream point of reference is TP = –2 inches of water and that at the downstream point TP = –4 inches, i.e., a pressure drop of two inches?

Solution. Note that Q, V, and VP will remain constant in the system. Only friction loss affects SP and TP. The answer is found

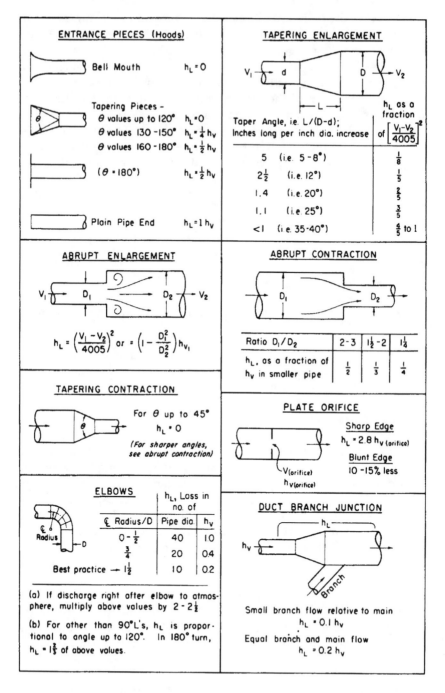

Figure 11-14. Losses in common fittings; rounded values (h_v = VP).

by reference to the friction loss chart, Figure 11-2, first converting the loss to the terms of 4 inches per 100 ft (the loss is 2 inches per 50 feet). Opposite the intersection of 4 inches loss and 10-inch duct we read the answer—approximately 2,900 cfm.

Example 7. What is the airflow through a duct branch consisting of 6-inch round duct with a plain open end (hood) when the static pressure 20 ft downstream from the open end is SP = –1.5 in w.g.? (Figure 11-15.)

Solution. The static pressure 20′ down the duct must consist of acceleration (1 VP), the entry loss (given in the figure as 1.0VP), and friction loss for 20 feet. For convenience and to avoid mixing the units of pressure loss, we shall consider the duct friction loss in terms of Table 11-1, where it is noted that there is a loss of 1.0 velocity head h_v (or, 1 VP), in 46 diameters. The length, 20 ft, is 40 diameters, therefore the friction loss is $(40/46)$VP, or $0.87h_v$. The total "loss" is, then, 2.87VP $= SP_{loss} = 2.87(V/4,005)^2 = 1.5$ inches w.g. Solving this gives V = 2,900 fpm. The area of a 6-inch duct is 0.1964 sq ft, so the airflow is Q = VA = 2,900 x 0.1964, or about Q = 570 cfm.

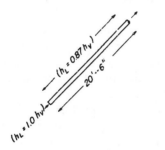

Figure 11-15. Example 7.   Figure 11-16. Example 8.

Example 8. What is the airflow through a duct branch 20 ft long when the diameter is 10 inches, the hood is a 45-degree tapering piece with negligible loss, and the downstream static pressure is SP = – 1.0 inches? (Figure 11-16.)

Solution. Since the entrance loss is negligible, and the static pressure includes the acceleration, 1VP, only friction loss is involved. Consider the duct friction loss in terms of Table 11-1, where it is noted that there is a loss of 1.0 velocity head h_v (or, 1 VP), in 52 diameters. The length, 20 ft, is 24 diameters, therefore the friction loss is $(24/52)$VP, or $0.46h_v$. The total "loss" is, then,

1.46VP = SP = 1.46(V/4,005)^2 = 1.0 inches. Solving this gives V = 3,315 fpm. The area of a 10-inch duct is 0.5454 sq ft, so the airflow is Q = VA = 3,315 x 0.5454, or about Q =1,800 cfm.

Example 9. What would the airflow be in the preceding duct branch, if the static pressure loss were recorded as 0.50 inch?

Solution. The calculation can be solved by reference to the relation that velocity is proportional to $h_L^{1/2}$. Therefore, for a duct system of fixed cross-sectional area, Q = $Kh_L^{1/2}$. In this instance the ratio of the pressure loss is 0.5/1 = 0.5 so that the reduced airflow is in proportion to $1/2^{1/2}$ = 0.71; that is, the airflow under the new conditions is Q = 0.71 x 1,800 = 1,300 cfm.

Adjustments for Proper Branch Distribution

It would be convenient in designing a system if, after selection of exhaust rates, Q, for each branch, and a suitable duct velocity, V, one could simply fix the necessary cross-sectional area of each branch by the relation A = Q/V. The preceding examples have demonstrated that airflow does not distribute itself between branches in accordance with fictitious branch velocities set down on paper without regard to losses in the different pathways to a common point, the junction of the two branches. The loss in the branch having lower losses inevitably must be increased to equal that in the other branch at the junction.

One could insert a restriction such as a blast gate in this branch and adjust it so that the head loss would effect the desired balance.

Instead of a restriction in the form of a single fitting, like the blast gate, one could modify the diameter in part or all of the entire branch so that increased losses would result. This can be done by trial and error until the reduced diameter is found that will pass the originally specified airflow at the resistance fixed by conditions in the other branch. Luckily, a mathematical relationship can be derived by from the basic friction loss formula, Equation (2).

$$h_L = fN\left(\frac{V}{4005}\right)^2 \tag{3}$$

Confining our consideration to the case where Q is a fixed value, it is seen that V and D are related. [Note that N = L/D.]

$$V = Q/A = Q/\frac{\pi D^2}{4} \tag{4}$$

Squaring both sides gives

$$V^2 = (\frac{4}{\pi})^2 \frac{Q^2}{D^4}$$

which can be simplified by grouping all constant factors, including Q, into one constant, K, and substituting in Equation (3) for V^2 to give

$$h_L = \frac{f\,K}{D^5}$$

If we ignore the variability of f with D, which is insignificant in this application, we note the useful fact that the friction loss of a duct varies inversely as the fifth power of diameter. A fifth power chart (Figure 11-17) has been constructed to facilitate adjustments in branch diameters.

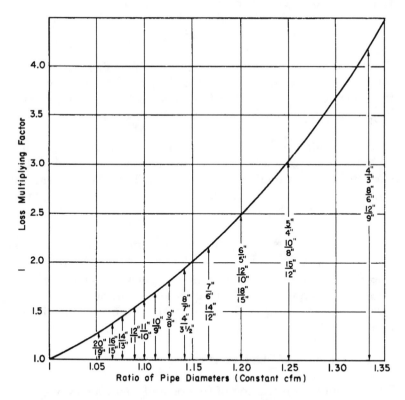

Figure 11-17. Fifth-power chart for estimating increased losses in a duct when diameter is reduced and airflow is unchanged.

In industrial practice, duct sizes above six inches are fabricated in intervals of whole inches and there is scarcely ever any useful

purpose served by specifications departing from this practice. Therefore the chart includes some specific duct diameter ratios for added convenience

Example 10. In the duct arrangement of the last two examples (see Figures 11-15 and 11-16), determine the diameter of the branches such that one branch will pass 570 cfm and the other 1,800 cfm through a common junction providing SP = -1.5" w.g.

Solution. The smaller branch has already been calculated to pass 570 cfm at 1.5" w.g. when its diameter is 6 inches. At a duct size of ten inches, the 10" branch carries 1,800 cfm at 1.0" w.g. The static pressure loss multiplying factor is 1.5/1 = 1.5. All of the loss was related to friction, so Figure 11-17 can be used. Entering the chart on the left side at 1.5, read to the 5th power line and then down to Diameter Ratio = 1.08. Thus

Useful diameter = 10/1.08 = 9.25"

Recalculating the loss with the nearest whole size duct, 9", the static pressure loss would be just slightly more than 1.5" w.g., or the flowrate must be slightly less than 1,800 cfm, demonstrated below using the following original estimating approach:

Since the entrance loss is negligible, and the static pressure includes the acceleration, 1VP, only friction loss is involved. Consider the duct friction loss in terms of Table 11-1, where it is noted that there is a loss of 1.0 velocity head h_v (or, 1 VP), in 52 diameters for the nine-inch duct.

The length, 20 ft, is 24 diameters, therefore the friction loss is (24/52)VP, or $0.46h_v$. The total "loss" is, still, 1.46VP = SP = $1.46(V/4,005)^2$ = 1.5 inches. Solving this gives V = 4,060 fpm. The area of a 9-inch duct is 0.4418 sq ft, so the airflow is Q = VA = 4,060 x 0.4418, or Q =1,790 cfm, well within normal design tolerances.

Accuracy of Head Loss Calculations

Throughout the description of friction and shock losses, we have tried to discourage any tendency to employ loss data more accurately than warranted by the facts. Handbooks in general report shock loss data in terms of more significant figures than needed by the engineer designing industrial exhaust systems, e.g., 0.02 h_v as the loss in a bell-mouth entrance, $0.11h_v$ for the loss in a taper hood of certain

proportion; and nearly all publications illustrate the summation of energy losses in three significant figures, such as "2.33 inches."

In these pages, head loss data have been rounded off, and their presentation simplified, in order to help the reader keep these numbers in perspective. The following discussion presents the basis for an independent evaluation by the reader of the quantitative significance of pressure drop estimates.

Consider an exhaust system wherein the design calculations have led to an overall head loss of 6 inches of water for a flow of 5,000 cfm. By noting that both friction and shock losses vary directly with (velocity)^2, and therefore, with (cfm)^2 for a given system (since cross-sectional areas are fixed values), it is possible to calculate the resistance for this particular system of any other airflow using

$$Q = K\sqrt{h_L}$$

and expresses the statement above. Q is the airflow, cfm; K is the constant of proportionality; h_L is the system head loss, inches of water. In the present problem, $Q = 5,000$ cfm and $h_L = 6$ inches.

Therefore

$$K = 5,000/(6)^{0.5} = 2040$$

from which the expression descriptive of this particular system,

$$Q = 2040\sqrt{h_L}$$

If flow through the system is caused to drop to 2,500 cfm, half its former value, the pressure drop will be one fourth of 6, or $1^1/2$ inches. Doubling the flow would increase the loss to 24 inches.

We may plot this relationship as shown in Figure 11-19. If now we assume that an error was made in the head loss estimate for this system, of 1/2 inch, that it should have been $6^1/2$ inches for 5,000 cfm, a corrected curve may be constructed in the preceding manner, the corrected equation being $Q = 1970h_L^{1/2}$.

The consequences of these types of errors are discussed after the figure.

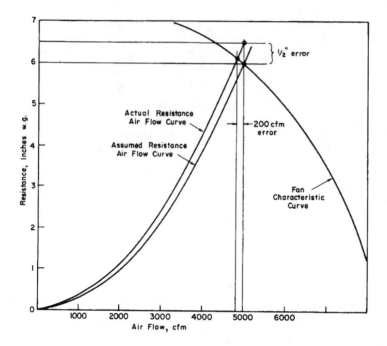

Figure 11-18. Graph illustrates the relationship between error in friction loss and the resulting variation from planned airflow.

Actual Flow; Fan Slippage

If a positive displacement pump having no slippage were employed, there would be no difference in flow between systems with differing resistance characteristics. Indeed, there would be no need to calculate duct resistance so far as airflow was concerned since it would be constant, determined only by the rate of displacement of the pump itself. However, the centrifugal fans that are used for ventilating and exhaust systems operate with relatively large clearance between the impeller and the housing and there is, therefore, slippage i.e., less air delivery than would be calculated from the rate of rotary displacement of the impeller blades. The greater the resistance against which they are required to operate, the less the flow of air, i.e., the greater the slippage.

The exhaust capacity of any particular fan against different degrees of resistance can be represented by one plot, for one speed of rotation, with head loss as ordinates, and airflow rate as abscissas. Characteristics of one fan at a given speed are drawn in Figure 11-18. Since the coordinates for this graph of fan performance are identical with those which describe the resistance-flow characteristics of a duct system, one may superimpose one graph upon the other. The intersection of the curves fixes the operating point of the combination

of the fan, at a designated speed, and the particular duct system; the intersection is the only point common to each and the only one satisfying the requirements of each.

We are now prepared to examine the magnitude of the error in system friction loss of 1/2 inch in the example described above, in terms of the difference in flowrate. Because the fan operating curve slopes at the points of intersection, it may be seen that the difference is even less than if the curve were horizontal at those points. However, it is possible to have the operating point occur at a nearly horizontal section of the fan curve; moreover some fans have characteristic curves having only a gentle slope. A horizontal section of fan characteristic curve represents maximum air slippage and one may therefore appraise the effect of error in calculating system resistance by assuming this worst condition. In the example above, the effect of pressure loss error between 6 and $6^{1}/2$ inches would represent a drop from planned 5,000 cfm to 4,800 cfm, a small change.

The relative magnitude of this error, on a percentage basis, is dependent on the slope of the system resistance curve at the point of consideration, and on the absolute value of the resistance.

[Editor's Note: This suggests that for local exhaust systems, it is important to choose a fan with a steep curve through the desired system operating point.]

Air Horsepower

The pressure variation to which air is subject in ventilating systems never exceeds 2 to 4% of one atmosphere and, therefore, it can be taken as subject to the same laws that apply to an incompressible fluid, like water.

The power required to overcome resistance to flow is equal to the product of pressure and volume flowrate. In common engineering units, it will be recalled it is

(pressure) x (flowrate)

$$\frac{lb}{ft^2} \times \frac{ft^3}{min} = ft \bullet lb/min$$

which becomes horsepower when divided by 33,000. The pressure unit needs to be changed, for convenience, from pounds per square foot to inches of water. To do so, it is noted that 12 inches of water is equivalent to 62.4 lb per sq ft and, therefore, 1 inch of water = 62.4/12 = 5.2 lb per sq ft. We can therefore derive

horsepower per 1 inch w.g. pressure per 1 cfm = 5.2/33,000 = 1/6356

Hence the general expression:

$$\text{air horsepower} = Q \bullet FTP / 6356$$

where Q is airflow in cfm and FTP is (generally) the total static pressure requirements at the fan in inches of water.

The product is termed *air* horsepower to distinguish it from the mechanical horsepower characteristic of the fan employed, efficiency of which is always less than 100%. Actual brake horsepower required is, then

$$\text{brake horsepower} = \frac{Q \bullet FTP}{6356 \bullet ME} \qquad\qquad \text{bhp at STP}$$

where ME, a decimal fraction, is the mechanical efficiency of the fan or air mover.

Example 11. Calculate the air horsepower required to draw Q = 10,000 cfm through a system having a FTP = 6 inches water.

$$\textit{Solution.} \quad \text{air horsepower} = \frac{Q \bullet FTP}{6356} = \frac{10000 \bullet 6}{6356} = 9.4$$

Pressure Relationships

In considering head or pressure losses due to friction and shock, one is really dealing in the final analysis with *power*, because a unit flowrate of 1 cfm is always implied. It is simply more convenient to drop the flowrate factor and power constant from consideration until a summation of all losses is effected, then the power requirement can be calculated for the final result.

Figure 11-19 is a reminder of the three forms of pressure or head and the types of tube connections that measure each one. We designate total head, static head, and velocity head by subscripts, thus, h_t, h_s, h_v, or TP, SP, and VP.

In a system completely free of friction or shock losses, the total pressure monitor (for TP or h_t) would register equal values at every point of the system without regard to velocity variations and, as will be recalled from Bernouilli's principle, static pressure and velocity pressure share the total pressure according to the velocity prevailing in any given section, TP = SP + VP.

The head losses that do occur are described as the difference between the total pressure at the two points of interest. Let h_L

represent the loss in head. It represents an irretrievable power loss which has been transformed into heat.

$$h_L = h_{t1} - h_{t2} \quad \text{(or, in the other format,} \quad SP_{loss} = TP_1 - TP_2)$$

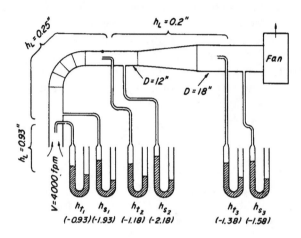

Figure 11-19. Pressure heads and their measurement.

Negative Pressures. In a blowing system, that is, one in which static pressures are positive at all locations, these relationships are easily understood. There is, on the other hand, frequent confusion in relationships when applied to the negative pressure of an exhaust system. The causes of confusion may be eliminated if the rules illustrated below are understood.

Study the simple exhaust system illustrated in Figure 11-19, in which all data are given or have been derived from friction or shock loss tables, with the exception of the values enclosed in parentheses below the U-tubes. (All measured values represent the average of the measurements at the point.)

Duct Velocities and Duct Sizing

The airflow through different hoods and branches of an exhaust system having been decided upon, the diameter of duct is selected next. Several considerations determine the duct air velocity, i.e., duct size, best suited to different situations. They are as follows: (1) obtaining the most economical size of duct; also duct size such that the total fan resistance matches the characteristics of commercially available fans; (2) minimizing airflow noise; (3) obtaining the maximum practical flexibility in hose when it constitutes a duct

branch, and (4) effecting a self-cleaning system where dusty air is to be handled.

Economics of Duct Sizing. The simplest industrial ventilating system of ducts for handling air contaminated by gases and vapors has no inherent duct velocity requirements. Therefore, its size, ideally, would be one which strikes a fair economic balance between first cost of the system, and cost of power to overcome friction in the system. The larger the area of the duct, the lower will be the air velocity and friction loss and, consequently, the cost of fan power. But, on the other hand, the larger the duct, the greater the material cost. There is an optimum range. (See Figure 11-20.)

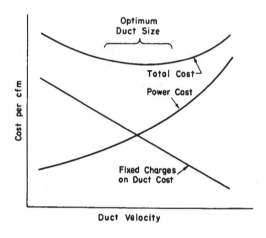

Figure 11-20. Economic optimum factors for duct velocity.

As a general rule, ducts sized for air velocities in the range 1,800 to 2,500 fpm meet the two conditions discussed. A figure of 2,000 fpm is often chosen by designers [circa 1998] as representing the optimum, but velocities well above and below the range cited are sometimes justified.

Air Noise. The greater the head loss of a system, the greater the tip speed required in a specific fan, thus creating greater noise. Also, the greater the velocity, the greater the noise created as air moves through the duct system. In a *non-industrial* ventilating system, these noise factors are usually a major consideration and, for this reason, velocities of less than one thousand feet per minute are not uncommon. Similar considerations apply to industrial ventilating systems but the permissible noise level is generally much higher.

Velocities and Maximum Flexibility. Floor vacuum cleaners of the type where the cleaning nozzle is connected to the bag filter assembly and fan by means of a flexible hose might be considered as exhaust systems in which minimum hose diameter is the foremost consideration because of the need for maximum flexibility. The dust-lifting capacity of the nozzle is determined only by the rate of flow of air, just as the performance of an exhaust hood is determined by the airflow through it. The velocity in the connecting tubing then is fixed at a high level in order to carry the desired airflow through as small a hose as possible. Limits are imposed because of the increased power required as the velocities are made greater. Commercial vacuum cleaners represent a compromise between the desire for maximum airflow, minimum hose diameter, noise control, and moderate size of motor. They are usually designed for hose velocities not exceeding 8,000 to 9,000 fpm.

Flexible Duct. Exactly the same consideration applies to an exhaust system where the branch consists of a flexible metal or rubber hose, which is repeatedly moved during working operation. This is illustrated in the use of flexible hose for fume removal in electric welding and in some of the arrangements employed for dust control in stone chiseling and in rock drilling. Inasmuch as the friction loss and consequently power requirement varies as the fifth power of velocity, for a fixed exhaust rate, the limits of practical velocities fall within narrow limits, 4,000 to 5,000 fpm where the fan and motor unit must be manually portable.

Velocities for Dust Systems

The most critical element in the design of an exhaust system for *dust* is the duct velocity. With most ducts, the ductwork can be made self-cleaning if high enough velocities are maintained to scour and transport the dry dust particles without interruption. If the air velocity is too low, settlement occurs in horizontal runs. When this occurs, the passage area is reduced and, for a given air flow, friction losses increase. This causes reduced airflow from the fan which causes a further reduction in air velocity in the duct line and thus, in time, encourages deposition of additional deposits of dust. A vicious cycle is apparent.

If the fan were a positive displacement air moving machine, then a stable condition would eventually be reached because, as the duct area is reduced by settled dust, air velocities would gradually increase until a point is reached where the air velocity exceeded the

transport requirements of the dust. This, however, does not occur where ordinary centrifugal fans are employed.

A duct velocity around 3,000 to 4,000 fpm is adequate for the great majority of industrial dusts but this requires extensive qualification and is discussed extensively in the following chapters. [Editor's note: check recent references for more information on appropriate scouring velocities for specific materials.]

Chapter 12

Exhaust Systems for Dust

Design of exhaust systems for dust control poses special problems stemming principally from the need to keep ducts free of settled dust and debris which would impede the flow or air and render the system ineffective.

The most common method of design is based on the fact that in most cases dust particles can be prevented from settling in either horizontal or vertical ducts if air velocities are maintained above certain values. Exceptions to this, as well as other bases of design, are discussed later in this chapter.

The magnitude of air velocities required for dust transport is a function partly of particle size and is best considered in two groups, fine dust and inertials. It will be noted in the following discussion that the particle size boundary between these two groups may be somewhat different from that described in Chapter 2, but the basic concepts are the same.

Fine Dust Transporting Velocities. The term *fine dust* as used herein means dust particles which are so small as to possess marked cohesive and adhesive properties. A handful of powder thrown at a vertical surface sticks there because of the physical properties of fine dust. Fine dust causes clogging of screens and impedes or completely stops passage of material even with vigorous vibration; this occurs in spite of the fact that the individual particle size is only a small fraction of the size of the screen openings. Particles below 10 to 20 microns certainly belong in the fine dust category, and it is probable that sizes several times this value should be included.

Transport of dust upward in a vertical run of duct is simply analyzed by reference to settling velocity characteristics for a particle of given weight. (See Chapter 6.) Any upward velocity in excess of the free settling terminal velocity of the particle would carry it along with the stream. In fact it has been shown in Chapter 2 that the settling velocity of fine dusts is so small that they may be regarded for practical purposes as having no weight. Similarly, in horizontal transport, the settling velocity of fine dust particles is insignificant.

Fine dust particles coming into contact with the wall of the duct may tend to adhere more or less tenaciously, and increased air velocity, in itself, will do little to prevent it. The viscous drag of air layers next to the walls is the only opposing tendency to adherence, but experience with very fine powders shows that such drag is often insufficient to keep the wall clean. Many fine dusts, therefore, cannot be satisfactorily handled in exhaust systems of conventional design.

Transporting Velocities for Clean Dry Inertials. Inertials are those solid particles having appreciable mass. In contrast to fine dust, they lack cohesive and adhesive properties. Fine beach sand typifies this class. Moist sand may be made to cohere into a mass, but only because of surface tension of the water film on the surface of the grains. Inertials in the present sense include any large particles that may be transported in an air stream—iron filings, sawdust, grain dust, or the like. Of more importance in the following discussion, inertials are the larger particles coexisting with fine dust in an industrial dust mixture.

DallaValle measured the velocities required to prevent settlement of dust particles in both vertical and horizontal duct lines. He employed particles of fine grain size (14 to 20 mesh or 1 to 5 millimeters) which in our classification would be considered as inertials. Moreover they were clean, that is unmixed with fine dust, a quality of great importance as will be shown in the following section. His tests for horizontal conveying indicated the following required air velocity for transport:

$$V = 105(\frac{Z}{Z+1})d_m^{0.4}$$

where V is the required transport velocity in fpm, Z is the specific gravity, and d_m the particle diameter in microns (1,000 microns = 1 millimeter; 25,400 microns = 1 inch).

It can be deduced from the same study that somewhat larger velocities are required for vertical than for horizontal transport, in the following ratio:

$$\frac{V_v}{V_h} = 0.27d_m^{0.2}$$

where V_v is the velocity required for transport upward in a vertical duct, and V_h is the velocity for horizontal transport.

Limitations of DallaValle Data. The formulas presented above must be viewed in the light of two limiting experimental conditions. They were obtained on a limited range of particles sizes (1 to 5 millimeters), and they were clean dry inertials. They doubtless provide a good guide within these limits. The importance of the second limitation is discussed in the following section.

Transporting Velocities for Mixtures of Inertials and Fine Dust. During transport through a straight duct, dust particles are constantly bombarding the walls of the duct because of the turbulent motion of the air stream and by gravitational effects in the case of inertials. By this mechanism the large inertials effect a scouring action tending to remove any fine particles previously deposited. The extent to which this action can maintain a clean interior surface is clearly dependent on the number and average weight of the inertials in relation to the proportion of very fine dust in the mixture, as well as the velocity imparted to them by the air stream.

It has been suggested that the coating of fine dust often observed in dust exhaust systems is due to electrostatic effects or to the hygroscopic nature of the dust. While it is possible that those qualities sometimes operate, they cannot effect coating of interior duct surfaces and eventual clogging of the system in the presence of the "sand blasting" action described.

This concept could serve as the basis for useful studies of required transport velocities for industrial dust mixtures in which a simple screen analysis, specific gravity, and nominal identity of the dust would constitute the essential data together with the critical velocity range.

A considerable number of practical data exist, but the dusts are often inadequately identified either as to particle size or process (not enabling experienced engineers to infer size characteristics).

Traps for Rejecting Inertials

Arrangements can be readily incorporated into the design of an exhaust system to reject very large inertials when velocities of 2,000 to 3,000 fpm are found adequate for transport of the resulting mixture of fine dust and small inertials. A margin of safety is desirable, however, and a figure of 3,000 fpm or a little larger is therefore indicated.

Rejection of large inertials where remaining material is conveyed by velocities of 3,000 fpm is accomplished very simply in two ways: one, by exhausting from a spacious enclosure with a baffle between the duct opening and the sources of flying inertials or, by similar means, preventing entrainment of inertials of excessive size and, two,

by providing a trap in the duct line as close to the hood opening as practical, for the case where large inertials cannot be prevented from entering the throat of the hood. Various hood forms and traps are illustrated in Figure 12-1a. It is clear from these illustrations that the trap can be simply constructed, even crude, for these purposes. In contrast, the arrangement in   Figure 12-1b permits no gravity separation of large inertials and would therefore require duct velocities at least 4,500 to 5,000 fpm.

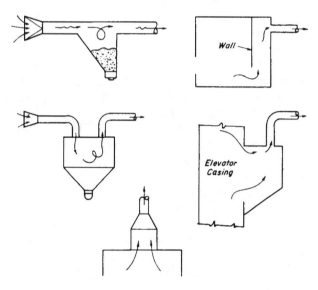

Figure 12-1a. Traps for rejection of inertials. Sizes passing are readily transported by duct velocities of 3,000 fpm.

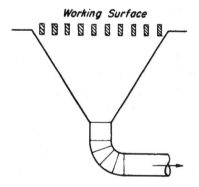

Figure 12-1b. Exhaust arrangement requiring high duct velocities.

There are two schools of thought concerning the advisability of designing exhaust systems for complete removal of all inertials, large and small, along with the fine dust. One attitude is represented in the standards of good practice for foundry dust control which advocates minimum duct velocities of 4,500 fpm in order to remove, pneumatically, the largest inertials encountered. Yet, experience indicates that by the simplest inertial-rejection traps, velocities of 3,000 fpm are adequate and liberal. It is argued that rejected material must be removed from traps by manual or mechanical means at suitable intervals, and this contributes to poor housekeeping during intervening periods; that use of a single system (the exhaust system) for both fine dust control and automatic removal of large inertials to a central disposal point is a simple and feasible procedure, especially desirable in mass production factories.

The pneumatic removal of large inertials represents an unnecessary economic waste in most instances and should be adopted only after careful consideration of the costs and advantages.

Power Cost of Higher Velocities. Since horsepower is proportional to the product of flow rate and pressure, and pressure losses are proportional to the square of velocity, the power requirement for a fixed exhaust rate, cfm, will be proportional to velocity squared. A comparison of system velocities of 4,500 fpm and 3,000 fpm as to power requirement indicates that use of the higher velocity would require $(4,500/3,000)^2 = 2.25$ times as much power as the lower velocity.

Abrasion of Duct Work. Conveying of inertials in a duct system subjects the walls to a continuous abrasive blasting action which is greater with increasing velocity, greater with increased size of particles, and greater with increasing quantities of material. These three influences increase the rate of wear on the duct material enormously. The wall thickness needs to be made greater to provide for a reasonable life of the system, at added initial expense, of course.

Multiple Branch Exhaust Systems

Because the different branch ducts of the typical exhaust system must inevitably be located at varying distances from the fan and be connected by main ducts that occasion different static pressure losses in the air stream, the flow of air through branches nearest the fan is facilitated and the flow through those more remote is impeded. Even more important, differences in resistance are occasioned by the use of varying size of duct in various branches. The result is that losses through the various branches must be adjusted in order to attain the

minimum airflow specified for the branch least advantageously located.

Method of Equalizing Friction

Such losses through each branch may be equalized (1) by restrictions in branch ducts (for example, dampers); (2) by systematic reduction of the diameter of some of the branch ducts; or (3) merely by exhausting greater than necessary volumes of air (which may be wasteful both in first cost of equipment and in power for operation). The second method is the best one and is developed in the following section.

Significance of Balanced Flow Design

The essence of good design method is in an orderly tabulation of data and calculations. Engineers having frequent occasion to design dust control systems will find it convenient to have calculation forms printed for ready use. [Editor's note: See for example, worksheets and computer/calculator programs available in the ACGIH *Vent Manual* and the editor's *Industrial Ventilation Workbook*.] Such forms are particularly valuable in the training of young engineers and facilitates checking of calculations.

Typically, calculations show significant figures more than warranted by the nature of the problem. Rounding off figures should be performed at the end of the calculations. In this connection, it is important that one recall the approximate basis on which original exhaust rate specifications are estimated. For example, if 1,200 cfm is specified, "1,200 cfm" could easily signify 1,100 to 1,300 cfm based on the state-of-the-art of design [which is that actual flowrates usually vary by about ± 10% design even under the best of circumstances]. Closer specification of exhaust rates is seldom warranted in dust control.

A problem will arise when the smallest branches have the highest friction losses and are farthest from the fan. The design is then penalized by the additional friction loss in sections of main connecting them to the fan. This suggests a general principle of considerable practical value enabling one, by simple inspection, to avoid grossly unfavorable duct arrangements, as follows:

Avoid where possible duct designs for dust systems where very small diameter branches and large diameter branches are connected to the same fan. Where large and small exhaust rates are indicated in a zone one may (a) attempt to connect the smallest branches to the section of main nearest the fan, (b) waste

exhaust capacity by arbitrarily increasing the exhaust rate specification from the smaller branches or (c) provide a separate exhaust system for the smallest branches. It will be clear from the foregoing why small diameter branches are so frequently found to be choked off by an accumulation of settled dust along their length.

Hood Static Pressure Specifications

In code rules administered by labor departments of several states [as of about 1950] one common provision specified that exhaust systems for grinding and buffing wheels and woodworking machinery be so designed that the static suction, measured in the branch duct just downstream from the throat of the hood, would not be less than 2 inches water gage. Diameters of branch ducts for various types and sizes of equipment were also specified. This specification originated many years ago in code rules of the state of New York [and had been adopted by a number of states]. Its significance in terms of dust transport as discussed above is of interest as brought out in the following problem.

Example 1. About what velocity in a duct will be found when a hood static suction of 2 inches w.g. prevails and the hood consists of a plain duct end?

Solution. The head loss at entrance to a plain end duct is approximately 1 velocity head ($h_e = 1VP$). Static suction [hood static pressure, SPh] is the sum of entry loss and velocity head.

$$h_s = h_L + h_v$$

and since the loss in this hood is 1 velocity head, $h_L = h_v$

$$h_s = 2h_v = 2 \text{ inches w.g.}$$

and

$$h_v = 1 \text{ inch w.g.}$$

from which

$$V = 4{,}005(1'')^{1/2}$$

$$= 4{,}000 \text{ fpm} \qquad\qquad \text{at STP}$$

The 2-inch suction rule is thus seen to specify transport velocities that are adequate to keep ducts continuously free of accumulations of debris and thus ensure effective dust control without the necessity of constant maintenance. Specifying branch duct size at the same time fixes the exhaust rate. Thus the two factors of prime importance to successful dust control are defined by a relatively simple formula easily checked by non-engineer contractors and state inspectors.

Blast Gates

A common method of balancing air flow between branches of a dust exhaust system involves sizing all ducts in proportion to the desired flow and providing each branch with a blast gate so that the flow can be adjusted after the system is in operation. This method should not be employed with dusts which would tend to collect on the gate projection, e.g., fibrous dusts, or highly abrasive materials such as those containing large inertials like stone chips. Very fine dusts that are free of any inertials and consequently "sticky" also would tend to accumulate around the gate and eventually plug the branch. If gates are to be used, they should be adjusted with the aid of water gage readings as soon as the system goes into operation and bolted or riveted into position.

Frequently blast gates are built into systems and left free for adjustment by the operators. This common practice has no virtue to recommend it and, unless there is an excess air velocity incorporated into the design, actually may be harmful by reducing velocities in some section of the main to a value permitting settling of debris. It is sometimes thought advantageous to be able to switch the exhaust capacity away from the unused equipment, but even this is largely fallacious. For one thing, if there are times when all equipment is used simultaneously, then the exhaust system should be designed for that condition and there is no need for switching additional exhaust capacity from an unused unit to another.

More interesting is the fact that the capacity of the branch that is cut off is not fully available to the other branches because of the interaction between increased friction loss (due to increased air flow) on the one hand, and the sensitivity of the fan to increased head loss, on the other.

Effect of Hood Entrance Losses

One other point is of interest in connection with the present discussion. It might be thought, at first glance, that advantages may accrue in designing exhaust hoods of such shape that they would contribute only a minimum entrance loss, in accordance with the data

in Chapter 11. Experience with the problems in this chapter, however, demonstrates that this would be true only for the key branch, the one having highest loss (usually the one most remote from the fan). For other branches it is clearly pointless to select hood shapes to reduce losses to low values when in the same flourish it is necessary to deliberately increase friction losses in the branch to restrict airflow to the specified value.

Summary of Design Principles for Balanced Flow

The following paragraphs set forth in a qualitative way the important points concerning the design of balanced exhaust systems.

(1) The branch ducts of smallest diameter in a multiple branch system having branches of widely varying size are likely to have lowest air velocities and, consequently, they are the ones most likely to plug due to settling of dust.

(2) The branches of smallest diameter are best connected to the section of main closest to the fan.

(3) The most economical arrangement from the standpoint of air power requirement for situations where dust sources have widely varying exhaust rate requirements (i.e., large and small diameter branch ducts), is to segregate the small diameter branches into an exhaust system separate from those of large diameter.

(4) An alternative arrangement to (3) that is advantageous where the small branches are closely clustered—combine them by short lengths of duct into a main branch of larger diameter, thus reducing the friction loss in that segment.

(5) Where a single small diameter branch is combined in a system with large diameter branches, it is often economical to increase the size of that branch to waste some exhaust capacity through it, for the sake of an over-all reduction in friction loss of the system.

(6) Adjustable blast gates in a dust exhaust system are seldom of value and are usually disadvantageous. When installed, they should be used for "shutoff" purposes only.

Vertical Duct Arrangement

The use of high duct velocities for dust transport and consequent need for attention to balancing, as in the foregoing discussion, can sometimes be avoided by the arrangement illustrated in Figure 12-2. All ducts are disposed at a steep angle, 60° or more, thus avoiding any horizontal duct in which dust could plug the duct. In vertical ducts, dust settles to the bottom where it can be collected and handled.

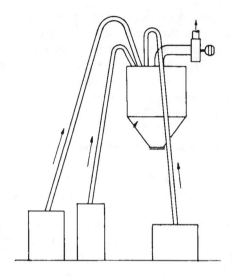

Figure 12-2. Where dust sources are clustered together, this vertical duct system may sometimes be employed. Steeply inclined ducts avoid the necessity for close attention to duct velocities and balance between branches.

Systems without Ducts

In Figure 12-3, a system is shown in which virtually no ducts are employed. Inertials from the operations drop into a hopper from which they are removed by mechanical conveyor, and fine dust is exhausted either to an overhead dust collector through a short vertical duct or directly to the exterior if quantities are not objectionable.

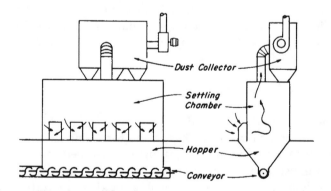

Figure 12-3. For operations where large quantities of heavy dust (inertials) are produced, removal by mechanical conveyor is usually indicated. Then there may be no need for ductwork as in this arrangement where all operations communicate directly to a central plenum or chamber.

Adjustable Dampers for Multiple Dust Sources

Situations are frequently encountered where dust is produced only intermittently at a number of different locations. If an exhaust system is designed for full and continuous capacity at all locations, a wasteful system results. If dampers are provided to block airflow from branches when not in use, yet with branch and main ducts proportioned conventionally with constantly increasing sizes as the fan or collector is approached, then velocities in the main duct will be decreased when flow through upstream branches is cut off and dust settlement may occur to clog the system.

If only a single branch will be in use at any one time, the entire duct system can be designed to have a uniform diameter for all branches and all sections of the main with exhaust capacity for a single branch. Aerodynamically this arrangement is satisfactory since it results in adequate velocities for dust transport in all parts of the system at all times. In practice, however, it is less than perfect because of the need to depend upon workers to open and close the dampers. This difficulty can be minimized by ingenious mechanical arrangements that greatly simplify the business of damper operation, but the only good solution is a completely automatic arrangement for opening and closing of dampers at the proper time.

Chapter 13

Evaluation and Control of Heat Exposures Using Ventilation

Terms and Units for Chapter 13

V = air velocity, fpm

vp_s = vapor pressure of water at skin temperature, millimeters of mercury

vp_{air} = vapor pressure of water in surrounding atmosphere, millimeters mercury

Alleviation of excessive heat in the workplace can be accomplished by one of the following ventilation-based methods:

(1) Supplying mechanically chilled air to the space, i.e., air conditioning. This method is generally not applicable to the hot and heavy industries with which this section is primarily concerned. Industrial applications of cooling in this chapter are primarily for process reasons, e.g., maintenance of a uniform temperature or humidity to facilitate the manufacturing process. Mechanical cooling applications in industry for worker comfort occur where there is no significant process heat. In these cases, the considerations are the same ones that apply to any other air conditioning installation.

(2) Flushing the space housing the workers and process heat sources with large volumes of outdoor air, i.e., general ventilation. The elementary heat capacity formula describes the relation between ventilation rate and temperature rise in the ventilating air.

Btu per min   = lbs air per min x (0.24) x temp. rise
                   = cfm x (0.075) (0.24) x temp. rise

or   cfm = 55 x (Btu per min) x (temp. rise)

231

In very warm weather, precisely the time when the need for thermal relief is most acute, the outdoor air has its lowest capacity for absorbing heat within acceptable limits for temperature rise. Clearly, therefore, this method has limited capacity for effecting *total* control. It will be seen that general ventilation is important as an adjunct to other measures. Furthermore, the discussion of this paragraph does not include the beneficial effects due to high air velocity, covered in a later item below.

It is interesting to conceive the practical economic limitations to mechanically induced ventilation in comparison with natural ventilation due to outdoor air motion. The enormous ventilation provided by natural wind forces is readily illustrated by imagining a small industrial building 100 feet long with open sides 20 feet high through which a "gentle" 10 mile-per-hour breeze passes freely. Since 10 mph is equivalent to about 900 fpm, there occurs—through the 2,000 square foot opening—natural ventilation at a rate of about 1,800,000 cfm. Ventilation of such magnitude is hardly duplicable by fans when costs are considered.

(3) Application of local exhaust hoods to all heat sources for direct removal of heated air, gases, or steam through duct systems to the exterior. Canopy hoods over such sources can be designed in accordance with principles set forth in Chapter 8. The air currents in this instance are of relatively small scale as compared with (4) below.

(4) In buildings where processes generate very large convection currents of hot air or gases which rise buoyantly, roof openings may be provided for their easy egress. The principle of operation is practically the same (except for dimensions of air current) as the hoods of (3) above, where the upper part of the building with its openings corresponds to the canopy hood of the smaller scale exhausting process. The principles have also been discussed in Chapters 8 and 15.

(5) The most common and most familiar method for alleviating the discomforting effects of heat is by blowing volumes of air at the exposed individual, as by fans or air jets, i.e., bathing the worker's body in high velocity air currents to accelerate the dissipation of body heat. In the temperature conditions usually prevailing, the main heat transfer mechanism is by evaporation of sweat from the skin wherein the latent heat, approximately 1,000 Btu per pound of water evaporated, is the significant factor. A great deal of research has been conducted to determine the rate of heat dissipation from the human body when subject to air currents of different velocity and humidity.

It is not possible to apply this knowledge to a certain calculation of heat dissipation due to many complicating factors, but an

approximate expression is available for estimating the maximum rate of body heat dissipation physiologically possible by evaporation of sweat and it can serve usefully in the analysis of hot environments, i.e.,

$$E_{max} \approx 10V^{0.37}(vp_s - vp_{air})$$

where E_{max} is the heat dissipation in Btu per hour, V is air velocity impinging on the body, feet per minute, vp_s is a vapor pressure of sweat (at skin temperature), and vp_{air} is the vapor pressure of water in the air, in millimeters of mercury. Skin temperatures vary in accordance with the thermal stress and physiological factors. For present purposes it is convenient to take it as 95°F. The corresponding vapor pressure is $vp_s = 42$ millimeters of mercury. Then

$$E_{max} \approx 10V^{0.37}(42 - vp_{air}) \tag{1}$$

There are certain physiological and environmental considerations which make it necessary to regard this expression as only a rough approximation, although correct as to order of magnitude. The value of vp_{air} can be obtained by taking the vapor pressure of water at the air temperature from steam tables, in millimeters of mercury, and multiplying by relative humidity expressed as a fraction. Figure 13-1 provides data for this purpose.

Example 13-1. What maximum heat dissipation might be expected when the temperature is 70° F, RH = 50%, and the airflow over a worker's body is about V = 700 fpm?

Solution. From Figure 13-1, vp_{H_2O} at saturated conditions = 18 mm Hg

$$E_{max} \approx 10V^{0.37}(42 - vp_{air}) \approx 10(700)^{0.37}(42 - [18 \times .50]) \approx$$
$$\approx 3,700 \text{ Btu/hr}$$

(6) Radiant heat is one of the most important factors in industrial hot environments. As is well known, it is transmitted in accordance with Stefan's fourth power law, and is unaffected by air motion. It can be impeded or intercepted only by shields placed between the source and the person. While these facts, as principles, are well known to all engineers, it is astonishing to observe how consistently they are ignored in industry. Radiant heat from hot surfaces is allowed to bombard the bodies of workmen unimpeded by shielding

screens, then cool air from a duct system, or from air circulating fans, is blown at the man to accelerate evaporation of sweat.

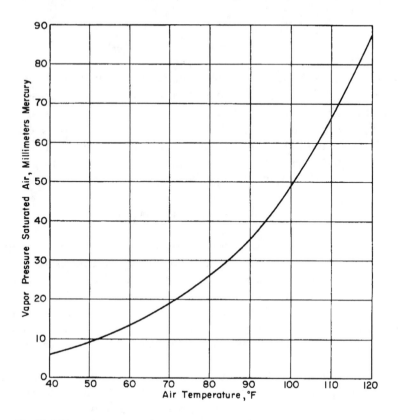

Fig 13-1 Vapor pressure of water in saturated air.

The design of shields for interception of radiant heat is peculiar to each industrial situation, but the problem is a tangible one involving common sense and some imagination to effect the improvement in a manner not to interfere with normal operation of the industrial process. Use of such shields has lagged mainly from insufficient appreciation of the magnitude of radiant heat transfer in relation to that by the convective process. Once its importance is grasped, a major incentive is apparent and improvements, often spectacular, can be effected at low cost.

[Editor's note: At this point in the original book, Hemeon provided methods for estimating heat stress that are more effectively covered in recent occupational health science texts. See for example, the AIHA "White Book."]

Chapter 14

Field Observations
for the Assembly of Design Data

The detailed analysis of an emission/exposure problem is an important phase of exhaust ventilation design. Preceding chapters must also be understood for effective assembly of the design data.

Location of Air-Contaminating Sources

The various points of origin or sources of dust, fume, vapor, etc., in an area requiring some form of ventilation will only sometimes be obvious. The more common problems are those in which the exact source(s) and the exact pulvation mechanism(s) are not apparent from a casual survey. It may be clear that individually small amounts of a highly toxic dust emanate from a large number of individual sources in a certain area, equipment assembly, or large machine. The temptation may be to design a large hood covering the entire machine, or to provide a large volume of general ventilation to the area, or to design a partial enclosure with exhaust. While any of these might be justified as the final design, they cannot be justified until all the particular sources have been located, the pulvation mechanism understood, and an attempt made to rate the several sources according to their probable relative contribution to the over-all contamination. With these data, the feasibility of designing local exhaust hoods can be appraised.

Situations requiring particular care by such an analysis are usually in one of the following categories:

(a) *A toxic substance like lead dust having a very low permissible atmosphere concentration.* In the manufacture of lead storage batteries a large number of handling and mechanical processes occur, most of them non-dusty by ordinary standards, but which do produce excessive quantities of dust when the result is compared with the atmospheric concentration considered acceptable to avoid lead poisoning [0.05 milligrams per cubic meter of air, 1998, OSHA PEL].

Maintenance of such low concentrations is one of the most difficult of all industrial ventilation problems. Most of the dust sources are invisible, and they are numerous and widespread. After initial exhaust of obvious dust sources the remainder must be identified by special effort.

(b) *Operations which produce such large volumes of dust as to visibly obscure the actual sources.* For example, an aggregation of crushing and grinding machinery within a small area may obscure visibility making it impossible to directly observe the individual emission sources contributing the dust.

(c) *A gas- or vapor-handling process in a series of equipment elements.* This is especially true where the acceptable atmospheric concentration is extremely low as would be the case for a highly malodorous substance like some of the mercaptans or the insecticide benzyl hexachloride. In this instance, the difficulty stems from the invisible character of the gas.

Identification of the source is subject to variability in interpretation of the term. It is intended to emphasize here the importance of tracking the source down to the actual pulvation or evaporation locale, thus separating this from the often confusing effect of *secondary air currents.* For example, in a process involving the continuous treatment of sheet material in a machine, one or two rolls may account for pulvation (or evaporation) of nearly all the contaminant, but, because of high velocity secondary air currents in the same vicinity induced by high speed roller elements, the over-all picture is confused, often to the disadvantage of a simple ventilation design. One observer may be satisfied to have identified the source in the general zone of a machine that includes both primary pulvation and the effect of secondary air currents. Another may succeed in narrowing it down to a single rolling element. In the latter case, it may be found possible to apply baffling and local exhaust of only moderate magnitude to obtain control of air contamination; or in some instances, discovery of the exact mechanism may suggest relatively simple modifications of the process or of the machine, to eliminate the contamination without any exhaust. The dividends from careful study of contamination sources are sometimes great.

Tyndall Beam Observations

One simple tool in the observation of dusty processes is some form of Tyndall beam illumination. In dark areas, observations will be improved by almost any kind of illumination. However, the Tyndall effect is the most valuable, and the tool emphasized in this discussion.

In the present application, it consists of the following simple elements: (a) a light source of any kind or intensity, (b) a shield interposed between the light source and the eyes of the observer, (c) arrangements to avoid or minimize reflection of light to the eyes of the observer from elements of the equipment being studied, and (d) a relatively dark background for the observer's eyes.

In its simplest form an ordinary pocket flashlight may be employed. The shield may be provided by one's hand, cupped around the flashlight lens to obscure all parts of the reflector. Reflection of light from nearby elements is minimized by the manner in which the light beam is disposed in the observer's hands; and the dark background can often be arranged by selecting the angle of sight so that it is lined up with some dimly lit wall of the room. It may also be realized by interposing a dark background in line with the observer's line of sight.

Simple arrangements like these may suffice for many problems. However, the value of this tool can never be fully appreciated until one has experienced the spectacular visual effects obtainable with more nearly ideal Tyndall illumination, in particular, high intensity illumination. (Figure 14-1.) Readily available equipment providing excellent results consists of a 110-volt extension cord, light socket, and a 150-watt bulb (or a short-life photographic photo-flood bulb) provided with a shield that prevents reflection of light directly to the observer's eyes. A dark background is readily provided by a panel of black cloth or paper; black velvet is nearly ideal.

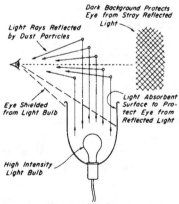

Figure 14-1. Diagram of high intensity local illumination to produce a Tyndall beam valuable in detecting dust sources, showing arrangements to minimize light reflection from sources other than dust particles.

A portable flashlight can be fitted with a light shield where a tool that can be independent of electric power lines is needed. Such a

flashlight should be, however, very bright and its limitations in comparison with more intense lighting should be recognized. The specialist in this work should be equipped with both this type and with one utilizing the power line circuit.

The limitation of the Tyndall light obviously is in its narrow beam. It creates the illusion that all the dust is concentrated within the area of intense illumination, much the way sunlight passing into a high mill building through the monitor creates the illusion that all dust is travelling upward along the beam of sunlight directly out through the monitor openings. The novice needs to be on guard against this; the Tyndall beam should be shifted into various positions in order to illuminate successively all portions of the zone of interest.

Measurement of Air Motion Outside Ducts

The appraisal of an existing ventilation system often can be made by direct observation without resorting to analysis by the techniques of occupational health. If the contaminant is a dust of low toxicity it may be that mere demonstration of the withdrawal of all visible dust by local exhaust suffices for the appraisal. Similarly, if it is invisible vapor, the effectiveness of local exhaust hoods could be demonstrated by observing the behavior of smoke streams. It is often desirable to measure low air velocities around local exhaust hoods especially at openings to partial enclosures like booths.

Use of Smoke Tubes. The generation of visible smoke to follow the travel of air currents incident to the operation of ventilation equipment is a useful device. Smoke tube kits are available from several different manufacturers.

One simple piece of equipment that can be assembled from glassware and reagents readily available in any chemical laboratory is illustrated in Figure 14-2 (top). Chloride smoke is formed by blowing air successively through bottles of ammonia and hydrochloric acid. This form of generator is best suited to the investigation of a particular problem where it is desired to assemble the unit quickly and conveniently. For general survey work, its portability and handling convenience is inferior to that of other forms.

The most convenient form of smoke tube is one in which air is blown through titanium tetrachloride (anhydrous), or stannic chloride (anhydrous). Figure 14-2 (bottom) illustrates a useful form of such a smoke tube. When air is forced through the mass of wetted granules, moisture in the air combines with chloride vapor, hydrolyzing it to a visible fume of hydrochloric and metal hydroxide. The smoke is somewhat irritating to breathe.

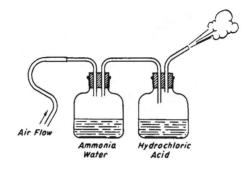

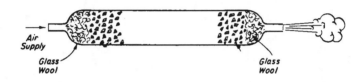

Figure 14-2. Simple smoke generators for studying air currents. (Top) Acid-ammonia smoke generator. (Bottom) Smoke tube generator using tin or titanium tetrachloride.

Testing Local Exhaust with Smoke. Among the many uses of the smoke tube in the study of ventilation air currents is the observation of local exhaust action where vapors or fumes do not permit direct observation of local air currents. A puff of smoke is released at various distances from the hood opening within the zone of contaminant generation and the resultant airflow observed. This technique is particularly useful as a follow-up to an air analysis survey which has indicated excessive contamination of the atmosphere even though an exhaust system is in operation, since it permits one to locate the zones of "leakage" from the various influence zones of the local exhaust hoods. Conversely, if in advance of an air analysis survey, smoke observations indicate some leakage, the significance of this cannot be inferred in quantitative terms; air analysis is necessary to determine its coverage effect in terms of average atmospheric concentration in the locations where workers are stationed.

Illustrative of the utility smoke observations, the performance of slot exhaust on a tank, for example, can be learned by observing the motion of a puff of smoke released near the center line of the tank, i.e., a point of contaminant release most remote from the exhaust openings. [Figure 14-3 (a).]

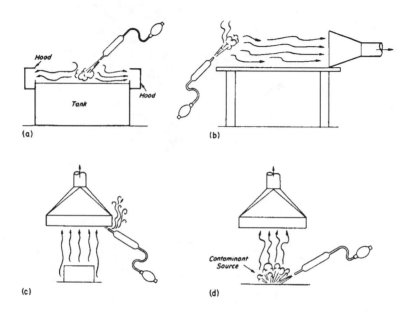

Figure 14-3. Use of smoke generator around exhaust hoods where the contaminant is invisible makes it possible to determine efficacy of hoods in control of contaminants. Smoke is ejected at the least favorable location in each case.

Lateral exhaust designed to withdraw contaminant from the total surface of a work bench would be tested by observing the travel of smoke release at the front edge of the bench, a point most remote from the exhaust at the rear (assuming, of course, that contaminant is also released near the front edge of the bench). [Figure 14-3(b).]

In the case of a canopy hood operating as a receiving hood above a hot process, smoke should be released at several points along the perimeter of the hood in order to establish whether there is any spill of hot, contaminated air outward into the open air. [Figure 14-3(c).]

Tetrachloride smoke tubes do not work well in currents of hot air, hence observation with them in such circumstances is inconvenient and difficult.

As has already been implied above, the escape of occasional wisps of air at points most remote from exhaust openings is not necessarily to be construed to indicate unsatisfactory operation of the exhaust hood. It may be that this represents, on the average, an insignificantly small percentage of the total contaminant. One can acquire, with some experience, a sense by which to mentally integrate observations of contaminated air quantities escaping from portions of the contamination zone, and draw conclusions as to the probably significance of the observed leakage.

We have seen, furthermore, how dilution can operate in combination with incomplete local exhaust coverage, especially in the case of solvent vapors, to result in satisfactory atmospheric concentrations.

The location of the most significant air velocity for various types of local exhaust installations is summarized in Table 14-1. Perusal of this summary will indicate or remind one of the points where, by the aid of smoke or otherwise, air currents should be observed in testing hood performance.

Table 14-1
Location of Critical Air Velocity in Various Hood Arrangements
(Important observation points for inspection by smoke tubes)

Type of Hood	Important Observation Point
Exterior hoods	At boundary of contamination area most remote from the exhaust hood
Booths (cold process)	At face, i.e., the plane of booth opening
Booths (hot process)	At face, and upper edge of opening
Enclosures (cold, no air induction)	At plane of openings
Enclosures (heated process)	At plane of those openings in topmost portion of enclosure
Enclosures (air induction)	At openings "downstream" of flow into enclosure. Process dust often provides visibility without smoke
Canopy hoods (cold process)	At boundary of contaminant area furthest below the hood
Canopy hoods (hot process)	At outer rim of hood

Air Velocities in Open Space

There is occasional need for the measurement of air velocities in free space such as in the zone adjacent to an exterior exhaust hood; more frequently it is desirable to measure air velocities at the face of a booth at openings into exhausted enclosures and at doorways leading into rooms ventilated at high rates. One traditionally employed instrument for this purpose was the swinging vane anemometer or velometer. A more recent addition is the thermal or hotwire anemometer or velometer which measures velocity in

accordance with the rate at which heat is lost from an electrically heated element. See your HVAC-equipment supplier for details.

Applications for Air Velocity Measurements

In many circumstances, the rate of ventilation of an enclosure, e.g., a paint spray booth, a ventilated room, or the like, can be estimated by measurement of the linear air velocity through the significant openings and the total area of openings, by $Q = VA$, i.e., the quantity rate of airflow is the product of the air velocity and the area of the openings. Care is needed in traversing the entire opening, particularly if there is some release of heat within the enclosure, for one may then observe strong inward velocities near the bottom of the opening, and weak air movement or even outward airflow near the top of the opening.

A velocity meter is sometimes useful as a supplement to the smoke tube, to demonstrate spillage of warm, contaminated air outward from under the perimeter of a canopy hood.

A number of miscellaneous uses for air velocity data will occur to the occupational health scientist, such as estimating the magnitude of air motion in a work space for which local exhaust is to be designed, measuring drafts in relation to the comfort of lightly clad workers during the heating season, and air velocities as they relate to comfort of workers in the hot industries during summer weather. It is useful in the development of experience to require frequent air velocity measurements.

Problem Definition

In earlier chapters, bases have been presented for reducing to its elements any ventilation problem that may be encountered, and in the present chapter, some techniques and tools have been described that will aid in accurate observation of the importance characteristics of the process to be ventilated. In the following section, elements of various processes and ventilation methods that have already been touched upon in several preceding chapters are summarized.

All processes, it will have been observed, can be broadly classified into cold processes (multidirectional and unidirectional pulvation) and hot processes. The data to be observed for each type of process are summarized in the following sections.

Cold Process Data

Cold, Multidirectional Processes. (1) With the aid of good Tyndall beam illumination, the pulvation distance, i.e., location of

the null point, concepts of which have been fully outlined in Chapter 6, can be determined. If dense clouds of dust obscure vision, some aid may be possible by arranging a temporary fan to partially clear the atmosphere. If this is not possible, it may be necessary for the observer to arrive at conclusions on the basis of inference. This implies that the pulvation mechanism has been located and identified.

(2) While impossible to define in exact quantitative terms, except in the case of solvent vapors, some attempt should be made to note the quantities of dust being dispersed into the atmosphere and as experience is acquired, this sense will assume value.

(3) In the case of dust, qualitative notes should be made of the behavior of inertials, whether the quantities are sufficient to be a housekeeping or handling nuisance, and in what direction they are projected, whether by mechanical high speed elements or by gravity.

(4) Note the nature, origin and magnitude of secondary air currents whether there are some that are caused by the process, whether they impinge directly on the primary pulvation *zone*, or whether they are air currents characteristic of the space and the weather. If the latter, it will be useful to measure their magnitude by one of the air velocity instruments described earlier in this chapter. The observer should visualize the location of a baffle that might be interposed between the pulvation zone and the source of secondary air currents, regardless of whether such a baffle is indicated in the final design of exhaust. These items are summarized in Table 14-2.

Table 14-2
Form for Observing and Reporting Design Data:
Exhaust Requirements of Cold Multidirectional Processes

1. Name of operation

2. Description of pulvation action: causes, locations

3. Recommended hood location and baffles; location of null point and X-distance

4. Control velocity recommended at null point

5. Exhaust rate recommended (Q=VA)

(5) Finally, a survey of the operational characteristics of the process should be made in relation to the kind of exhaust hood that may be judged feasible. While an exhausted enclosure is, of course,

the most efficient, an exterior hood is simplest, and, if exhaust requirements are low, may be most acceptable and practicable.

Cold Unidirectional Projection. The following elements describe the data needed to fully understand the pertinent characteristics of this type of pulvation process.

(1) The rate of generation of dust or inertials in pounds per minute, or like units. This might sometimes be inferred roughly without measurements but a careful estimation is desirable. In the discussion of Chapter 7, an unanswered question was raised covering the particle sizes significant in the air induction process. With this question in mind, the author in one instance measured the weight of material that could be collected by gravity in a deep receptacle held in line with the axis of a stream of dust and inertials from a rotating cut-off wheel cutting clay tile. (See Figure 14-4.) While this was only a fraction of the total dust formed in the process it was clearly the factor most significant in the air induction process.

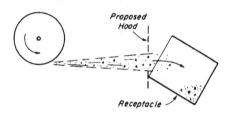

Figure 14-4. Arrangement employed for measuring the rate of flow, R, or inertials incident to an exploration of exhaust requirements for a receiving hood.

(2) The initial velocity with which inertials are projected needs to be noted, usually by inference, from the known speed of the mechanical element causing projection, i.e., the same speed can be assumed for practical purposes.

(3) The projection distance is to be noted. This distance is usually that between the point of origin of the dust and the location of a proposed receiving hood. The hood location depends largely on mechanical feasibility. In this observation, note should also be made of the maximum distance within which all fine dust appears to be efficiently carried without deviation.

(4) The angle of spread of the cone of the dust-inertial trail should be observed to permit determination of the cross-sectional area of the stream at points corresponding to those noted under item (3).

(5) Some information on particle size should be obtained. A sample of the dust-inertial mixture is collected for this purpose, perhaps in conjunction with the determination of weight rate described in item (1).

(6) A notation of the mineral character of the dust will generally suffice for determining the specific gravity of the particles, by reference to handbooks.

Falling Materials (Materials Handling)

The following data are required for complete description of dusty processes in the gravity fall of crushed material. (See Chapter 7.)

(1) The weight rate of flow of material, e.g., pounds per minute, a figure which is generally available from operating data.

(2) The falling distance, measured from the topmost point to the level *where the column of material enters* the enclosure to be exhausted. It should not include the falling distance within the enclosure except when, as in the case of enclosed chutes, the proposed point of exhaust is from an enclosure at the bottom of the chute and where the *entire* travel of the falling column is within an enclosure that is to be exhausted.

(3) The particle size distribution of the material must be obtained (see discussion of preceding section), and also (4) the specific gravity.

(5) The cross-sectional area of the falling particle stream at the elevation where it enters the enclosure to be exhausted. This is the point corresponding to the one determining falling distance as discussed under item (2) above.

(6) The splash characteristics are a significant element in the description of falling material. If the process is already enclosed (elevator housing, bin, etc.), and there are cracks or other openings in the housing at the same level where the fall of material is arrested, one needs to observe the extent to which dusty air streams outward through such orifices. It is often difficult in the absence of exhaust from the enclosure to determine whether the outward flow is due to the over-all induction of air or is simply the local splash effect resulting from close proximity between the wall and the point of material arrest. Inferences can be drawn and, also, comparison of observations at different points, of the housing. If the operation is not enclosed, the observation is easily made.

Heated Processes

The prime data required to describe aerodynamic characteristics of heated processes producing air contamination are five in number.

(1) The rate of heat release or generation available for heating the buoyant column of air that carries the contaminant upward away from the point of origin. Note that this excludes that fraction of heat that is radiated from the source and which, therefore, does not contribute to the heating of the air. This data can usually be obtained from the operating characteristics of the process although assumption may be necessary for estimating the fraction radiated.

In the case where heat originates at hot surfaces, satisfactory estimates can be made by observation of the surface temperature and of the total surface area corresponding to this temperature. (Then see Chapter 8.) If the source of heat is steam, the rate of evaporation provides a basis for estimating thermal transfer rate.

(2) The horizontal cross section of the rising hot air column is inferred from the horizontal cross section of the heat source. It is not to be confused with the hot surface area referred to in the previous paragraph, even though they may sometimes be the same.

(3) The vertical dimension of the hot body contributing its heat to the rising air column, if there be such a surface, or the horizontal dimension if it be a horizontal surface.

(4) The vertical distance between the source of heat and the contaminated air column, and the elevation of the bottom plane of the proposed canopy hood.

(5) Having in mind the possibility of applying a lateral exhaust hood to the process, some observations should be made on the apparent "strength" or deflectability of the rising hot air column. In the case of tiny heat sources like soldering operations, the rising column is easily deflected by cross drafts and this is readily demonstrated.

For large processes, on the other hand, the demonstration is not simple and, in any case, we do not know how to describe this factor in our present state of knowledge. The observer is advised, however, to note the behavior of the column as it is subjected to prevalent drafts of the particular space.

Measurement of Induced Airflow

Chapters 3 and 7 discuss the airflow induced by inertials projected at high velocity or falling by gravity. Since this phenomenon often determines the rate at which air must be exhausted from such apparatus as crushed materials handling systems, it is important to have available a technique for estimating the rate of such airflow.

The direct method involves the temporary installation of a test exhaust system with facilities for varying airflow. Then the exhaust rate can be increased until observations at other openings to the

enclosure indicate that all air leakage is inward, and the proper rate of exhaust is determined.

An indirect method utilizes the simple principle of dilution wherein a test gas-like carbon dioxide is metered from a supply cylinder at a fixed rate into the system at the point where induced air enters the system. (See Figure 14-5.) At a point downstream where the air is being discharged into the open, or into succeeding equipment, a sample of the dusty air is aspirated at a metered rate and the concentration of the test gas is thus determined, and simple arithmetic may be applied to determine the rate of airflow induced through the system.

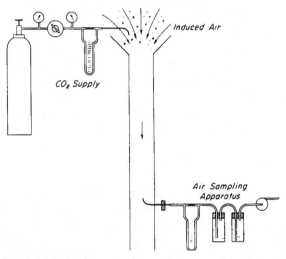

Figure 14-5. Diagrammatic layout of apparatus employed for airflow measurement by the dilution method using tracer gas.

In the initial consideration of a given problem of this character, painstaking visual observations should be made of the general character of air motion into and out of various openings using the strong illumination Tyndall beam technique detailed earlier, aided by smoke-trail observations with a smoke tube. Where the magnitude of airflow is large, special measurements to determine its rate may be warranted. Having established the flow pattern through the system in a qualitative manner, that is locating the points where induced air enters the system and where it leaves, the desirable location of exhaust connections will be apparent and the test problem then reduced to one of determining the rate of airflow between points of entrance and exit.

A cylinder of test gas is placed adjacent to one of the points of entrance and fed at a metered rate into the entering air stream. After a suitable time lapse for establishment of equilibrium, and while the

test gas stream flow is continued, air is continuously sampled at one of the exit points for a few minutes, or for a period determined by the requirements of chemical analysis, to determine the concentration of gas in the exit air stream.

The gas to be used is selected for analytical convenience or because it does not occur naturally in the process being studied. Carbon dioxide can be used but the background concentration must be taken into account.

It is important to insure thorough mixing of test gas with the induced air stream, but this constitutes no difficulty in most circumstances. Where there are, however, numerous points of air entrance and air discharge, more than ordinary care may be required to insure a truly representative sample of test gas in the analyzed air stream. It may be necessary to introduce the CO_2 into two or more secondary air streams before they join to form the main stream. Proof of uniform mixing is demonstrated when the concentration of the CO_2 is found to be the same at the several points of discharge.

A number of other test gases will suggest themselves for use in special circumstances. Convenience dictates the necessity that they be commercially available in cylinders.

Establishing Sources by Air Analysis

In the development of control systems for substances like lead dust or fume with very low permissible atmospheric concentrations, a point is reached where all obvious sources of air contamination have been taken care of by local exhaust hoods, yet air analyses in the working location may indicate concentrations exceeding permissible levels.

At this stage, intensive study of all possible remaining sources is required. Air sampling in the vicinity of suspected sources may be of assistance, but it is a laborious procedure and therefore of limited utility.

Where it is done, it should involve simultaneous sampling at two and preferably three locations, one near the source, and the other at a little distance reflecting background concentrations. During such sampling, close observations of air currents should be made and Tyndall beam tests, for example, made at the center of interest.

Measurement of Airflow by Static Pressure

The pressure loss at the entry, through a hood to a duct, can be regarded as analogous to the pressure drop across an orifice meter. The pressure is atmospheric in the open and is reduced in the duct to the value of the negative static pressure, which is readily measured.

If the coefficient of entry, C_e, is known, such a measurement can serve usefully for the determination of airflow rate into the hood. Fortunately, adequate information on these entry coefficients is becoming known, so that the technique to be described constitutes a useful and simple way for measuring airflow.

This technique consists in measuring the static suction or static pressure at one or two small holes in the wall of the duct a little downstream (2 to 4 duct diameters downstream from the hood entrance), as shown in Figure 14-6.

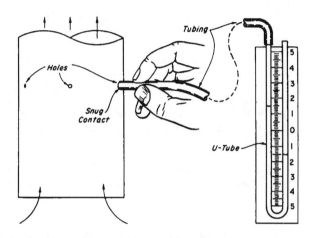

Figure 14-6. Measuring static pressure at a hood (plain end of a duct illustration) for determination of airflow.

The reading in inches of water is a measure of the velocity pressure in the duct plus entry loss at the entrance, i.e., through the hood into the duct. This reading of static head, h_s, taken in conjunction with the cross-sectional area of the duct A, and the coefficient of entry, C_e, which is dependent on the shape and proportions of the hood, enables a direct calculation of the airflow quantity, Q, in cubic feet per minute by means of the following formula (at STP):

$$Q = 4005AC_e\sqrt{h}_s$$

where Q = cubic feet air per minute
A = cross-sectional area of the duct, in square feet
C_e = coefficient of entry (values for different types of hoods are given in Figure 14-14)

h_s = height of water column measured with a manometer, in inches; static pressure in duct near the duct, SPh.

Missing a manometer? A U-tube, even though crudely constructed, can be used for these purposes in a pinch. It can be made by bending glass tubing into the shape of a U and attaching it to a strip of wood or metal having a scale marked in inches and tenths.

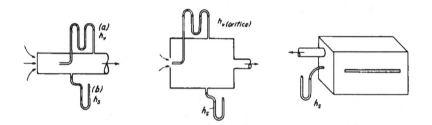

Figure 14-7 (left). The nature of the coefficient, C_e, is illustrated. When comparison is made between the Q equations using data of (a) and (b), $C_e = \sqrt{1/(K+1)}$.

Figure 14-8 (center). Where air flows through an orifice into a chamber, C_e is described differently than in 14-7, due to zero velocity in the chamber or plenum. In this case, $C_e = \sqrt{1/K}$.

Figure 14-9 (right). The pressure loss when air flows through a slot-shaped sharp-edged orifice is different from that of a circular orifice, for though the width of vena contracta is the same, the area is different.

Entry Coefficient. The entry coefficient is defined as that factor by which $h_s^{1/2}$ must be multiplied to obtain the equivalent value of $h_v^{1/2}$ in the ordinary flow formula, or

$$C_e = \sqrt{\frac{h_v}{h_s}} = \sqrt{\frac{VP}{SP_h}}$$

Thus, the flow might be determined as in Figure 14-7 from a measurement of velocity pressure, h_v. Then airflow is Q = 4,000A$(h_v)^{1/2}$. Alternatively, the static suction pressure, h_s, taken in conjunction with a suitable value of C_e could be employed to calculate the flow rate. Then

$$Q = 4005 A C_e \sqrt{h_s} = 4005 A \sqrt{h_v}$$

from which we obtain $C_e = h_v^{1/2}/h_s^{1/2}$. The value of h_s is the result of two pressure changes, one due to the establishment of velocity, hence it includes one velocity pressure, h_v, and the other due to energy irretrievably lost in turbulence, the head loss, h_L. That is, $h_s = h_v + h_L$. Therefore,

$$C_e = \sqrt{\frac{h_v}{h_v + h_L}} \qquad (1)$$

In the foregoing equations all h values are expressed in units of pressure, e.g., inches of water. If head loss be expressed in terms of velocity heads, i.e., $h_L = Kh_v$, then

$$C_e = \sqrt{\frac{1}{1+K}} \qquad (2)$$

where K is the number of duct velocity pressures lost in the entry, i.e., $SP_{loss} = K \bullet VP$.

Example 1. If a coefficient of entry for a given hood is $C_e = 0.90$, what is the head loss in numbers of VP?

Solution. Rearranging equation (2) and solving for K gives

$$K = (\frac{1}{C_e^2} - 1) = (\frac{1}{0.9^2} - 1) = 0.24$$

That is, a C_e value of 0.90 corresponds to an entry loss of about 1/4 velocity head (velocity pressure).

Example 2. If a hood has loss of $h_L = 1$ velocity head = 1VP, what is the entry coefficient?

Solution. The number of VP is one, so

$$C_e = \sqrt{\frac{1}{1+K}} = \sqrt{\frac{1}{1+1}} = 0.71$$

Some confusion may arise in applying these relationships to sharp-edged orifices and similar shapes where the orifice is followed by a chamber or large duct of different area from that of the

orifice, as in Figure 14-8. The following discussion illustrates the
correct development. As before,

$$Q = 4005A_{orifice}\sqrt{h_{v(orifice)}} = 4005A_{orifice}C_e\sqrt{h_{s(chamber)}}$$

from which

$$C_e = \sqrt{\frac{h_{v(orifice)}}{h_{s(chamber)}}}$$

The value, h_s, as in the preceding case, equals $h_v + h_L$, but since
velocity in the chamber is zero, $h_v = 0$, after the jet has expended its
energy , we get

$$C_e = \sqrt{\frac{h_{v(orifice)}}{h_L}} = \sqrt{1/K} \qquad\qquad (3, 4)$$

Head Loss in Round Sharp-Edged Orifice. The value for h_L is
obtainable from experimental data which has demonstrated that the
velocity at the vena contracta is about 1.65 times that through the
area of the orifice. (The reciprocal is 0.60, the well-known value for
coefficient of contraction in such an orifice.) Also, it is clear that the
head loss is precisely the value of one velocity head based on vena
contracta velocity since the stream energy is allowed to expend itself
turbulently to near zero velocity in the chamber. Then,

$$h_L = 2.8h_{v(orifice)} \quad or \quad SPh = 2.8VP_{orifice}$$

It will be noted that placement of this value for h_L in Equation
(4) results in the expected C_e value of 0.60.

Coefficient and Loss in Slot Sharp-Edged Orifices. One can
readily compute the head loss and entry coefficient for the slot-
shaped orifice shown in Figure 14-9 by application of the principles
and data discussed in the preceding section.

Synthesis of Entry Coefficients. An understanding of the
relationship between C_e and static pressure loss provided in the
preceding discussion enables one to synthesize a series of pressure
losses into a single coefficient as illustrated in the following.

Example 3. Consider a hood having an entry loss of 1 velocity
head, followed by an elbow having a loss of 1/4 velocity head

(Figure 14-10). Static suction measurements can be made only after the elbow due to space limitations. What entry coefficient, C_e, should be employed in airflow calculations?

Solution. The total loss factor $K = 1 + 1/4 = 1.25$. Hence

$$C_e = \sqrt{\frac{1}{1.25 + 1}} = 0.67$$

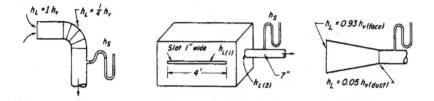

Figure 14-10 (left). Addition of losses and synthesis of entry coefficient, as in Example 3.

Figure 14-11 (center). Synthesis of an entry coefficient from individual losses, allowing for differing flow areas, as in Example 4.

Figure 14-12 (right). Entry coefficient for a simple shaped hood is deduced as in Example 5.

Example 4. A sharp-edged slot, 1 inch wide and 4 feet long, communicates with a chamber as shown in Figure 14-11. A 7-inch duct exhausts from the chamber. Compute the entry coefficient for the combination of orifices that can be applied to the static suction measured in the 7-inch duct.

Solution. Since areas of slot and duct are different, a little additional arithmetic is required to convert them to the same area basis before adding them together.

The loss through the slot is 1.3 velocity heads, based on orifice dimensions, $4 \times 1/12 = 0.33$ sq ft. The loss at entry from chamber to 7-inch duct is 0.5 velocity head, based on duct area, i.e., 0.267 sq ft. Convert the 1.3 velocity head loss to a loss in terms of duct velocity by which , added to 0.5 gives 1.35 duct velocity heads as the total loss, K. The loss summation may also be performed if desired by assuming a definite slot velocity and computing the losses in inches of water at both locations, adding them and converting back to duct velocity heads.

The value of C_e is then found by

$$C_e = \sqrt{\frac{1}{1.35 + 1}} = 0.65$$

Example 5. Consider the conical hood illustrated in Figure 14-12 (right) and estimate the magnitude of the entry coefficient on the assumption that two energy losses occur in series, first at entrance through the wide face and, second, at the throat. Assume the first head loss to be 93% of a velocity head based on velocity through the flare of the face, and the second loss to be 5% of velocity head in the duct (See Chapter 11).

Solution. An easy way to follow the calculations is to assume a velocity in the duct and compute losses in inches of water although the velocity factor actually cancels out and the calculation could be performed as in Example 4. If the duct velocity is 4,000 fpm, then the velocity head in the duct will be $VP = (V/4,000)^2 = 1$ inch. Velocity through the face will be $4,000/1.8 = 2,220$ fpm, and the corresponding velocity head, $(2,220/4,000)^2 = 0.31$ inch. Loss at the face is then $0.93 \times 0.31 = 0.29$. The loss at the throat is 5% of 1 inch, or 0.05 inch. Total loss is 0.34 inch and K is therefore 0.34 duct velocity heads. The corresponding value of C_e, from Equation 2, is

$$C_e = \sqrt{\frac{1}{0.34 + 1}} = 0.86$$

Experimental Hood Coefficients

[Editor's note: The loss factors presented by Hemeon, while still valid, have been expanded and/or refined in more recent texts.]

Experimental observations on entry coefficients for taper hoods of various shapes and dimensions are summarized in Figure 14-13.

The small angle, long taper hood (Figure 14-13) appears to experience its major loss at the face of the hood, and calculations like the last example, above, come close to accounting for the values found experimentally. The large angle hood (Figure 14-13B) are thought to have no significant loss at the face but appreciable loss at the throat, i.e., the junction of taper and duct.

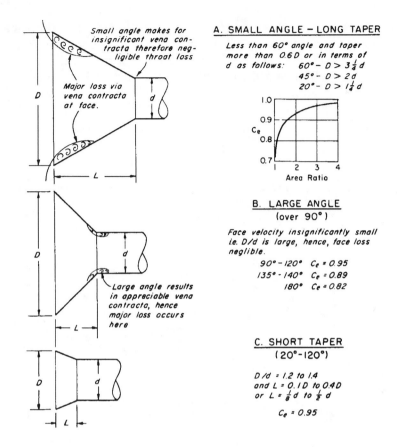

Figure 14-13. Effect of entry shape.

Hoods with abnormally short tapers (Figure 14-13C) minimize entrance energy losses and consequently have high entry coefficients by a happy combination of dimensions that results in a conjunction of the vena contracta formed by entrance through the hood face with the connecting duct section. This would nearly eliminate losses due to abrupt expansion from a vena contracta to a duct of larger diameter.

[Editor's note: In studies of similar hoods, DallaValle and others have reported C_e values ± 10% those of Figure 14-13. The difference can probably be explained by slight differences in hood construction.]

Exhaust from Chambers

The methods of determining airflow by measurement of hood static pressure are also applicable where a duct exhausts from a plenum or chamber, provided that the size of the plenum is large

relative to the duct diameter, thus resulting in no abnormal constriction to free flow in the vicinity of the duct entrance. In this case, it is necessary to correct for negative pressure prevailing in the chamber by subtracting the measured static suction in the chamber from the static pressure in the duct. (Figure 14-14J.)

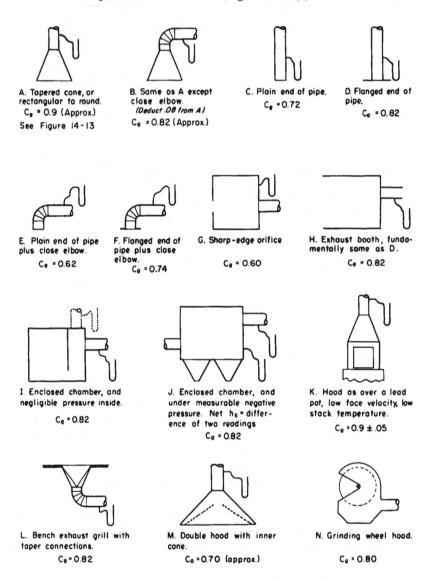

Figure 14-14. Summary of hood entry coefficients.

The most common example of this case is where a duct leads from a cloth dust collector under negative pressure. The method described

is not applicable, however, where, as often occurs, the length of duct from chamber to exhaust fan is less than about one diameter.

A fair approximation of the airflow entering some part of an exhaust system through a sharp-edged orifice or a fitting resembling such an orifice may also be obtained by static pressure measurements. As indicated in Figure 14-14G, the coefficient of entry for such an orifice is 0.60. Seldom will one encounter the laboratory sharp-edged orifice in industry, but occasionally there are similar pieces such as bench grilles with relatively small holes in the steel plate.

Example 6. There is a rectangular hole 3 inches by 2 inches in the side wall of a large duct. The static suction in the duct at this point is -2 inches (SP). How much air is leaking through the hole into the duct? (Assume STP.)

Solution. Substitute the above data in the general formula, using the entry coefficient of 0.60.

$$Q = 4005 A C_e \sqrt{SP} = 4005(\frac{3 \times 2}{144})(0.60) \sqrt{2} = 140 \text{ cfm}$$

Pitot Tube

The Pitot tube with manometer is a basic instrument for measurement of velocity pressures and will give accurate results if properly used. The Pitot tube, for best results, should be used in conjunction with an inclined gage [or calibrated electronic manometer] and by making a duct traverse, although in some circumstances passable results may be obtained with only the center line reading. The possible combinations of the instrument and number of readings are indicated below in increasing order of their accuracy.

  (a) With a vertical U-tube, single center line reading only
  (b) With a vertical U-tube, making a traverse
  (c) With an inclined gage, single center line reading only
  (d) With an inclined gage [or calibrated electronic manometer], making a traverse

In general, a vertical U-tube may be used only if the velocity pressure is more than 0.5 inch of water. An error of 0.1 inch in reading the gage in this range will result in an error in rate of airflow of about 10%. If the velocity pressure is 1.0 inch, an error of 0.1 inch in reading would amount to about 5% error.

Table 14-4
Approximate Locations of Pitot Tube in Six-Point Traverse
for Obtaining Average Air Velocity

Reading Number	Distance of tip from duct wall,%
1	5
2	15
3	30
4	70
5	85
6	95

Example 7. Airflow is to be measured by means of Pitot tube in a twelve-inch duct by traversing the diameter for six readings. How far in from the side should one measure? Measured values of VP are shown below. What is the approximate flowrate, Q?

Solution. The proper location of each of the readings is obtained from the data given in Table 14-4. Those locations are given in column (2) (below). Velocity pressure readings observed with an inclined gage are in column (3), computed velocities or those from Table 14-5, and average velocity and airflow under the tabulation. (Assumes STP.)

Reading Number	Distance from wall, inches	Measured VP inches w.g.	Velocity fpm
1	0.6	0.74	3440
2	1.8	0.79	3560
3	3.6	0.85	3690
4	8.4	0.80	3580
5	10.4	0.81	3600
6	11.4	0.72	3390

Average velocity = 3,540 fpm.

Airflow,   Q=VA

$\qquad$ =3,540 x 0.7854

$\qquad$ = 2,780 cfm

Table 14-5
Velocity Pressure to Velocity Conversions
At standard conditions rounded for US units.
For non-standard conditions, multiply $V \cdot d^{-0.5}$. For example, if VP = 1.00"w.g. and
d = 0.86 then V = 4,005 $\cdot$ $(0.86)^{-0.5}$ = 4,005 $\cdot$ 1.078 = 4,320 actual fpm.

VP inch/mm wg	V fpm mps		VP inch/mm wg	V fpm mps		VP inch/mm wg	V fpm mps	
0.01 0.25	400	2.04	0.39 9.91	2501	12.7	0.76 19.3	3490	17.8
0.02 0.51	566	2.88	0.40 10.16	2530	12.9	0.77 19.6	3514	17.9
0.03 0.76	694	3.53				0.78 19.8	3535	18.0
0.04 1.02	801	4.08	0.41 10.41	2564	13.0	0.79 20.1	3560	18.1
0.05 1.27	896	4.56	0.42 10.67	2600	13.2	0.80 20.3	3580	18.2
			0.43 10.92	2626	13.4			
0.06 1.52	981	4.99	0.44 11.18	2660	13.5	0.82 20.8	3627	18.5
0.07 1.78	1060	5.39	0.45 11.43	2687	13.7	0.84 21.3	3671	18.7
0.08 2.03	1132	5.76				0.85 21.6	3690	18.8
0.09 2.29	1202	6.11	0.46 11.68	2720	13.8	0.86 21.8	3714	18.9
0.10 2.54	1266	6.44	0.47 11.94	2746	14.0	0.88 22.4	3757	19.1
			0.48 12.19	2775	14.1			
0.11 2.79	1328	6.76	0.49 12.45	2803	14.3	0.90 22.9	3800	19.3
0.12 3.05	1387	7.06	0.50 12.7	2832	14.4	0.92 23.4	3841	19.5
0.13 3.30	1444	7.35				0.94 23.9	3883	19.8
0.14 3.56	1499	7.62	0.51 13.0	2860	14.5	0.95 24.1	3905	19.9
0.15 3.81	1551	7.89	0.52 13.2	2890	14.7	0.96 24.4	3924	20.0
			0.53 13.5	2916	14.8			
0.16 4.06	1602	8.15	0.54 13.7	2945	15.0	0.98 24.9	3965	20.2
0.17 4.31	1651	8.40	0.55 14.0	2970	15.1	1.00 25.4	4005	20.4
0.18 4.57	1699	8.64				1.02 25.9	4045	20.6
0.19 4.83	1746	8.88	0.56 14.2	2995	15.2	1.04 26.4	4084	20.8
0.20 5.08	1791	9.11	0.57 14.5	3024	15.4	1.05 26.7	4105	20.9
			0.58 14.7	3050	15.5			
0.21 5.33	1835	9.33	0.59 15.0	3076	15.6	1.10 27.9	4200	21.4
0.22 5.59	1879	9.56	0.60 15.2	3100	15.8	1.15 29.2	4295	21.9
0.23 5.84	1921	9.77				1.20 30.5	4385	22.3
0.24 6.10	1962	9.98	0.61 15.5	3128	15.9	1.25 31.8	4480	22.8
0.25 6.35	2003	10.2	0.62 15.7	3155	16.0	1.30 33.0	4565	23.2
			0.63 16.0	3179	16.2			
0.26 6.60	2040	10.4	0.64 16.3	3205	16.3	1.35 34.3	4655	23.7
0.27 6.86	2081	10.6	0.65 16.5	3229	16.4	1.40 35.6	4740	24.1
0.28 7.11	2120	10.8				1.45 36.8	4825	24.5
0.29 7.37	2157	11.0	0.66 16.8	3255	16.6	1.50 38.1	4905	25.0
0.30 7.62	2190	11.2	0.67 17.0	3278	16.7	1.60 40.6	5065	25.8
			0.68 17.3	3300	16.8			
0.31 7.87	2230	11.3	0.69 17.5	3327	16.9	1.70 43.2	5220	26.6
0.32 8.13	2270	11.5	0.70 17.8	3350	17.0	1.80 45.7	5375	27.3
0.33 8.38	2301	11.7				1.90 48.3	5520	28.1
0.34 8.64	2330	11.9	0.71 18.0	3375	17.2	2.00 50.8	5665	28.8
0.35 8.89	2369	12.1	0.72 18.3	3400	17.3	2.50 63.5	6330	32.2
			0.73 18.5	3422	17.4			
0.36 9.14	2400	12.2	0.74 18.8	3445	17.5			
0.37 9.40	2436	12.4	0.75 19.1	3468	17.6			
0.38 9.65	2470	12.6						

Table 14-5
Duct and Pipe Sizes Commonly Available

Diameter, Inches	Area Sq. Feet	Diameter, Inches	Area Sq. Feet
3	0.0491	14	1.069
3.5	0.0668	16	1.396
4	0.0873	18	1.767
4.5	0.1104	20	2.182
5	0.1364	22	2.640
5.5	0.1650	24	3.142
6	0.1963	28	4.276
7	0.2673	30	4.909
8	0.3491	36	7.069
9	0.4418	42	9.621
10	0.5454	48	12.57
11	0.6600	54	15.90
12	0.7854	60	19.63

Velocity Pressure Measurements

Where the pressure exerted by a free stream of air of known area can be conveniently measured, it affords a direct measure of airflow. Such a measurement may be made conveniently, for example, with a Pitot tube in the stream, pointing parallel and opposite to the flow. The other ends of the tube, connected to a water manometer, register the velocity pressure directly, in inches of water. The velocity of the air stream at this point may then be calculated by the formula

$$V = 4005 \sqrt{VP} \quad \text{at STP (or, } V = 1096 \sqrt{VP/\rho} \text{)}$$

where V = velocity in feet per minute
VP (h_v) = velocity pressure in inches of water.

The total air flow, cubic feet per minute, is then given by the average velocity in feet per minute over the whole area multiplied by the area in square feet.

$$Q = VA$$

The accuracy of such a measurement depends largely on the time and care devoted to it. Thus one may make a careful ten-point or six-point traverse with an inclined gage or electronic manometer and get accurate results (less than 5% error).

If a preliminary exploration of pressures along one diameter of the duct indicates them to be distributed reasonably symmetrically, one can then measure the maximum center line velocity. The airflow for the duct of the area A square feet and center line velocity pressure VP inches is as follows:

$$Q \;=\; (0.9) \cdot 4{,}005A\,\sqrt{VP} \;=\; 3{,}600A\,\sqrt{VP} \qquad \text{at STP}$$

The figure, "0.9," known as the "duct factor," represents the ratio of average to centerline velocities usually found in long straight sections of duct, 6-8 duct diameters downstream and 2 to 3 duct diameters upstream of any fitting.

The "duct factor" varies considerably in practice, depending upon the nature of upstream and downstream disturbances.

Where discharge of air follows immediately after the fan or shortly after passing an elbow, the distortion of velocity pressures across the duct diameter is too great for a single reading to be of even approximate value. Where this is true, another straight-duct location should be found or a complete traverse of the duct should be made.

Chapter 15

Fans, Injectors, Natural Draft Chimneys and Roof Ventilators

A consideration of the engineering factors that describe the functioning of the air-moving mechanism (e.g., fan) used with the hoods and connecting duct system completes the assembly of a basic exhaust system. The air-moving mechanism is usually a power-driven exhaust fan, either of the centrifugal type or, occasionally, the axial flow type. It may, however, consist of an air ejector (or injector) using a centrifugal blower or compressed air for jet air, or, if the contaminated air is hot, a natural draft chimney may serve the purpose. Roof ventilators operating on natural draft or by power-driven fans also properly belong in this discussion.

Fan Performance

One of the purposes of calculating resistance to airflow (i.e., finding static pressure needs) through a duct system as previously outlined, is to enable the calculation of power requirement for the air moving equipment. If power were no object and if the air pumping equipment had positive displacement characteristics like a bellows or a piston pump, then one could dispense entirely with duct loss calculations. Design would consist of noting the total exhaust rate required and this would fix the capacity of the positive displacement exhauster.

Slippage, however, is characteristic of all exhaust fans, and the amount of slippage is dependent on the pressure the fan is required to develop. When the outlet of the fan is closed, reducing flow of air to zero, the pressure developed is, for practical purposes, close to its maximum. This pressure is called the *static no delivery* (SND) pressure. At zero pressure, that is, against no resistance as would be the case if there were no connecting duct work, the fan is described as operating at *free delivery* or *wide open*.

A graph of pressure (resistance to airflow) as ordinates and airflow rate as abscissae including these two limits, and intermediate points describe the slippage characteristics of a fan. Usually referred to as the *pressure-volume characteristic curve* or *P-V characteristic curve*, or simply *characteristic curve*, it is a universal device for describing fan performance. Different fan types vary in performance, mainly in the extent and manner of their response to increased or decreased resistance and this is summarized in the characteristic curve. Other factors are mentioned below.

Axial/Propeller Fans

Propeller fans in use today are of two principal types: (a) those with blades which are stamped out of sheet metal, and (b) those with blades which have thickness and are shaped like the propellers of an airplane, the "airfoil blade."

Fans of the first type are like the common air fan seen in homes and offices. Industrial exhaust fans of this type have the same type of blading, differing principally in ruggedness of construction and in mounting of blades and motor. The blades are usually fashioned of sheet metal curved helically in the manner of a screw. Their capacity is materially reduced when required to operate against static pressures as little as 0.1 inches of water. Propeller fans having airfoil type blades can operate at resistances up to 1 to 2 inches of water. Operation at these higher resistances is accomplished, however, at the cost of materially higher noise level.

Where the system resistance is low, propeller fans are by far the lowest in cost for a given exhaust capacity of any in the fan family.

The limitation of propeller fans in their capacity to move air through duct systems is evident from the pressure limits cited. They are commonly used for exhausting air through short lengths of duct serving paint spray booths and the like. Low air velocities and short duct lengths make this type of application feasible. The airfoil type can serve duct systems of great lengths where large diameter ducts are used to maintain low frictional resistance. Then it becomes, in effect, a tubeaxial fan.

Tubeaxial Fans

Propeller fans of the second type are available mounted in a short cylinder to facilitate their connection to duct work. These are termed *tubeaxial fans* but they are not actually changed basically in their operating limitations by the mere addition of the cylinder section. This is provided solely for convenience of the buyer who intends to connect the fan to duct work and requires a type of propeller blade

that is adapted to operate against higher resistances than the stamped sheet metal type.

Propeller fans should not be used for exhausting dust-laden air through horizontal duct sections because the high duct air velocities required for dust transport lead to excessive resistance.

Generally, duct diameters for use with propeller fans should be selected for air velocities that do not exceed about 1,500 fpm. Velocities around 1,000 fpm are more suitable.

Estimating Axial/Propeller Fan Capacity

While there is considerable choice in speed of the direct-connected motor employed for driving fan blades available to the manufacturer, they are generally assembled for capacities most commonly in the range of 1,000 to 1,500 cfm per square foot of circular area within the blade ring—roughly, the circular area corresponding to blade radius. Belt-driven fans are sometimes assembled for especially quiet operation that accompanies low speed, where the capacity ratio is still less, but this is not common in industrial applications.

These figures provide a convenient basis for rough estimation of the free delivery capacity of a propeller fan. If the circular area of the opening is about 7 square feet, the fan could reasonably be assumed to have a capacity within the range of 7.000 to 10,000 cfm.

Power for Propeller Fans

Power requirements for centrifugal fans are based on "fan static pressure" and "fan total pressure," (usually the latter) as defined by AMCA. (See References.) Propeller fan power requirements are usually based on total pressure, h_t, which is about equal to the velocity pressure, h_v, through the ring. The *mechanical* efficiency of propeller fans seldom rises above 30% and may be lower.

Thus the power required for operation of a 36-inch propeller fan designed to operate at a free delivery capacity of 1,500 cfm per square foot, i.e., 1,500 fpm through the total ring area, and at a mechanical efficiency of 30% would require power calculated as follows:

$$h_v = (1,500/4,005)^2 = 0.14 \text{ inches w.g.}$$

$$hp = \frac{cfm \times h_v}{6356 \times ME} = \frac{10500 \times 0.14}{6356 \times 0.3} = 0.77 \text{ hp}$$

A motor would be selected for this fan at some figure in excess of 0.77 rated horsepower (rhp).

Considerations of power requirement for propeller fans do not often engage the attention of the user because he purchases a fan and motor combination—the selection has already been made by the manufacturer.

It is important for the user to recognize that as resistance is increased, the power demand of the fan motor increases steadily to the point of no flow, or "shut-off." This is opposite to the behavior of centrifugal fans where power decreases to a minimum at no air delivery.

Vaneaxial Fans

A type of axial flow fan, the vaneaxial fan, is usually dealt with in fan engineering texts along with propeller fans (including tubeaxial fans) but beyond the fact of airflow direction paralleling the fan axis, the similarity ceases. The fact is that it is no more like the propeller fan in its characteristics and operation than it is like a centrifugal fan.

It is characterized by short blades designed with an airfoil cross section operating in a narrow annular space bounded on the outside by the housing and with an inner boundary formed by a large hub blocking off something like two-thirds of the total diameter. The whirling motion imparted to the air stream in passage through the rotor is neutralized by stationary vanes at inlet or outlet, or both.

The vaneaxial fan is competitive with the centrifugal fan in many applications. It is adaptable to the development of higher pressures than required in industrial ventilation applications. Its efficiency is as high or higher than that of centrifugal fans. An outstanding characteristic is its compactness. Hence space saving is a primary advantage over centrifugal fans. On a first-cost basis, it will seldom compete with an ordinary propeller fan except in large applications where superior efficiency is important or where it is called upon to operate against some resistance which it can do at a much lower noise level.

Centrifugal Fans

The common types of centrifugal fans are characterized by inclination of the blades either forward, backward, or radial with respect to rotation. While they differ in their respective efficiencies, this is not an important consideration in many applications. Where power consumption is a minor factor, the principle differences are in their change in capacity with varying pressure (described by the

characteristic curve), the physical form of the fan wheel (of importance in relation to solids in the air stream), the magnitude of resistances they can effectively operate against, and the rotational speed required for a particular duty. The last named factor is of practical importance from the standpoint of mechanical maintenance and fan noise.

Air streams heavily loaded with dust are usually handled by the common *radial,* or *paddle wheel fan,* the blade and housing of which are ruggedly fabricated to withstand the sand blasting action of dust particles passing through. The characteristic curve of such fans is well adapted to provide the relatively high static pressure usually required for dust-exhaust systems. Dust transport velocities of 3,000 to 4,000 fpm or higher commonly result in resistances of 5 to 10 inches of water, or even greater with the resistance of an added dust collector.

The static efficiency of this type of fan *in the most favorable range of operation* can be expected to lie between 55 to 65% or better. In practice, lower efficiencies are more common because of the need to choose a system operating point (SOP) on a steep—but less efficient—point on the fan characteristic curve.

The paddle wheel fan has radial blades and may have either a steel plate housing or a cast iron housing although these do not distinguish this particular fan from others. It is the oldest type of industrial fan; it first saw extensive use in exhaust systems for handling of wood and leather waste in woodworking and shoe industries in the 19th century.

The *forward curved-blade fan* has a fan wheel suggestive of a rotating squirrel cage and the latter term is often used to designate this family. Its shallow blades are cusp-shaped and point forward in the direction of rotation. In comparison with the paddle wheel fan, it is characterized by a wide housing, and by an inlet opening of large diameter, almost as great as the housing itself.

The pressure-volume curve of this type of fan dips and rises to a peak in the range from zero flow to about one-half wide-open capacity, beyond which it curves downward toward full wide-open capacity. Over the latter range the slope is pronouncedly less than for the paddle wheel fan when compared on a basis of equal capacity, wide open. This means that a relatively small change in resistance of the duct system results in a relatively large change in the rate of air-flow.

The forward-curved multivane fan is designed to handle large volumes of air at fairly low resistance. That is, on the basis of equal rotational speed, the multivane fan has greater exhaust capacity than other types of centrifugal fans. The static pressure developed in

proper application is low, i.e., most commonly in the range 2 to 4 inches of water more or less.

The static efficiency of the multivane fan lies usually between 40 to 60%.

This fan is adapted primarily to the handling of clean air at low temperatures. The blades readily accumulate dust or fume so it is particularly unsuited to applications where buildup may occur and injure its functioning. A good rule to follow in this connection is that the forward-curved fan should not be used if the task is one in which one might expect some solid or liquid accumulation on the interior surface of the duct work, since deposits might accumulate on its blades at an even more rapid rate.

High volume fans other than the multivane are manufactured, having shallow but wide steel plate blades, but having housing proportions similar to the multivane, that is, wide housing and large diameter inlet. In fact, this dimensional characteristic of high volume, low pressure is useful to bear in mind to aid in the identity of fan types.

Backward curved blade fans have characteristics quite different from those of the forward curved blade fans. They operate at higher speed to develop equal pressures and, at equal speed, have lower airflow capacities. They are not only the most efficient (60 to 70% static efficiency) of the three types of centrifugal fans that have been discussed herein but also have a nearly flat power characteristic curve, an important advantage in some applications. Seen superficially, their housing proportions are not greatly different from those of paddle wheel fans.

Selection of Fans

Ordinarily a fan can best be selected for a given duty with help from a representative of the manufacturer (who is skilled in the field). If, however, selection is to be made on a competitive basis it is necessary to insure that comparisons are made on the basis of the same type of fan, since the boundaries of application indicated in this discussion are by no means hard and fast.

Where the type of fan needed is obvious and no specific advice is required of the supplier, a selection can be made by reference to rating tables and figures of the manufacturers. Methods for reporting fan capacities are illustrated in Table 15-1 and Figure 15-1.

Fans should always be selected that have an AMCA certification because standardized tests have been used to determine performance characteristics. This makes it easier to compare fans between suppliers and provides some assurance that performance will be met.

Table 15-1
Fan Rating Table

Capacity cfm	1″ FTP		2″ FTP		3″ FTP		4″ FTP	
	rpm	bhp	rpm	bhp	rpm	bhp	rpm	bhp
4870	287	1.21	395	2.44	482	3.76		
5570	292	1.39	398	2.79	484	4.17	558	5.78
6270	297	1.58	401	3.13	485	4.67	559	6.33
6970	302	1.77	405	3.48	489	5.15	560	6.96
7660	309	2.01	412	3.83	493	5.72	563	7.59
8350	316	2.24	416	4.24	496	6.27	566	8.28
9050	325	2.49	421	4.59	501	6.75	570	8.98
9750	335	2.77	426	5.01	505	7.31	573	9.75
10430	345	3.07	433	5.43	511	7.94	576	10.4
11120			441	5.92	516	8.49	585	11.1
11820			448	6.32	522	9.05	590	12.0
12520			457	6.89	528	9.74	597	12.65
13220			467	7.38	535	10.4	599	13.5

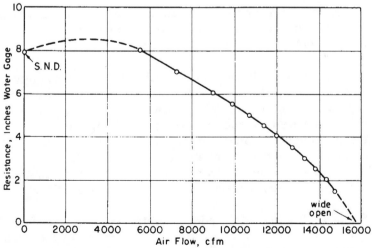

Figure 15-1. Fan Characteristic Curve as developed from AMCA standard rating test.

   Each point of rating given in Table 15-1 is located on a separate
pressure-volume characteristic curve as illustrated in Figure 15-2.
This has been indicated by the fragment of a curve drawn through
each point. While some extrapolation from the Q-Pressure values at
one speed to another by application of the fan laws permits some
extension of a given "curve," it is a laborious procedure. Where the
characteristic curve is needed for calculations of the type illustrated
in problems which follow, one can obtain it from the manufacturer.

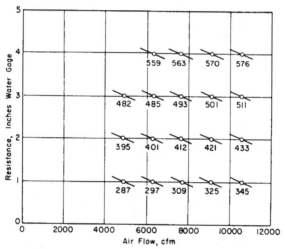

Figure 15-2. Graphical representation of fan rating tables of the type
shown in Table 15-1. Each point in the table is a single point on a
characteristic curve for one particular fan at various rpms (as designated
by the numbers shown).

   The following examples illustrate the use of fan characteristic
curves. These are problems that cannot be solved in the absence of
such information.

   Example 1. A fan having the characteristic curve of Figure 15-3
serves an exhaust system having a resistance of 4 inches water
gage when handling 12,000 cfm. These conditions are changed in
cold weather when, due to lack of properly heated air supply to
the building with closed windows, the room housing the exhaust
system is under a negative pressure of 1 inch w.g. (the fan
discharges outdoors). Determine the extent to which the capacity
of the fan is reduced in such conditions of air "starvation."

   Solution. The effect of the added negative pressure is as though a
restriction were attached directly to all inlets to the exhaust
system, and the original system resistance curve no longer
applies. A new system operating point (SOP) has been created. Its

point of intersection with the fan characteristic is found by moving upward along the fan curve to the new resistance, 5 inches w.g. The new airflow is then indicated on the abscissae—about 10,800 cfm, a decrease of about 10%.

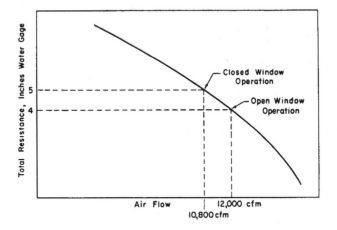

Figure 15-3. Data and solution for Example 1 showing the basis for estimating the reduction in airflow rate due to an increase in the system static pressure.

Such a decrease might actually be insignificant to the operating effectiveness of some exhaust systems operating at already high static pressures. One objectionable feature of such a situation is in the discomfort due to cold drafts in the peripheral areas of the room, as well as that on entering and leaving the space through doorways. It should be noted that the practical result on the operation of an exhaust system is dependent on the shape of fan characteristic curve and the location of the system operating point (SOP) on it.

Fans of the kind used in dust exhaust systems would seldom be materially affected by an "air bound" condition but multiple blade fans might well be choked down considerably and propeller fans very seriously so.

Example 2. A fan having the characteristic curve shown in Figure 15-4 is to be employed to exhaust air through several 40-ft lengths of flexible hose in parallel arrangement. The total pressure loss, h_L, for one hose when handling 200 cfm has been estimated at 5.1 inches. Estimate the airflow rate through each hose when one, two, three, and four hoses are connected in parallel to the inlet of the fan.

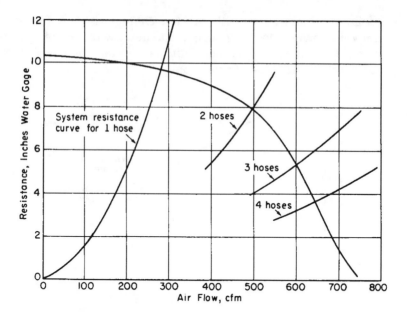

Figure 15-4. Data and solution for Example 2 illustrating use of fan characteristic curve for determining airflow through multiple branches in parallel arrangement.

Solution. Construct the system resistance curve for a single hose by calculating several points from the relation

$$Q = K(h_L)^{1/2}$$

$$200 = K(5.1)^{1/2}$$

where

$$K=88.5$$

and

$$Q = 88.5(h_L)^{1/2}$$

The resulting curve is drawn in Figure 15-4 which intercepts the fan characteristic curve at Q = 275 cfm. This will therefore be the actual flowrate for a single connected hose attached to this fan.

Next consider two hoses in parallel. In combination with each other, a new system has resulted, one point of which will be Q = 200 • 2 = 400 cfm at a resistance of 5.1; the airflow will be about

doubled at the same resistance. (The result is as though the two hoses had been replaced by a single hose or duct of large diameter.)

Then, using the evaluation technique shown above,

$$Q = K(h_L)^{1/2}$$

$$400 = K(5.1)^{1/2}$$

whence

$$Q = 177(h_L)^{1/2}$$

The intercept of this new system curve on the actual fan curve is seen to be about 500 cfm for the two hoses, or 250 cfm each.

The same procedure applied to 3 and 4 hoses indicates that three hoses will pass 600 cfm or 200 cfm each, and about 175 cfm in each of four hoses.

Estimating Capacities of Used Fans

Occasionally it is necessary to estimate the capacity of a used fan for a new application when either the identification name plate is lacking or rating tables for the particular fan are not available.

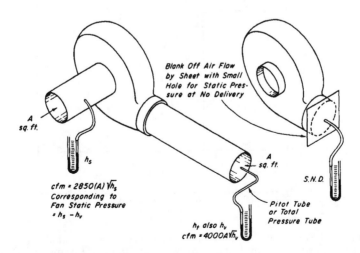

Figure 15-5. Method for approximate measurement of characteristics of a fan. Static-no-delivery pressure and airflow nearly wide open.

It is possible to approximate the performance by the simple tests illustrated in Figure 15-5 and described below.

The shut-off no-delivery static pressure (SND pressure) is easily measured by blanking off the inlet or outlet and observing the reading of a water manometer connected to a small hole in the baffle plate. Next, a value close to the wide-open airflow capacity is measured by attaching a short length of duct to the outlet or to the inlet (a flange is desirable for the latter to minimize inlet shock loss). The airflow can be measured at the outlet by means of a simple bent tube (or a Pitot tube) that will register total pressure, about equal in this case to velocity pressure. The static pressure can be measured in the wall of the inlet duct as described for measurement of airflow at the inlet to exhaust hoods. Both measurements serving as a check on each other are desirable.

These two points are placed on a graph of pressure against airflow rate and connected by two straight lines to the point where an airflow rate is 50% of the wide open value, and static pressure equal to, or a little less than, the static no-delivery value. Then a freehand curve is drawn outside of the straight lines but intersecting the three points. It can be demonstrated by reference to typical characteristic curves of various types of industrial fans that the right-hand segment of the curve is a fair approximation of the actual characteristics. The shape of the left segment varies with the type of fan but need not be drawn, since it is not in the proper or desired operating region.

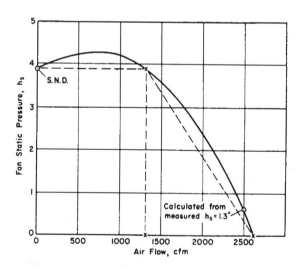

Figure 15-6. Plotting the measurements obtained by the method illustrated in Figure 15-5. A horizontal line is drawn from the SND point to 50% of wide-open capacity, then a sloping line is drawn from that point to zero pressure, wide-open capacity. (Calculations of Example 3.)

The wide-open airflow rate is not the value actually obtained by the method illustrated in Figure 15-5 because the inlet occasions a loss of about 1 velocity head. This fact is treated in the following example.

Example 3. An unidentified paddle wheel fan is tested by the Figure 15-5 method giving an SND value of 3.9 inches water gage. The velocity pressure with inlet and outlet duct in place, but no other restrictions, is 0.64 inches and the duct area (both equal) is 0.7854 sq ft (12-inch diameter). Construct the characteristic curve. Assume STP.

Solution. The static no-delivery value is placed on the graph directly (Figure 15-6). The velocity is V = 4,005(0.64)$^{1/2}$ = 3,200 fpm. Since there is a loss at the entrance to a plain duct end of about one velocity head (loss = 1VP), then the static pressure at inlet represents about 2 velocity heads, one for the loss and one for the actual velocity head prevailing. The fan static pressure therefore is about 0.64 inch, the magnitude of the head loss. This value is then used for plotting the lower point together with airflow (Q = VA = 3,200 • 0.7854 = 2,500 cfm). We estimate that the sloping curve when constructed will intercept the zero pressure line at about 2,600 cfm. At half this value (Q = 1,300 cfm) draw a vertical line which will intersect a horizontal line from the SND point. A free-hand curve is then drawn as illustrated in Figure 15-6.

Fan Capacity Estimation by Inspection

It is possible to estimate order-of-magnitude installed exhaust capacity by reference to some of the facts that have been set forth in this and the preceding chapter.

A radial-bladed fan installed for dust control can be assumed to have been designed for inlet linear air velocities of 3,000 to 4,000 fpm, in general, hence the diameter of fan inlet, or of the main duct connected to the fan if different, is a clue to the probable air flow. Thus, a 24-inch duct main, having a circular area of 3.124 sq ft could be judged to be operating at a probable capacity of 9,000 to 12,000 cfm. The velocity could be as low as 2,000 to 2,500 fpm, or as high as 4,500 fpm, or greater, and while they are less likely values, they are recognized as possibilities in making the estimate. In the case of systems that utilize the high-volume type of fan wherein dust transport is not a factor, the best guess as to prevailing air velocities is in the range 1,500 to 2,500 fpm. The lower values are most likely for the smaller systems (smaller diameter ducts) and the higher values

for the larger systems, because, it will be recalled, friction losses are less the larger the duct diameter.

With a little experience, the engineer or hygienist will recognize the signs indicating duct air velocities in the higher range, e.g., noticeably greater than normal air noise, or complete transport of large inertial particles, observed visually or by the sound of their bouncing passage through the duct.

This type of observation is often useful in discovering the probability of inadequate transport velocities in a duct system as in the following case.

Example 4. A dust exhaust system is connected to partial enclosures at the openings of which air velocities are judged to be slightly more than 100 but probably not in excess of 200 fpm. (Your hand begins to become sensitive to air movement at about 100 fpm. Here some experience is necessary, such as will be obtained, for example, by continued use of an air velocity meter at booths and other exhausted enclosures.) The total area of enclosure openings is 24 sq ft. The ductwork leads to a main duct having a diameter of 30 inches. What conclusions can be drawn concerning the design and operation of the system?

Solution. The probable airflow through the opening is judged to be not in excess of Q = 200 • 24 = 4,800 cfm; possibly one half as much. The circular area of the 30-inch main is A = 4.91 sq ft. Therefore if the appraisal of airflow into enclosure openings is anywhere near correct, the linear air velocity in the main would not be greater than 4,800/4.91, or approximately V = 1,000 fpm. This is far below the requirement for adequate dust transport and one concludes, therefore, that there is a good likelihood that the duct system is not self-cleaning. This, of course, is only a preliminary conclusion, needing to be checked.

These ideas concerning design velocities make it possible to draw some preliminary conclusions concerning the probability of a given used fan being applied to a newly designed exhaust system, even without any pressure and volume testing. Radial, straight-bladed exhaust fans are designed for most efficient capacity in the velocity ranges cited above, 2,500 to 4,500 fpm, and high volume fans in the range 1,500 to 2,500 fpm. These statements are generally valid only when tempered by consideration of the static pressure the fan will be required to develop. A multivane forward-curved blade fan, for example, would not in general be adaptable to a high resistance system.

Estimating Required Speed. A rough rule for estimating the speed required of a centrifugal fan is given by the following. The velocity pressure corresponding to the linear velocity of the tip of the wheel can be an indicator of the static pressure that might potentially be developed. The application of the following rules-of-thumb varies widely with different types of fans.

- In the case of a high volume, *forward-curved blade fan*, the static no-delivery pressure (which can be considered about the same at 50% of wide open capacity) is not likely to exceed 2.0 times the velocity pressure corresponding to the tip speed of the wheel.

- For a *backward-curved blade* fan the SND pressure is unlikely to exceed 0.7 times the tip speed velocity pressure.

- A *straight-blade* (paddle wheel) fan cannot be expected to develop more than about 1.25 times the peripheral velocity pressure.

Design And Use Of Injectors (or, air ejectors)

There are occasionally important specific applications where an air injector has particular advantages over a fan for moving air. The injector is highly inefficient compared with fans, and may require from 5 to 10 times as much power for the same duty. In some circumstances, however, the lower first cost of the injector may be attractive. Some of the following situations offer possibilities for advantageous use of injectors, for which another name is ejectors.

- The total power required even by an injector in a particular case may be so low as to make the efficiency factor of no concern compared with the simplicity of an injector installation.

- Possibilities of considerable economic advantage for the injector lay in circumstances where there is a highly intermittent demand for exhaust capacity. One can imagine a process generating large quantities of air contaminants for 20 to 30 seconds every five minutes, i.e., 10% of total time. The low first cost of an injector installation combined with arrangements for intermittent operation of the injector exhaust provides an attractive possibility for an economical design.

- Where an exhaust air stream carries particles of sticky material like tar mist or slow drying paint mist, or highly corrosive materials which would quickly foul the moving parts of ordinary fans, the air injector may have greatly superior advantages; it would also become fouled and its operational characteristics modified, but at a much slower rate. Furthermore, it could be much more easily cleaned and maintained.

- Another field of application is in some circumstances where, for example, it is desirable to experiment with an exhaust installation to determine the most effective rate of exhaust, circumstances wherein theoretical calculations cannot be applied effectively.

• An injector can often be applied to use an existing compressed air supply, which still further simplifies the installation.

Cylindrical Injectors

G.E. McElroy ("Design of Injectors for Low Pressure Air Flow," *Technical Paper 678,* U.S. Bureau of Mines, 1945) has written an excellent treatise on this subject which is recommended to the reader desiring more detailed knowledge of this subject than is provided herein. The operating characteristics of a cylindrical injector illustrate most simply the basic factors determining the performance of both cylindrical and venturi types.

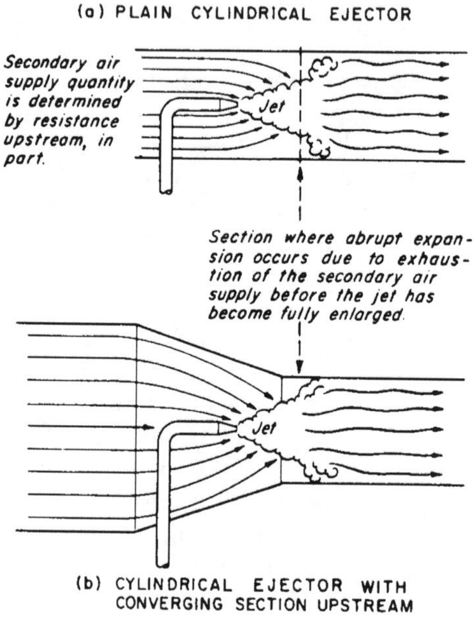

(a) PLAIN CYLINDRICAL EJECTOR

Secondary air supply quantity is determined by resistance upstream, in part.

Jet

Section where abrupt expansion occurs due to exhaustion of the secondary air supply before the jet has become fully enlarged.

Jet

(b) CYLINDRICAL EJECTOR WITH CONVERGING SECTION UPSTREAM

Figure 15-7. Cylindrical injector: (a) diagrammatic illustration of the relationship between primary air from the nozzle, jet expansion, and secondary air supply. (b) Addition of a tapering section does not change performance materially. Note, however, that it would become a Venturi injector by the simple addition of a diverging section downstream.

The air supply nozzle creates an air jet which expands inside the duct just about as it would in open air, a subject treated in Chapter 9. It entrains, and mixes with surrounding air; in the present case air moving forward through the duct. If, due to resistance characteristics of the duct system in relation to the energy characteristics of the jet stream, the secondary air supply is insufficient to permit the jet mixing to continue expanding at the conical angles prevailing

initially, the final expansion required to fill the duct is visualized as occurring abruptly (Figure 15-7). That is, velocity of the air stream of the mixing jet during the first stage of distance decreases regularly as it accepts the quota available secondary air moving forward from the rear. When this has been used up, the air velocity then decreases abruptly in accordance with the dimensions of the duct cross-sectional area, and the total air flow. The abrupt velocity decrease occurring in the present illustration is accompanied by a transformation of velocity pressure to static pressure and the latter provides the motive power which causes movement of the secondary air up to the jet and discharges the air mixture forward. The quantitative relationships prevailing in this expansion are those that are applied to the simple abrupt expansion in a duct described in Chapter 11.

The greater the duct resistance at the prevailing flowrate, the greater the degree of expansion of the jet from the depicted stream diameter of maximum entrainment to the diameter of the duct. As the duct resistance approaches zero, the expanding cone of jet air and secondary air approaches the point where there is no abrupt expansion, i.e., the cone angle continues regularly until it intercepts the duct walls.

If, now, the diameter of the upstream duct section is increased (Figure 15-7b), requiring a converging duct section to bring it down to the original diameter at the jet, we find that the operating characteristics are unchanged (ignoring the decrease in frictional loss for the secondary air, upstream, and, also, the insignificantly small increased energy loss in the converging section). It will be noted that the new arrangement of duct work represents one half of a venturi shape; the expander or diffuser section is missing. A typical pressure-volume characteristic curve for this shape is given in Figure 15-8.

Venturi Injectors

The velocity pressure in the stream, after mixing of jet and secondary air, can be transformed to static pressure by attaching a tapering expansion piece to the assembly. The additional static head thus provided results in an increase in the flow of secondary air thus modifying the relationship described in the preceding section. It is now possible for the jet to expand completely to fill the total duct area without any abrupt expansion, even though there is appreciable frictional duct resistance to be overcome. (Note that with the simple cylindrical injector such continuous expansion could occur only when duct resistance became zero.) In this instance the static pressure made available by the diverging expansion is available to overcome duct resistance.

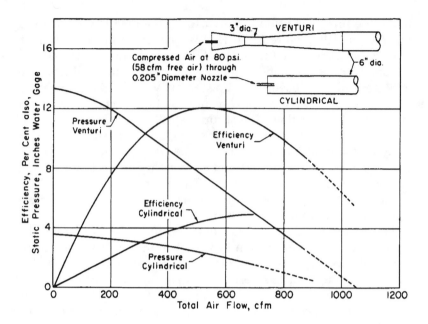

Figure 15-8. Typical characteristic curve for a cylindrical injector and corresponding Venturi injector. (Weeks experiments cited by McElroy.)

For any desired ratio of jet airflow to induced secondary airflow, the maximum efficiency is obtained when all elements of the design are such that the jet stream does entrain the secondary air without abrupt expansion. This point is called the point of rating; its efficiency, the design efficiency. In this condition, the velocity pressure, h_v, after complete mixing, equals this total pressure, h_t, at the discharge for a blowing system.

Among other factors, the ratio of jet diameter to duct or throat diameter is an important element of design. To render this throat diameter independent of the diameter of the preceding duct, the converging duct element is added and a complete Venturi ejector results.

Design of Injectors

The complete derivation of formulas by McElroy sets forth the relationships between various factors in injector design. The ones needed in most ordinary engineering work are presented below. The point of rating is the most favorable design and the following formulas apply to that specific design.

(1) The first step in the design is to decide upon a ratio of total airflow to jet airflow, the quantity ratio, noting that the efficiency which determines the power requirement is the inverse of that ratio.

$$\text{Efficiency} = \frac{Q_{jet}}{Q_{total}} \times 100$$

or

$$\frac{\text{Jet diameter}}{\text{Throat diameter}} \times 100 \qquad (1)$$

A quantity ratio of 20 to 1 means an efficiency of 5%, one of 4 to 1 means 25%. The conditions of a given problem will determine the ratio selected. If the power requirement is low due to low resistance or low airflow, then low efficiency may be acceptable for the sake of a high airflow ratio. Similarly, highly intermittent use of an air injector may dictate a high airflow ratio (small jet blower) with its accompanying low efficiency.

Following this decision note the relationship,

$$\frac{Q_{total}}{Q_{jet}} \simeq \frac{Q_{throat}}{Q_{jet}} \times \sqrt{\frac{\text{Density jet air}}{\text{Density total air}}} \qquad (2)$$

wherein the term involving density ratio disappears when there is no important difference in their temperatures. This relation fixes the size of throat and jet.

Where the injector is to operate as an exhauster, the total Q is derived.

$$Q_{total} = Q_{exhaust} \times \frac{Q_{total}/Q_{jet}}{Q_{total}/Q_{jet} - 1} \qquad (3)$$

(2) The next step is to calculate the total duct, h_t, of the system to be equipped, allowing for the increased airflow due to the jet, on the blowing side of the injector. The outlet velocity pressure, h_v, is added to the total h_L to give h_t.

(3) The size of the tapering enlargement in terms of throat area and downstream duct area for "design efficiency" is next calculated from the data of (2) above.

$$\frac{\text{duct or pipe area}}{\text{throat area}} = \frac{A_{duct\ or\ pipe}}{A_{throat}} = \sqrt{\frac{h_{t(system)}}{h_{v(discharge)}}} \qquad (4)$$

(4) The jet velocity is next fixed by an expression which describes the design velocity head of the jet, $h_{v(jet)}$, in terms of jet area, system resistance or pressure, h_t, and throat area.

$$h_{v(jet)} = h_t \frac{A_{throat}}{A_{jet}} = h_t \left(\frac{D_{throat}}{D_{jet}} \right)^2 \tag{5}$$

The ratios of throat/jet and duct/throat combined with the duct diameter have already fixed the diameter of the jet. The jet diameter is equal to the inside diameter of the nozzle in low pressure airflow where the jet air is not caused to contract as it would, for instance, in passing through a sharp-edged orifice.

However, when compressed air supplies the jet, this is no longer true. In that case the jet area to be employed in the design is obtained from

$$\text{Jet area} = CKA_n \tag{6}$$

where

C = the coefficient of orifice discharge; 0.60 for a sharp-edged orifice; and 1.0 for a well-rounded nozzle.
K = the compressibility coefficient for air, obtainable from Figure 15-9.
A_n = area of the nozzle itself.

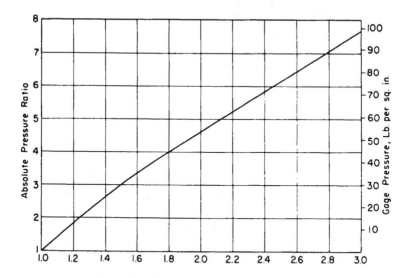

Figure 15.9. Chart for estimating K for use in equation 6.
K = (Jet area)/(C)• (Nozzle area)
[Note: X-axis is Compressibility Coefficient, K]

The preceding equations apply to one set of design conditions that are determined by the resistance of the duct system.

Characteristic Curve for Injectors

Following the design or selection of an injector it may sometimes be desirable to analyze its performance with system resistance different from that originally contemplated. A characteristic curve of pressure vs. volume drawn from three points (maximum capacity, i.e., wide-open or zero pressure; static no-delivery pressure or zero capacity; and the point of rating) can be constructed as follows:

The *maximum capacity* of an air injector is given by

$$\frac{Q_{total}}{Q_{jet}} \approx \frac{Q_{duct\ or\ pipe}}{Q_{jet}} \times \sqrt{\frac{Density_{jet\ air}}{Density_{mixed\ air}}} \qquad (7)$$

which for equal air densities becomes

$$\frac{Q_{total}}{Q_{jet}} \approx \frac{Q_{duct\ or\ pipe}}{Q_{jet}} \qquad (7a)$$

and in another form describing maximum *exhausting* capacity

$$Q_{exhaust} = Q_{jet}\left(\frac{D_{duct\ or\ pipe}}{D_{jet}}\right) - Q_{jet} \qquad (8)$$

The *maximum pressure* developed by an injector, i.e., *static no-delivery* pressure is given by

$$h_{t(max)} = 2h_{v(jet)} \times \frac{A_{jet}}{A_{throat}} \qquad (9)$$

From these two points plus the point of rating for a particular injector, a characteristic curve can be drawn. The point of rating is that given by the design of Q of original specifications, and h_t from (4) above, i.e.,

$$h_t = h_{v(jet)} \times \left(\frac{D_{jet}}{D_{throat}}\right)^2 \qquad (10)$$

Example 5. A system is being designed to handle 5,000 cfm of air at room temperature and it is decided to employ an air ejector as an exhauster. The duct is 17 inches in diameter corresponding to a

velocity of 3,180 fpm, and velocity pressure, h_v, of 0.63 inches. The friction and shock loss, h_L, in the duct work for the flow of 5,000 cfm totals 4.5 inches, calculated by the methods outlined in Chapter 13. Bearing in mind that a high ratio of air exhausted to air used for the ejector nozzle results in lower primary air requirements but, conversely, lower power efficiency, it has been decided to design for an efficiency of 10%, from which it follows that the ratio of total airflow (including nozzle or jet airflow) to jet airflow will be 10 to 1.

Calculate (a) the required diameter of the throat of the venturi, (b) the diameter of the nozzle, and (c) nozzle airflow.

Solution. First calculate total airflow from which jet airflow is obtained, knowing that net exhaust airflow is to be 500 cfm, by

$$\frac{Q_{total}}{Q_{jet}} = \frac{10}{1}$$

and

$$Q_{total} - Q_{jet} = 5,000$$

$$Q_{total} - \frac{Q_{total}}{10} = 5,000$$

$$Q_{total} = 5550 \text{ cfm}$$

therefore

$$Q_{jet} = 550 \text{ cfm}$$

Then calculate throat size which, in relation to the duct size, describes the proportions of the tapers of the venturi.

$$\frac{\text{duct or pipe area}}{\text{throat area}} = \frac{D^2_{duct \ or \ pipe}}{D^2_{throat}} = \sqrt{\frac{h_{t(system)}}{h_{v(discharge)}}}$$

$$\frac{17^2}{(\text{Throat Diameter})^2} = \sqrt{\frac{4.5 + 0.63}{0.78}}$$

where 4.5 and 0.63 are h_L and h_v, as stated by the conditions of the problem and 0.78 is the discharge velocity pressure corresponding to the velocity of 5,550 cfm passing through a 17-inch duct outlet (area, 1.576 sq ft).

Throat diameter = 10.6 inches

Next the jet diameter can be determined from Equation (1) or (2).

Jet diameter = 10.6/10 = 1.06 inches

Then jet velocity is calculated from Equation (5).

$$h_{v(jet)} = h_t \frac{A_{throat}}{A_{jet}} = h_t \left(\frac{D_{throat}}{D_{jet}}\right)^2 = 5.13\left(\frac{10.6^2}{1.06^2}\right) = 513$$

$$V = 4005\sqrt{h_v} = 4005\sqrt{513} = 90,000 \text{ fpm}$$

If this value had not exceeded 15,000 to 20,000 fpm, one could have proceeded with the design based on low pressure airflow, i.e., insignificant compression of nozzle air. The value of 90,000 fpm exceeds even the velocity of sound, 66,000 fpm, hence it is clear that the compression factor must be taken into account, as follows.

The velocity within the nozzle will be velocity of sound (i.e., the velocity corresponding to all pressures above critical pressure), or 66,000 fpm. The ratio

91,000/(c)66,000 = 1.38

where c = 1 as for a well-rounded nozzle. This (1.38) is the value of K in Figure 15-9. Expressed in other words, the area of the jet will be 1.38 times the area of the nozzle. Referring to Figure 15-9, find gage pressure required for K = 1.38; = 25 lb per sq inch.

The nozzle area will be (jet area)/(1 x 1.38), therefore, nozzle diameter, d = $1.06^2/1.38$ = 0.81 inch.

In summary, the ejector will consist of a well-rounded nozzle with airflow of 550 cfm at gage pressure of 25 lb per sq inch; the orifice diameter is 0.81 inches discharging into a venturi throat 10.6 inches in diameter, the latter connected, by venturi tapering pieces to the 17-inch diameter duct.

Symbols and Units Used in Natural Draft Formulas, Chapter 15

A_i = cross-sectional area of stack, $\pi D^2/4$, sq ft
A_w = stack wall area, πDL, sq ft
D = stack diameter, ft
D' = stack diameter, inches
f = duct friction factor for stack, $= 1/N$
G = mass flowrate through stack, lb per (sq ft) (hr)
H = overall coefficient of heat transmission through stack wall, Btu per (hr) (sq ft) (F)
H' = rate of sensible heat contributed by process to gas stream, Btu per min
H_i = coefficient of heat transmission through inside gas film
h_L = friction loss in stack, inches water gage
h_v = velocity head of stack gas stream, inches water gage
L = stack height, ft
N = number of stack diameters for loss of 1 velocity head $= 1/f$
q_c = volume flowrate through stack at room temperature, cfm
t_c, T_c = temperature of room air, deg. F and absolute, respectively
t_{ch}, T_{ch} = temperature of stack gas, deg. F and absolute, respectively
t_e = temperature of stack gas at entrance, deg. F
t_f = temperature of stack gas at exit, deg. F
Δt = average temperature differential between stack gas and room
V = stack gas velocity, fpm
ρ_{ch} = density of stack gas, lb per cu ft
ρ_c = density of room air, 0.075 lb per cu ft

Natural Draft Chimneys for Hot Gases

Occasionally a canopy hood may be disposed to receive a stream of hot gases from below. If there is sufficient heat in relation to total air or gas flow to be handled by the hood, a natural draft steel stack extending upward through the roof may be sufficient, obviating the need for a fan. This is an advantageous arrangement where gases passing through a duct are at such elevated temperatures as would require a fan with special features for its protection against heat.

Design of a natural draft stack consists essentially in the selection of a stack area for a given height that will accept all the hot gases delivered to it by the hood. The hood exhaust rate required for the particular process determines the stack duty. A relationship that permits stack diameter calculation is outlined below. It is based on establishing an equality between the force due to buoyant gases inside the stack, and the force (head loss, h_L) required to overcome friction losses due to gas flow and force to establish velocity in the stack, (velocity head, h_v), as follows:

$$\frac{L}{5.2} \left(\rho_c - \rho_{ch} \right) = h_L + h_v$$

where L is the height of the chimney in feet; ρ_{ch} and ρ_c are, respectively, densities of hot chimney gases and of the cold surrounding air in pounds per cubic foot; h_L and h_v are the resistance head and velocity head, in inches of water.

Substitutions and transformations, as will be outlined, result in the following expression for design stack velocity:

$$V = 21 \sqrt{\frac{LDN(\Delta t)}{L + DN}}$$

where V is stack velocity, fpm; L is stack height, feet; D is stack diameter, feet; N is reciprocal of duct (stack) friction factor, i.e., number of diameters for a loss of 1 velocity head; and Δt is the elevation of chimney temperature, t_{ch}, above room temperature, t_c, i.e., $t_{ch} - t_c$. Room temperature can usually be taken to be 70°F.

Average *chimney temperature* is a function of the entering temperature of gases and heat losses through the chimney wall. Estimation of the latter is a function in part of V and D, hence the above equation cannot be solved directly. However, calculations can be made and corrected with the aid of Table 15-3. The entering gas temperature, t_e, is calculated from the following equation, the ordinary heat capacity formula

$$t_c = \frac{55H'}{q_c} + 70$$

where H' is the heat release rate in Btu per min., and q_c is the rate of air or gas flow in cubic feet per minute based on room temperature (70°F).

From Table 15-3, a stack diameter is then selected such that its area will accept the flow of gases specified by the problem, assuming the chimney temperature to be the same or a little less than entrance temperature.

The real chimney temperature is next estimated for this size of chimney by

$$t_{ch} = \frac{200H'}{DL} + 70$$

If this value of chimney temperature is much different from the value originally assumed, then a different selection and a recalculation, or estimate, from Table 15-3 is made to arrive at a final selection of stack diameter and velocity.

Example 6. A canopy hood is to be erected over some furnaces in a brass foundry which it has been decided should exhaust hot air and fumes at a rate of 7,500 cfm by natural draft. The furnaces burn oil at a rate of 10 gallons per hour; the heating value is 150,000 Btu per gallon. Calculate the necessary chimney dimensions.

Solution. The heat generated is

150,000/60 x 10 = 25,000 btu/min

The entrance temperature will therefore be

$$t_c = \frac{55H'}{q_c} + 70 = 250° \, F$$

Next proceed with a preliminary estimate from Table 15-3. It indicates that for a 20-ft stack, velocities will be of the order, 1,000 fpm at t_{ch} a little above 200°F over a considerable range of diameters.

Taking this velocity, calculate the diameter required to handle 7,500 cfm.

$0.7854D^2 = 7,500/1,000$

Solving for D:

D = 3.1 ft.

Inspection of the tabulated data indicates that no correction is required. The fact that velocity values change so little with varying diameter greatly simplifies the calculations, as the above problem illustrates.

Table 15-3
Chimney Gas Velocities

Linear gas velocities at chimney temperature for chimneys or stacks of various sizes and temperatures calculated from $V = 21\sqrt{(LDN(t_{ch}-70)/L + DN)}$

Diam. ft	Area sq ft	Flue	Gas	Temp.,	°F			
		150	200	400	600	800	1000	1200
		Flue	Gas	Vel.,	fpm			
L = 10 ft								
0.5	0.196	490	625	1010	1270	1500	1690	1860
1.0	0.785	540	690	1100	1380	1640	1840	2020
2.0	3.14	560	710	1140	1440	1700	1900	2100
3.0	7.07	570	730	1160	1470	1730	1960	2140
4.0	12.5	580	740	1170	1480	1750	1980	2160
L = 20 ft								
0.5	0.196	620	800	1280	1610	1900	2040	2360
1.0	0.785	710	900	1440	1820	2150	2420	2680
2.0	3.14	760	970	1550	1960	2320	2600	2870
3.0	7.07	780	1000	1590	2020	2380	2670	2940
4.0	12.5	790	1020	1620	2050	2420	2720	3000
L = 30 ft								
0.5	0.196	685	880	1400	1780	2090	2350	2590
1.0	0.785	800	1030	1520	2080	2460	2750	3050
2.0	3.14	890	1150	1820	2300	2720	3040	3360
3.0	7.07	930	1190	1900	2400	2820	3180	3500
4.0	12.5	940	1200	1920	2440	2880	3220	3540
L = 40 ft								
1.0	0.785	880	1110	1780	2260	2660	3000	3300
4.0	12.5	1070	1370	2190	2760	3440	3580	4050

Derivation of Natural Draft Formulas

The following equates the positive buoyant force of a stationary column of hot gases with the force of friction required to establish and maintain flow.

$$\frac{L}{5.2}\left(\rho_c - \rho_{ch}\right) = h_L + h_v$$

The value 5.2 will be recognized as the quotient 62.4/12, the density of water and inches per foot, respectively, so that both sides of the equation are expressed in inches of water. $\rho_c = 0.075$ at ordinary temperatures.

The density term can be expressed in terms of absolute temperatures,

$$\rho_{ch} = 0.075 \frac{T_c}{T_{ch}}$$

hence the left side of the original equation becomes

$$\frac{0.0145L(t_{ch} - t_c)}{T_{ch}}$$

Velocity head and friction loss head are as follows:

$$h_v = \left(\frac{V}{1096}\right)^2 \cdot \rho_{ch} \quad \text{and} \quad h_L = f\frac{L}{D}\left(\frac{V}{1096}\right)^2 \cdot \rho_{ch}$$

Substitution in the original equation and rearrangement gives

$$V = 21\sqrt{\frac{LDN(t_{ch}-70)}{L + DN}}$$

where V is stack velocity, fpm; L is stack height, ft; D is stack diameter, ft; t_{ch} is stack temperature, F; and $N = 1/f$ is the number of stack diameters for friction loss of 1 velocity head, 1VP.

The addition of h_v to the right-hand side of the equation is subject to the following qualification. If the stream of hot gas coincides with the duct cross section at the point of entrance then the stack need not be charged with an energy requirement for velocity head; the velocity has already been established by the heat of the process. In that case the term, h_v, would be omitted from the equation. Where on the other hand, a substantial part of the velocity is dissipated within a hood vestibule as in Figure 15-10(a), velocity must be reestablished by the stack. Also, in the former case, there would be no entrance loss because the air stream conditions at this point preclude the formation of a vena contracta.

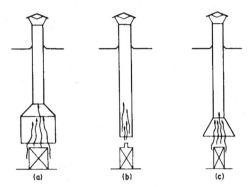

Figure 15-10. Different arrangements of natural draft stack for removal of hot gases. Since hot gas stream is delivered to the duct interior, there is assumed to be little or no vena contracta, therefore, little or no entry loss.

In the case illustrated by Figure 15-10(a), a vena contracta would tend to form, but since a portion of the hot gas enters directly without intermediate dissipation of its velocity, the vena contracta would probably be considerably repressed, hence the entrance loss would be expected to be much less than in ordinary hood entrances.

In this treatment, the equations are developed for the less favorable situation where the hot gas velocity is dissipated within a hood enclosure but entrance loss is deemed insignificant.

Value of t_{ch}

Heat losses from the chimney determine the value of t_{ch}, and this quantity can be estimated from the overall coefficient of heat transmission, H. If we equate heat loss through the chimney walls (in terms of H and the mean temperature differential between stack and surroundings) to that described by the gas temperature drop, we get

$$HA_w(\Delta t) = 0.24 GA_i (t_e - t_f) \qquad \text{(a)}$$

where

H is the over-all coefficient of heat loss; A_w is the wall surface area of the chimney; A_i is the inside cross-sectional area of the chimney; t_e and t_f are, respectively, the entrance and exit temperature of the gases; G is the mass flowrate in pounds per (square foot) (hour); and Δt is the mean of the temperature differences between flue gas and ambient air entrance and at exit.

Magnitude of H

The transmission of heat from the hot gases through the chimney wall which reduces the average gas temperature, t_{ch}, is the resultant of two principal resistances to heat flow, that due to the inside gas film and another due to corresponding film on the stack exterior.

Heat transmission through the inside gas film may be found from a simplified expression, derived from a well-known formula for air flowing inside ducts,

$$H_i = 0.024(0.24)\frac{G^{0.8}}{D^{0.2}}$$

and

$$H_i = 0.0034G^{0.8}$$

where H_i is in Btu per (hr) (sq ft) (F) and G is the mass flow rate of gases in lb per (hr) (sq ft). The diameter term appearing in the original expressions has been eliminated by assuming average diameter of 16 inches, since it is found that between limits of 4 to 60 inches there is an insignificant change in the constant.

The value of G usually encountered will correspond to a lower limit of about 40 and upper limit of around 80; these values correspond to velocity of 700 fpm at 200°F, and 3,000 fpm at 1,000°F. The values of H_i corresponding to these limits are 0.06 to 0.11.

The magnitude of the coefficient of heat transmission for the exterior of the stack would range from 5 to 10, hence, in comparison with the interior film coefficient, may be disregarded for practical purposes.

We may take as a conservative value, $H_i = 0.1$, and obtain a simplified expression by which to estimate the average chimney temperature, t_{ch}, from the entrance temperature, t_e, and the final chimney temperature, t_f.

The term Δt is properly the logarithmic mean temperature difference between hot gases and the surroundings. However, the arithmetic mean gives a close enough estimate for present purposes and is more convenient.

$$\Delta t = \frac{t_e - t_c + (t_f - t_c)}{2}$$

where t_e is temperature at entrance to the chimney, t_f the final temperature, and t_c the cooler temperature of the surroundings which will be taken as 70°F.

Substituting into Equation (a), $H_i = 0.1$, $t_c = 70°F$, and the above expression for Δt, and noting that $A_i/A_w = D/4L$, gives the following:

$$\frac{t_e + t_f - 140}{t_e - t_f} \cong \frac{10GD}{8L} \tag{b}$$

Values of t_e and of G are available from the conditions of the problem as follows. From

$$H' = q_c(0.075)(t_c - 70)(0.24)$$

we get

$$t_e = \frac{55.5H'}{q_c} + 70$$

where H' is the rate of that contribution from the process to the stream of air going to the chimney in Btu per min.; q_c is rate of airflow in cubic feet per minute, based on the room temperature, 70°F.

From

$$60H' = 0.24GA(t_e - t_f)$$

we obtain the value of G:

$$G = \frac{60H'}{0.24A(t_e - t_f)} = \frac{250H'}{A(t_e - t_f)}$$

Finally, one obtains $t_{ch} = (t_e + t_f)/2$ by substituting the above expression for G in Equation (b) which gives

$$t_e + t_f - 140 = \frac{2500DH'}{8LA}$$

Note that

$$\frac{D}{A} = \frac{4}{\pi D} \qquad \text{and}$$

$$t_{ch} = \frac{t_e + t_f}{2} = \frac{200H'}{DL} + 70$$

Roof Ventilators

Natural draft roof ventilators function mainly as devices to keep rain from passing downward through a hole cut in a roof top and to prevent reverse airflow due to wind currents. If rain and wind were not influencing factors, then ventilation just as good would be obtained through unadorned roof openings.

Clearly, therefore, the rate at which air passes outward through such ventilators is a function of a difference in temperature between air inside the building and that outside as well as the coefficient of discharge through the roof opening and the ventilator structure. In fact, the principles applicable here are identical with those discussed in Chapter 8 describing the effect of the thermostatic head on air leakage through cracks or orifices in the upper part of a heated enclosure. In that discussion, a formula was developed to describe the velocity of air passage outward through orifices that may also be applied to roof ventilation openings.

$$V = 20 \left(\frac{LH'}{A_f} \right)^{1/3}$$

where V = air velocity through the ventilator opening, fpm

L = effective height of the hotter air column, i.e., inside the building, ft.

H' = rate of heat released to the interior air, Btu per min

A_f = area of roof ventilator (assuming no restriction to air supply into lower part of building), sq ft.

In actual application to building ventilation problems, an important question arises as to the value of L to be employed. Simplifying assumptions can be made as in the following example.

Example 7. A 30-foot high building is capped by a roof ventilator having total opening area of 512 sq ft. Total heat release inside is 10,000 Btu per min. Estimate the maximum rate of ventilation assuming free access of outdoor air into the lower part of the building.

Solution. Assume the effective height, L, of the hot air column to be a little less than the total building height, say 27 ft. Then the velocity outward through the monitor opening will be

$$V = 20 \left(\frac{LH'}{A_f} \right)^{1/3} = 20 \left(\frac{27 \times 10000}{512} \right)^{1/3} = 160 \text{ fpm}$$

The ventilation rate will therefore be

$$Q = VA = 512 \times 160 = 82,000 \text{ cfm}$$

Equation (15) of Chapter 8, from which the preceding equation was derived, leads to the following form permitting calculation of ventilation rate from data on inside and outside temperatures, taking orifice coefficient $C = 0.6$

$$V = 8.0C \sqrt{\frac{L\Delta t}{T_h}} = 8.0 \times 60 \times 0.6 \sqrt{\frac{L\Delta t}{T_h}} = 290 \sqrt{\frac{L\Delta t}{T_h}}$$

If T_h is taken to be about 600-700°R, the expression becomes, in round numbers,

$$V = 10\sqrt{L \bullet (\Delta t)} \tag{c}$$

which is essentially the equation cited in some texts (e.g., ASHRAE *Fundamentals Handbook*).

Value of L with Restrictions

In applying the equations illustrated in the preceding discussion, it is seldom correct to take the height of roof opening to represent the value of L, because this implies that all of the interior is subject to a thermal pressure outward whereas, in practice, there are negative pressures at lower levels, causing inward motion of air and, conversely, positive pressures are outward air motion at upper levels.

The effective value of L is the distance downward from the roof opening to that elevation where pressures are changing from positive to negative, i.e., the point of neutral pressure. Emswiler (Emswiler, J.E., "The Neutral Zone in Ventilation," *Trans. Am. Soc. Heating and Ventilating Engineers,* Vol. 32, p. 59 [1926]) discussed this question and suggested that ventilation rates be derived by calculating inward air motion at various lower levels and then, similarly, outward airflow through upper level openings (Figure 15-11).

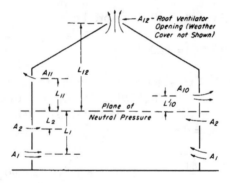

Figure 15-11. Inward and outward air motion through a building due to thermal pressures generated by release of process heat, illustrating neutral pressure point in relation to direction of airflow, and the identity of values of A and L for Equation (c).

Since L is not known, Emswiler described a unique trial-and-error method with various assumed values of L until inward airflow is found to equal outward flow. This can be done using the preceding equation. Since air temperatures are usually variable with elevation, it is necessary to take average values of Δt corresponding to values of L. Such a method of estimating ventilation rate is represented by the following equality where subscripts represent various elevations, the left side series being inward airflow at low levels and the right side those outward at upper levels.

$$10A_1\sqrt{L\Delta t_1} + 10A_2\sqrt{L\Delta t_2} + etc = 10A_{10}\sqrt{L\Delta t_{10}} + 10A_{11}\sqrt{L\Delta t_{11}} + etc$$

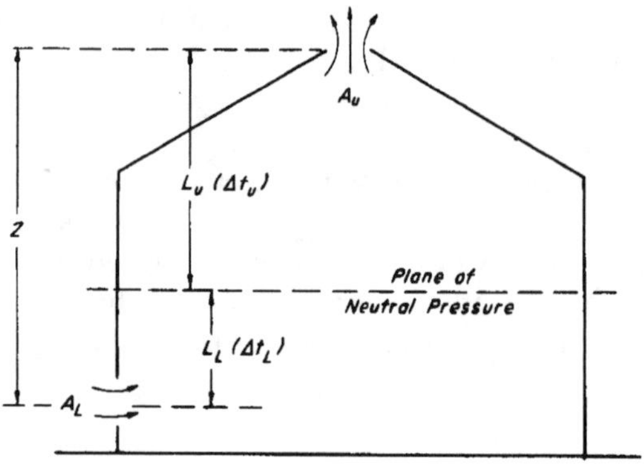

Figure 15-12. Where all inlet openings are at one lower level and the roof ventilator is the principle outlet, the simplified relationship of Equation (d) applies.

For the simple and common case where all lower level and upper level openings are at substantially the same elevation (Figure 15-12), the following simpler equation can be written:

$$10A_L \sqrt{L \triangle t_L} = 10A_u \sqrt{L \triangle t_u} \tag{d}$$

where subscripts L and u represent lower and upper levels, respectively. Letting Z equal the distance in feet between lower level and upper level openings, $L_L = Z - L_u$. When substituted in the preceding equation and rearranged, we get an expression for the distance, L_u, from the point of neutral pressure to the upper level openings, i.e., a value of effective L.

$$L_u = \frac{Z \left(\dfrac{A_L}{A_u}\right)^2 \times \left(\dfrac{\triangle t_L}{\triangle t_u}\right)}{1 + \left(\dfrac{A_L}{A_u}\right)^2 \times \left(\dfrac{\triangle t_L}{\triangle t_u}\right)}$$

This value is then employed in the following general expression.

$$Q = 480CA_u \left(\frac{L_u \triangle t_u}{T_u}\right)^{0.5}$$

It will be noted that all conditions are those prevailing in the upper zone above the point of neutral pressure. Substitution of L_u into this equation gives

$$Q = \frac{480CA_L}{\sqrt{T_u}} \left[\frac{Z \triangle t_L}{1 + \left(\dfrac{A_L}{A_u}\right)^2 \times \dfrac{\triangle t_L}{\triangle t_u}} \right]^{0.5}$$

Motion of Hot Air Masses

The most important application for roof ventilators is in those industries where large amounts of heat are released to ambient air as in glass plants and metal smelting and refining industries. When some of the equipment gives off an air contaminant in addition to heat, the considerations discussed in the following paragraph are of importance.

Hot air heated by contact with hot equipment rises due to its buoyancy in a column which, in size and velocity, is a function of the area of the heated surfaces and the temperature of the heated air. Roof ventilators have no influence whatever on the air columns at these lower levels. All such hot air columns rise toward the roof and join a pool of hot air which, while it has no well-defined lower boundary, is a real situation of some importance to an understanding of various phenomena (Figure 15-13).

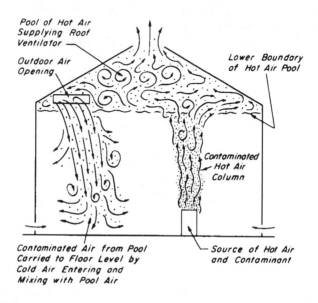

Figure 15-13. Behavior of contaminated air column and hot air pool and mechanism of recontamination of floor-level areas.

As a column of hot air rises toward the pool, mixing turbulently with surrounding air, which is ever hotter with increasing elevation, it finally reaches a point where the temperature is equal to that of the surrounding air and buoyancy is lost. This level may be said to define the location of the lower boundary of the pool so far as that particular hot air column is concerned. Obviously, it will vary with the thermal characteristics of the particular air column being considered. Thus it will be seen there are two phenomena operating to destroy the buoyancy of the hot air column before it reaches the roof level; one, the existence of a hot air pool of some depth under the roof; and two, the fact of a progressive reduction in air temperature of the hot column by reason of its progressive dilution with surrounding air. So that even in a situation where no roof and therefore no pool exists, the air column will eventually lose its buoyancy.

Disposal of Smoke or Gases Through Roof Ventilators

If any hot air column transports a smoke or objectionable gas, the extent to which it is successfully removed from the building without contamination of the lower, inhabited levels near the floor is dependent on the elevation of the lower boundary of the pool in relation to disturbing air currents. If that boundary is lower than the neutral pressure zone and there are outside wall openings, then air entering the building will mix turbulently with the contaminated air of the pool and transport it downward to the occupied levels near the floor (see Figure 15-13). The advantage of high roof level is evident in this connection. The pool of contaminated air being well elevated, there is less likelihood of its accidental transport to floor level.

There are two methods available for correcting such an air contamination problem. The first is to increase the rate of ventilation through the building, which will have the effect of raising the lower boundary of the pool. This is effected by the fact that the temperature of air in the hot air pool is lowered by the increased ventilation, hence the buoyancy of the hot air column is enhanced and it can travel to a higher elevation before its travel is neutralized. Increased ventilation, it should be noted, can be effected only by increasing the area of supply openings as well as that of roof openings. This method is seldom a practical solution due to the magnitude of increased ventilation required.

The other method is to isolate the contaminated air column from the pool by providing a separate conduit for it leading directly out through a separate roof ventilator. This is a more positive method and, therefore, generally much more satisfactory than the one previously described.

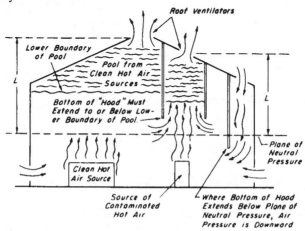

Figure 15-14. Considerations when designing natural draft hoods.

The conduit will generally take the form of curtain walls hanging from the roof to enclose a space above the particular process, forming in effect a huge canopy hood connected to a roof opening with a ventilator for protection from rain and wind (Figure 15-14). Success of this arrangement will depend on the following conditions.

First, the curtain wall enclosure must extend downward to a point below the lower boundary of the hot air pool. Second, it must be at a higher elevation than the point of neutral pressure.

The former condition stems from the basic requirement that the hot air column have sufficient buoyancy to attain the level of the hood. The second condition is necessary to ensure an upward direction of airflow through the hood and ventilator. If the point of neutral pressure were above the hood opening, one could expect a reverse down draft of cooler outdoor air.

A third condition is that the capacity of the roof ventilator serving the contaminated hot air hood must be sufficient, not only to accommodate the volume of contaminated air, q, arriving at hood level, but also an additional quantity to ensure positive indraft through the inactive areas, A, of the hood opening, i.e., capacity = AV, where we may assume that V would be in the neighborhood of at least 50 to 100 fpm. If there are no heat sources other than those releasing the contaminant, then the required ventilator capacity can best be estimated by

$$V = 20 \left(\frac{LH'}{A_f} \right)^{1/3}$$

where, as before, A_f is the area of the roof ventilator, in square feet, H' is the heat release rate, Btu per min, L is the height from the point of neutral pressure to the ventilator opening, and V is in feet per minute. It is to be noted that this result describes only the air quantity represented by the product AV. In addition, the ventilator must have the capacity to handle the quantity of contaminated hot air, q.

Where the building houses other major heat sources, hot air from which passes outward through a separate roof ventilator, the ventilator capacities for each are best estimated together as a single problem.

Ventilator Capacity in Relation to Convectional Air Flow

Another set of circumstances that will affect the behavior of contaminated hot air masses is related to the principles discussed in Chapter 8 wherein it was explained that a canopy hood operating as a receiving hood must accommodate the total flow of hot air

delivered to it. In the present instance the lower boundary of the hot air pool can be regarded as the equivalent of the bottom of a large hood. If a massive source of heat causes upward flow of hot air at a rate greater than the total outflow through the roof ventilator, then a vertical circulation of air will occur between the hot air pool and the air layers nearer the floor. This will result in unnecessary contamination of air at working level by air which, having once reached the pool above, should not have been permitted to move downward again. This condition is illustrated in the following example, which, citing some dimensions from an actual industrial problem, also serves to demonstrate the large scale of air motion and heat release characteristic of hot industries.

Example 8. A glass products plant consists of a building 50 ft high from floor level to roof openings, topped by roof ventilators, totalling 1,800 sq ft in area. Glass tanks and auxiliary equipment emit heat to the interior at a rate of 3 million Btu per min, of which 1 million Btu per min is estimated to pass outward through walls and roof top as radiation that does not heat the interior air. The remainder escapes as heated air which passes upward and out through the roof ventilators. By the relationships embodied in the equation discussed in Chapter 8,

$$q_o = 29(H'A^2m)^{1/3}$$

it is estimated that the air set in motion by the hot equipment surfaces approximates 2 million cfm.

Air travel outward through the roof ventilator in the most favorable circumstances is that corresponding to an effective height, W, of something a little less than the building height of 50 ft, say 40 ft. The outward air velocity through roof openings is estimated by

$$V = 20 \times \left(\frac{40 \times 2 \times 10^6}{1.8 \times 10^3} \right)^{0.33} = 700 \text{ fpm}$$

The area opening is 1,800 sq ft, hence the air flow, Q is

$$Q = AV = 1,800 \times 700$$

$$= 1,260,000 \text{ cfm.}$$

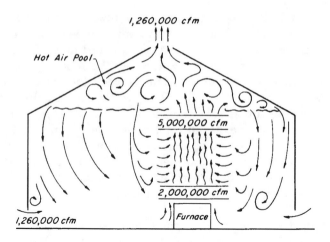

Figure 15-15. Illustration of air motion where roof ventilator capacity is lower than airflow rates sent into motion by heated processes. Effect of conditions of this example is to force a downward flow of 5 million minus 1.26 million cubic feet of air per minute thus contaminating lower levels with air from the overhead pool.

We may now consider the behavior of the 2 million cfm air stream rising as convection above the hot furnaces, in relation to the total ventilation of 1.26 million cfm. (See Figure 15-15.) The difference between these two quantities represents an airflow increment that can be supplied to the air column only by a process of recirculation. Actually the increment is larger because of the turbulent mixing that occurs with surrounding air layers as the column travels upward toward the lower boundary of the overhead pool.

As indicated in Figure 15-15, the total air motion at that elevation is estimated to be 5 million cfm, hence the total increment above that supplied by the outdoor air of 1.26 million is 3.74 million cfm. These requirements can be met only by a downward motion of air from the hot air pool to replace that adjacent to the hot air column that was carried upward.

Thus a constant recirculation occurs and if there is any contaminant in the rising air columns, the pool air is contaminated and subsequently contaminated air is carried downward from the pool to floor level.

Chapter 16

Particle Separators and Dust Collection

Exhaust air streams, having served their process or ventilating function, must eventually be discharged to the outdoor atmosphere. They will have entrained a gaseous or particulate contaminant which will have to be diluted, neutralized, or removed.

Methods for neutralization of gaseous contaminants, where required, can be readily analyzed and easily applied on the basis of well-developed engineering principles. In contrast, the engineering of particulate separation is relatively complex because of their heterogeneous character re: particle size and characteristics, not to mention the wide variety of air/gas stream qualities that need to be considered. Not surprisingly, therefore, this chapter will deal almost exclusively with particle separation principles and devices.

The separating efficiency of a dust collector is the principal characteristic (a) in comparing competing devices and (b) considering its applicability to a particular air/gas stream. Unfortunately, it is not a simple matter to derive performance figures describing particle separation efficiency and this makes it difficult to compare different devices on the basis of efficiency figures that are derived from different sources. Therefore it is important to use standard particle sizing tests, e.g., ASHRAE 52.2, when testing for collection efficiency.

We have been discriminate in citing comparative efficiency figures, except where a single worker using unique techniques has reported some comparative values. We cite percentage dust "penetration" in order to emphasize the significance of this figure in appraising particle separation performance.

The other question involving separation efficiency is performance required for a particular air/gas stream. Where maximum concentration of solids in the effluent air stream is regulated by regulations or ordinances, the required efficiency would be based on the quantity of dust in the air stream as defined by the code agency.

Where no local regulations exists, one may be at a loss for a suitable criterion. It is common practice to consider separation efficiencies above 90 percent as inherently good but this is obviously a fallacious basis for judging adequacy of a dust collector installation.

In another aspect pertaining to separation efficiency, some fail to consider the distinction between "hourly," "monthly," and "yearly efficiency," terms which remind us that efficiency tests on a device carefully "conditioned" beforehand may not be fully representative of average performance over a period of time. Some collectors are inherently "steady" in their performance, others tend to be "unsteady"; and, of course, in all types, the quality of maintenance has a bearing on the same question.

Physical Characteristics of Process Particulate Materials

[Editor's note: The literature on particle physics has widened considerably over the past three decades. The fundamental equations, approaches, simplifications, and data presented by Hemeon in this chapter appears to be valid for overview, conceptual, and historical values. Check the current literature for the latest technical information and product descriptions.]

Fumes. Fumes are those solid particles generally resulting from a thermal process of vaporization, followed by condensation to a solid state either by cooling or by oxidation. An example is the formation of zinc oxide fumes which occurs by evaporation of zinc, as from molten brass, which is oxidized immediately as the zinc molecules meet oxygen of the atmosphere. Similar action occurs with all oxidizable metals; variations between them depend on the vapor pressure of the molten metal and its characteristic rate of oxidation. Metal fumes are also formed when an oxyacetylene torch is applied to metals. In electric welding, the coating of the welding rod is volatilized to a fume.

The physical characteristic of fumes that is important in air cleaning problems is the extremely small particle size. In comparison with dust, there is a very large number of particles per unit weight, and the inertia of each particle is very small. As a consequence they are not effectively removed from a gas stream by any type of dust collector which relies on a particle inertia, such as cyclones, devices which include mechanical rotating elements, low energy water scrubbers, and the like. Fume particles can be removed by electrical precipitation and by filtration through cloth filters and, to some extent, by high energy water scrubbers.

Flocculation. Immediately following their formation, the small particles comprising the fume are subject to the process of flocculation wherein individual particles, in violent Brownian motion, collide with each other and form aggregates or flocs having effectively higher inertia. This phenomenon is most rapid for the smallest particles in the typically heterogeneous size mixture, and also more rapid in dense concentrations of particulates. Thus it is a fast reaction at the site of fume formation, gradually decaying with time, and it is self-limiting as the process itself reduces particle concentrations and the relative population of the smallest particles.

Collectibility. The flocs having effectively higher inertia tend to be more readily collected in inertial types of particle separators. This explains in part the observation, for example, that 50 percent by weight of the fume from the open hearth steel process was readily removed in scrubbers and cyclones not having high separation efficiencies on particles smaller than 5 microns.

The shortcomings of common particle sizing techniques discussed in a following section (Separation Mechanisms, Particle Size, and Size Measurement) as an index of particle collectibility are exaggerated in the case of fumes because of their pronounced flocculation tendencies.

In cloth filtration (e.g., baghouses), the effect of the small particle size of fumes is in a relatively high pressure drop for a given weight, per unit area, and at a fixed gas flow rate.

Mists. Mists are liquid droplets which behave in air cleaning devices like fumes or dust, depending on their particle size, i.e., inertia. They warrant separate classification, however, because of the character of the mass after it has been agglomerated by the particle separation process. So-called tar fume emanating from hot molten tar is actually a mist because the individual particles are liquid, a fact which becomes evident when they are brought together in a collector. Cigarette smoke is a tarry mist. Oil mists are formed when oil is subjected to elevated temperatures. Oil mist is also formed in certain high speed machining operations, but in this case the misting action is mechanical and the individual droplets much coarser than in the thermal process.

As has been suggested, collection of mists could be effected as though they were solid particles except for their liquid characteristics in condensed form. Thus a tar or oil mist could be filtered through cloth, but the liquid would soon coalesce to saturate the filtering element with a sticky liquid and prevent further functioning. Successful application of an electric precipitator to mists would depend not so much on precipitating action but on the character

of the deposit and on practicality of measures for its periodic removal. Dusts and fumes are dislodged by a vibratory rapping on the collected electrodes. Some liquid mists will flow under gravity as deposited. Others require special measures such as flushing with water or other liquid or washing by hot steam jets.

Mists that are water solutions of soluble solids (such as salts in electroplating) may become dry solids by evaporation of water in transit to the collector. In some instances, therefore, it may be possible to collect the contaminant in a cloth bag filter, if dry. Obviously corrosive constituents in relation to bag material would need to be considered in any practical application. Similarly, the over-spray incident to spray painting is a mist during its early life but the evaporation of its thinner may transform it to a dust which could be filtered through cloth.

Dusts. The dust classification is the broadest one of those cited in this discussion. As a general thing, it results from mechanical processes like crushing or abrading of solids, or by disturbance of already pulverized material. Particle sizes may range from very coarse to very fine; the former is typified by sawdust and other wood waste, collection of which can often be effected to a satisfactory degree of efficiency in relatively simple devices like cyclone collectors. Fine dust includes higher proportions of small particles of a size characteristic of fumes, requiring collection devices of superior efficiency.

Separation Mechanisms, Particle Size, and Size Measurement

Except for cloth bag filters and electric precipitators, the particle separation mechanisms operative in all other types of commercial dust collectors are (a) gravity sedimentation, (b) centrifugal force, and (c) impaction.

In gravity sedimentation, the velocity of separation is indicated by the terminal velocity, u_t. In inertial separators, e.g., cyclones, the separating force is magnified, and separation velocity is some multiple of terminal velocity. In the impaction mechanism, as will be shown later, the separation velocity is also some multiple of terminal velocity u_t; the effectiveness of separation by this mechanism is directly proportional to the product of terminal velocity and initial velocity,

$$\frac{u_t \cdot u_o}{g}.$$

Thus in all these cases terminal velocity of the particle is an index of its collectibility in these various types of particle separators.

Unfortunately, it is not true, as widely assumed, that particle sizes as commonly reported in the technical literature are a fairly good basis for calculating terminal velocities. This would be true only if the particles comprising industrial dusts and fumes were completely dispersed and discrete. In fact, however, as was particularly observed in the preceding discussion of fumes, they may sometimes be flocculated to an extreme degree. One [older] particle size measurement technique required the collected sample of particulates to be dispersed in water or other fluid preparatory to observation and measurement of particle size, either by microscope examination or by sedimentation in a liquid. The liquid dispersion step destroyed the flocs and hence the effective inertial size of the fume dust mixture as it existed in the gas stream. This is illustrated in Figure 16-1.

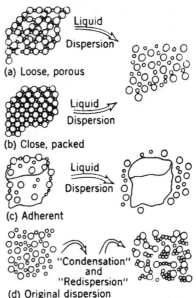

Figure 16-1. Problems in particle sizing techniques. Various forms of particle aggregation indicated at the left; and at right, the breakup of aggregates due to dispersion in liquid during preparation for size analysis using wet methods. (a) through (c) represent typical condensed dust masses; d) represents the air elutriator sizing method in which particles were originally well-dispersed but were imperfectly re-dispersed.

(a) Loose, porous

(b) Close, packed

(c) Adherent

(d) Original dispersion

Techniques involving re-dispersion of a collected dust in a stream of air for gravity or centrifugal elutriation are also subject to serious errors because the resulting test dust cloud may be less completely dispersed than it was originally (Figure 16-1, d), and in that case the measured particle size, or settling velocity, might be greater than in the original gas stream from which the dust sample was collected.

Thus predictions of collection efficiency in the one case would be on the high side and in the other case too low.

These measurement errors are probably less serious in the case of dust whose particles are predominantly greater than 10 microns.

The practical importance of this discussion is in the warning it sounds against too ready acceptance of particle size data in connection with estimates of dust collector efficiency.

Theory of Impaction

The collision of dust or mist particles with droplets of water in various types of scrubbers or with fibers in some types of dust filters due to relative velocity of the two is one of the important mechanisms of dust collection.

The theory of the process of impaction of a dust particle onto a water droplet target, fiber target, or the like is of great value to an understanding of dust separation mechanisms even though the experimental values available from research results are not practically applicable in a quantitative way to the design or the appraisal of such equipment.

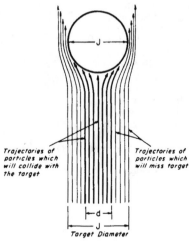

Figure 16-2. Particles in a gas stream flowing around a target (e.g., droplet, fiber) may collide therewith (i.e., impact) depending on their position in the approach path.

Figure 16-2 depicts a target (water droplet, fiber, etc.) having diameter, J, which lies in a gas stream in which numerous particles are suspended. As the gas flows around this obstacle, the particles within a path leading to this target are subject to two influences affecting their motion. Their inertia tends to attract them to the target, and this tendency is opposed by the viscous drag of the gas

streaming around it. Particles in a central core will succeed in colliding; those in outer streams may elude the target.

The efficiency of collection is obvious from the geometry of stream lines and target. If the target is a cylinder (fiber, for example) the efficiency will be d/J. If the target is a sphere (e.g., water droplet), the efficiency will be represented by the ratio d^2/J^2.

Various theoretical and experimental investigations have shown that target efficiency is directly related to a ratio, the impaction parameter

$$\frac{\text{stopping distance}}{\text{target "diameter"}}$$

Stopping distance of a particle is the distance required for the velocity of a particle projected through a fluid with some initial velocity, to decay to zero due to viscous drag forces. The projection of particles discussion in Chapter 2 briefly described the relationships. In the area of our interest the motion of particles in the impaction process is in the streamline flow region and, therefore, stopping distance, S_t is given by the following equation, where u = o. (Viscosity at 70°F.)

$$S_t = 0.29 \times 10^{-5} u_o \cdot Z \cdot d_m^2$$

In more general terms it can be shown that

$$S_t = \frac{u_o \cdot u_t}{g}$$

Thus, the impaction parameter is approximately

$$\frac{0.29 \cdot 10^{-5} u_o \cdot Z \cdot d_m^2}{\text{Target diameter (ft)}}$$

Transforming the units of target diameter from feet to microns, we get

$$K = \frac{0.88 u_o \cdot Z \cdot d_m^2}{j_m}$$

where u_o is gas velocity relative to the target, fps; d_m is particle diameter, microns; j_m is target diameter, microns; and Z is specific gravity.

The relationship between K and target efficiency according to two particular research investigations is summarized in Table 16-1 for spherical targets (e.g., water droplets) and in Table 16-2 for cylindrical targets (e.g., filter fibers).

Table 16-1
Fractional Efficiency in Impaction of Particles onto
Spherical Targets (Ranz and Wong)

$K = \dfrac{0.88 u_o \cdot Z \cdot d_m^2}{j_m}$	Efficiency, fractional
0.04	0
0.09	0.08
0.16	0.4 (0.17 - 0.42)*
0.25	0.55 (0.3 - 0.6)*
0.49	0.74

Table 16-2
Fractional Efficiency in Impaction of Particles onto
Cylindrical Targets (Ranz and Wong)

$K = \dfrac{0.88 u_o \cdot Z \cdot d_m^2}{j_m}$	Efficiency, fractional
0.06	0
0.16	0.03
0.36	0.13 (0.08 - 0.18)*
0.64	0.32 (0.3 - 0.4)*
1.0	0.13 (0.35 - 0.55)*
1.4	0.13 (0.5 - 0.7)*
2.5	0.13 (0. - 0.9)*
3.2	0.13 (0.08 - 0.93)*
4.0	0.13

*Range of experimental results

Types and Classes Of Particle Collectors

The wide variety of particle separators available [circa 1963] for cleaning of industrial gas effluents is merely testimony to the fact

that no one type is possessed of all engineering virtues, nor free of some disadvantage.

The classifications in Table 16-3 are based on the different mechanisms by which particles are separated from the gas stream; the scrubber group has been provided with several sub-classes for better understanding of their operating characteristics.

Low Loss, Large Particle, Dry Collectors

The simplest dust collectors are the low-pressure drop, large particle arrestors which depend on force of gravity, or on a 90 to 180 degree change of direction of the gas stream. Pressure drops do not generally greatly exceed 0.25 to 1.0 inch w.g. Although lowest in cost and relatively free from problems of metal abrasion (with some obvious exceptions), their field of application is quite narrow.

These simple collectors include those employed for the purpose of reducing required transport velocities and to reduce abrasion on downstream duct work and other equipment.

They find application also for fly ash removal from small coal burning furnaces where it is desired to avoid the cost of a fan, required for operation of more efficient types of collectors. Their low pressure drop requirements are adequately met by the natural draft provided by the hot flue gases in the chimney.

The economy of this low efficiency arrestor over collectors of higher efficiency is not so much in the relative cost of the collectors themselves, but relates to the need for an exhaust fan, and also the need for cooling the gas to levels the fan can tolerate.

Gravity types. The particle collection characteristics of the simple particle trap illustrated at the upper left on Figure 12-1 and of the shelf type gravity dust separator ("Howard dust collector") is determined by the trajectory of the particle in relation to the dimensions of the hopper or of the horizontal collecting shelf or plate.

Considering the horizontal passage between two plates of the shelf-type separator, if the gas were in streamline motion the trajectory would be the result of the particle settling velocity vector and the horizontal gas velocity vector (Figure 16-3a). In turbulent gas flow, however, smaller particles are affected by the gaseous eddies which will have the effect of delaying the arrival at the shelf of a portion of the particles (Figure 16-3b) and the efficiency of the trap will not correspond therefore to that estimated by the idealized streamline flow pattern.

Table 16-3
Historical Classification of Particle Separators

I. *Large particle, low loss, dry collectors:* Embraces those depending on gas direction change of 90 to 180 degrees, gravity sedimentation; pressure losses range from 0.25 to 1.0 inch w.g.

II. *Cyclone dust collectors, dry:* Includes large diameter units and small diameter tubes of various types.

III. *Wet wall cyclones:* Vertical or horizontal axis; pressure nozzles or swirl-induced wetting.

IV. *Inertial collectors, dry:* Stationary and rotating.

V. *Water scrubbers or washers:*
  a. Spray towers; zone adjacent pressure nozzle
  b. Spray towers; force of gravity
  c. Wet impingement; shear atomization and simultaneous impaction on stationary targets.
  d. High gas velocity atomizers; shear atomization of water; mutual impaction and mist separator.

VI. *Cloth filters*

VII. *Fiber-bed filters:* Impaction, agglomeration, filtration and wash renewal.

VIII: *Electrostatic precipitators*

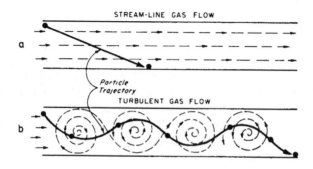

Figure 16-3. Trajectory of particles in a gas stream flowing between horizontal plates. (a) Laminar flow; and (b) turbulent flow.

Change-of-Direction Types. One type of large particle trap that [had] been widely used on iron cupolas is illustrated in Figure 16-4. It was commonly referred to as a scrubber but the principal dust collection mechanism would seem to be the change-of-direction feature and that the water stream functions mainly to cool the steel cone and, secondly, to transport collected grains of dust to ground level.

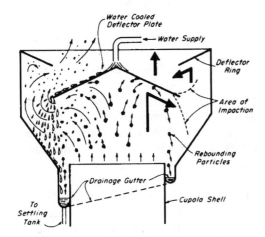

Figure 16-4. Large particle separator of the change-of-direction type. Water flow cooled the conical baffle and transports separated dust to ground level.

Another type of dust trap in this category, not illustrated here, conducts the gas stream into a converging passage the walls of which are louvered. The gas passes through the louver openings leaving the largest particles to continue along the axis, due mainly to their momentum, to the small opening at the passage apex where they are deposited in a collection hopper. A fan-induced flow of gas through the apex dust discharge port may or may not be provided. Pressure drop through the louvers and hence the overall pressure loss is less than 0.2 inch. (This collector is not to be confused with the louvered cone collector referred to later whose pressure drop is 1 to 3 inches w.g.)

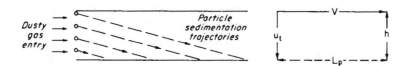

Figure 16-5. Trajectory of particle in a stacked plate dust collector is the result of horizontal gas velocity (streamline flow) and its settling velocity.

Dust collection characteristics of this type of device in terms of Percent Dust Penetration is probably fairly represented in the following tabulation:

Particle Size	Dust Penetration,%
10 micrometers	80-90
20	50-70
30	30-40
50	10-20
70	5-10

Efficiency-size relationship in design of shelf dust collectors

An interesting and illuminating demonstration of the relative "collectibility" of large particles vs. small particles and the increasing cost of collection equipment with decreasing particle size is presented in the discussion which follows (*Hemeon, 1958*) and which also provides an appropriate preface to the later discussion of cyclonic type collectors.

The shelf type collector finds little or no industrial application these days, but it frequently appears in essays by various writers on the subject because it so conveniently illustrates the principles of gravitational sedimentation dust collection. We also use it here for these purposes. In the following development we outline the design of a series of such collectors in order to arrive at some figures demonstrating the objectives mentioned in the preceding paragraph.

In order to avoid the uncertainties in performance due to the turbulence characteristics described in the preceding section and illustrated by Figure 16-3, we shall design the collectors so that gas flow between the plates is in streamline, laminar flow motion. (Re < 2000).

Design parameters

A simple vectorial analysis as depicted in Figure 16-5 provides the basis for describing the ability of a stacked plate dust collector to remove particles of various "sedimentation sizes" (larger particles). It shows the trajectories of several particles of a single size whose positions as they enter the channel are uniformly distributed from top to bottom. We can characterize the capturing ability of such a channel in terms of the particle most favorably located to travel the greatest distance, h, and reach the floor after traveling the horizontal distance, L_p. The actual value of L_p can be determined if the terminal settling velocity, u_t, of the particle is known and where a particular value of horizontal gas velocity, V, is specified. As is

apparent from Figure 16-5, the significant relationship may be described by

$$\frac{u_t}{h} = \frac{V}{L_p}$$

We can now proceed to the derivation of an expression that will approximately describe the size of a full-scale stacked plate dust collector in terms of its efficiency for capture of particles of various sedimentation sizes. Obviously, the size is adequately described by a statement of the total area of all shelves, at least in terms of required material for construction.

If we let N represent the total number of plates, or of channels, W_p the width of the plate assembly and L_p its length, we see that

$$\text{Size (sq ft)} = N \bullet W_p \bullet L_p$$

Let the total height of the plate assembly be represented by H_p, then

$$N = H_p / h$$

and

$$\text{Shelf area} = \frac{H_p}{h} \bullet W_p \bullet L_p$$

We now substitute for h its equivalent

$$h = \frac{L_p u_t}{V}$$

and we obtain

$$\text{Shelf area} = \frac{H_p \bullet W_p \bullet L_p \bullet V}{L_p \bullet u_t} = \frac{H_p \bullet W_p \bullet V}{u_t}$$

In this general expression it is seen that the product of H_p and W_p is face area and the total product of the numerator is gas flow rate. Hence the dust collector size is adequately described by the ratio of gas flow rate and particle terminal velocity; in the units feet and feet per minute the size is the ratio of

$$\frac{\text{gas flow rate, cfm}}{\text{terminal velocity, fpm}}$$

There is need for another qualification. The foregoing analysis applies only to a condition of streamline flow in the channels which means that the Reynolds number must not exceed a value of about 2,000. This fixes the product (diameter)x(velocity) in the Reynolds number which, by substitution of a value for kinematic viscosity (1.6 X 10^{-4}, air, 70°F) and simplification, noting that diameter = h in the present notation, we get

$$hV = 230$$

where h is in inches and V in feet per minute.

Example 1. Let us now apply these relationships to the design of a collector to handle 25,000 cfm of air with dust at a particle size of about 20 microns and of specific gravity = 3. From Equation (7) [see Chapter 2 appendix; Stokes Law], the settling velocities for these particle sizes (converted to feet per minute) are estimated to be:

Particle Size, Micrometer	Terminal Velocity, fpm
20	6.6

The required shelf area then for 20 micron particles will be

Area = 25,000/6.63 = 3,770 sq. ft, or 150 sq. ft per 1,000 cfm.

Although not required for our present purposes it is interesting to see what the dimensions of such a collector might be. If we specify a face velocity of 600 ft per min, then by hV = 230, the value of h is fixed at 0.383 in. The face area would be

A_f = 25,000/600 = 41.7 sq. ft

which could be represented by face dimensions of 8.35 ft high by 5 ft wide. The length of the plates, from $L_p = Vh/u_t$, would be 2.88 ft and 262 plates would be required. For other particle sizes:

Particle Size, Micrometer	Terminal Velocity, fpm
10	1.7
5	0.4
2.5	0.1

The corresponding shelf areas required for 10, 5, and 2-1/2 micron particle collection would be 600, 2400, and 9600 sq. ft per 1,000 cfm, respectively. The practical significance of these numbers is

dramatically evident if we transform these areas to the weight of steel plate required and then note the cost. This is set forth in the following tabulation.

Particle Size Collected, micrometers	u_t fps	Shelf Area Required, ft^2	Total Wt. of 16-Gage Steel Plate, tons	Cost at $260 per ton
20	6.6	3,770	4.7	1220
10	1.7	15,100	18.8	4880
5	0.4	60,000	75	19,500
2.5	0.1	240,000	300	78,000

It will be noted that those particles which account for the dominant dustfall nuisance, 20 microns and larger, are easily collected at low cost; removal of the smaller particles are extracted in such an apparatus only at a high cost.

For those of us who subconsciously distort the dust-micron scale the tabulation is useful in reminding us that there is as great a difference from an engineering standpoint between 2 and 4 micron particles as there is between 20 and 40 micron particles.

Cyclones (Dry)

The limitations of gravitational force in the separation of particles from a gas are extended to some degree by equipment that subjects the particles to centrifugal force.

The inertia of dust particles as exemplified by their settling velocity also determines their behavior in cyclones, and the relative sizes of a series of such collectors for the collection of particles of decreasing size. For example, if a cyclone dust collector of a certain diameter were able to collect nearly all particles of 20 microns diameter it would require four smaller diameter cyclones of similar proportions to attain the same efficiency for 10 micron particles, the same proportions we found for the shelf dust collector.

Cyclone Performance. Cyclone dust collectors have been the subject of extensive research for one interested in design research and the "principles" governing chaotic motion of *small* particles. A perusal of the literature will provide considerable intellectual stimulation. For the engineer concerned with applications all this knowledge can be concisely summarized, for the cyclone has up to now [1963] successfully defied all attempts to reduce its performance in the collection of dry dust particles to a rational basis.

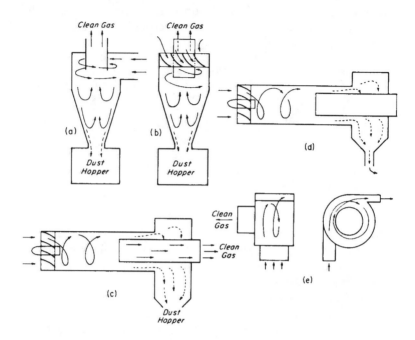

Figure 16-6. Various types of cyclone dust collectors.

Cyclone Types. In the original cyclone design, air enters the body tangentially and, inside the cyclone, assumes two spiral forms, one circulating adjacent to the outer wall, subjecting dust particles to centrifugal force and causing the larger of them to migrate radially toward the wall, and a concentric inner gas spiral containing residual finer particles, moving upward toward the exit duct passage at the top. Dust concentrated in the outer spiral moves down the cone wall and is projected into the dust collection hopper below the cyclone.

The original cyclones having diameters measured in feet were successfully applied to collection of debris in the woodworking industries and the like. The knowledge that the centrifugal forces to which the particles are subjected is a function of the radius of rotation of the dusty gas stream, eventually led to cyclones of small diameter, measured in inches, whose efficiency is materially greater. In order to accommodate larger gas flows, these small diameter units are assembled into multiple cyclone tubes operating in parallel. A variety of forms has been developed over the years (see Figure 16-6) almost all of which can be described in terms of gas flow directions and handling of the collected duct, as shown in the following information:

<u>**Gas Flow Direction**</u>

Gas Entry	Gas Exit	Collected Dust Disposition
Tangential	Axial, 180°	Projection to hopper
Annular	Axial, 180°	Projection to hopper
Axial	Axial, straight through	Projection to hopper
Annular	Axial, straight through	Concentration in peripheral gas fraction
Tangential	Axial, 180°	Concentration in peripheral gas fraction

In the simplest case, dust concentrated along the outer wall of the cyclone migrates spirally downward and is precipitated into a collection hopper by its momentum and by gravity and the partially cleaned gas passes in the opposite direction upward and out through the discharge flue.

A small diameter cyclone causes the dust gas to enter the cyclone body from the top through swirl vanes in the annular space surrounding the exit duct, and the dust concentrated near the wall is projected as before into the collection hopper; the gas passes outward to the exit duct in a direction 180 degrees from its point of entry after passing through the typical double spiral pathways.

In still another design, the dusty gas enters axially through swirl vanes, the central core of relatively cleaner gas passing out the opposite end in the same direction, while dust concentrated in the annular periphery is projected into the dust collection hopper.

Dust which is concentrated in the whirling outer layer of dust in all the preceding types of cyclones is not effectively separated unless it is actually caused to reach the outer wall, and considerable proportions of dust adjacent to that wall, therefore, may be re-entrained in the inner spiral moving upward toward the exit. This is improved in other designs wherein the dust so concentrated in the peripheral annular gas stream is removed in a small fraction of the total gas flow and treated separately in an auxiliary small dust collector of high efficiency. Effluence from this small collector is then recirculated to the main dusty gas stream.

In another form, the gas flow is straight through from one end to the other end of the cyclone body, the cleaner gas stream in the central core passing directly outward and the peripheral annular gas stream containing the higher concentration of dust removed for subsequent treatment in the auxiliary collector. The dusty gas stream enters tangentially in another form of this same type.

Cyclone Design. The relative dimensions which affect the dust separating efficiency of cyclone type dust collectors are well known but are usefully considered only in qualitative terms.

(1) The smaller the diameter of the cyclone body, the greater the centrifugal force to which the particles are subjected and consequently the higher the efficiency. This, however, is subject to the principle of diminishing returns and it is doubtful that diameters less than 3 to 4 inches accomplish much more in dust separating performance.

(2) Increased gas inlet velocity, other things being equal, tends to result in increasing efficiency and accompanying pressure drop. The limitation relates to the fact that proper functioning of the two internal gas spirals is eventually impaired due to increasing tendency to intermingling and consequent premature re-entrainment from the outer spirals to the inner spiral traveling outward toward the exit.

(3) Increased length of the cyclone body tends to increase separation because of the additional number of turns afforded the dust-laden gas stream in the outer spiral, but this is also limited for approximately the same reasons as cited above, because, as the length is increased beyond a point, turbulent mixing between the outer, dust-laden gas spiral and the relatively cleaner gas spiral in the core occurs to an increasing extent, thus nullifying the advantages.

(4) A smaller diameter of gas outlet duct relative to the body diameter is also advantageous to increase efficiency within limitations; but the pressure drop increases more rapidly than the benefits in increased efficiency when these proportions pass beyond a given point.

(5) The parallel arrangement of multiple small-diameter tubes necessary for large gas-handling capacities is accompanied by lowered efficiency compared with those found with a single cyclone tube, due to the tendency for uneven flow to the various tubes which may lead to reverse flow through some of them upward through the collection hopper. This tendency is aggravated if the dust being handled has adhesive characteristics tending to reduce the dimensions of the opening at the bottom of individual tubes.

(6) One of the most common factors reducing efficiency of all types of cyclone collectors is in leaking dust valves between the cyclone proper and the dust collection hopper. Where the cyclone is operated under negative pressure, as is usually the case, and upward flow of air from the hopper into the cyclone is permitted and this may seriously impair the operating efficiency of the cyclone. This fault is inherent in the design of many dust valves even in the products of some reputable manufacturers, and in other cases it is due to faulty maintenance.

Cyclone Efficiencies. Inadequately developed particle size measurement techniques and lack of standardization [circa 1963] [made] it difficult to offer sound comparisons of dust collection efficiency of different types of cyclones in inertial dust collectors. One author (Caplan) avoided this problem by citing efficiency figures for all types of so-called high-efficiency multiple-tube cyclones in broad particle size categories (Table 16-4).

Table 16-4
Dust Collection Performance of
"High Efficiency" Cyclones (after Caplan)

Particle size micrometer	Efficiency, percent	Penetration percent
< 5	50-80	20-50
5-20	80-95	5-20
15-40	95-99	1-5

Difficulties and limitations in significance of comparative test data were partially overcome in cases where tests were made by an identical technique and by the same worker. Table 16-5 presents the results of comparative tests of this nature and, although the cyclone types tested are not exactly described or identified, they do provide a useful picture of spread in performance of various types of cyclones.

Table 16-5
Relative Dust Separation Performance of 5-Micrometer Dust Particles in Several Types of Commercial Cyclones (after Stairmand)

	Collection Efficiency %	Penetration, %	Relative Penetration
Multi-tubes, 6", high efficiency	89	11	1
Long cone, high efficiency	73	27	2.5
Multi-tube, 12", low pressure (about 1")	42	58	5.3
Medium Efficiency, high throughput, commercial, four feet diameter	27	73	6.5

Comparisons must be made by reference to the percent of dust penetrating the device which, as can be seen, shows a range from 11% to 73%, or a sevenfold difference in performances as between the most efficient and least efficient types.

[Editor's note: Those interested in the latest information and the various design aspects of dust collectors are referred to the considerable body of technical literature that has been published in the last four decades. An excellent resource is the McIlvaine Company's various scrubber and filter manuals. See References.]

Wet Cyclones

In all dry type cyclones there is some tendency for reentrainment of dust which has actually made contact with the outer wall due to the highly turbulent gas conditions within the apparatus. Wetted wall cyclones are another variety of inertial dust collector which tend to have higher efficiencies since all dust that has once reached the outer wall of the cyclone is entrapped in the film of water and is not therefore reentrained. One type is represented essentially by a conventional cyclone which has been equipped with water sprays supplied from low pressure nozzles.

Another type, resting on a horizontal axis, has a reservoir of water at the bottom and the swirl of dust gases induces a flow of water upward and around the inside wall surface, paralleling the spiraling flow of gas.

The enhancement of efficiency due to the wetted wall is illustrated by the following test results (after Stairmand), the wet wall showing double the effectiveness (one half the penetration) of the dry cyclone, for 5-micrometer particles:

	Efficiency	Penetration
Dry Cyclone	73	27
Same, wet wall	87	13

Johnson also reported relative efficiencies of a cyclone, dry vs. wet wall, with a particular test dust. The difference in penetration was about 25 percent less.

	Efficiency	Penetration
Dry cyclone	59	41
Same, wet wall	67	33

Other Types of Inertial Dust Collectors

Louvered Cone. Another type of dry dust collector is a louvered cone device wherein the larger particles in the gas stream are concentrated in a fraction of the total gas flow for treatment in small high-efficiency dust collectors. Although quite different in form, it can be seen that its operating principle is similar to that of cyclone

dust collectors, differing in that the main gas stream in passing through the small louver openings, is subjected to only one-half turn or 180 degrees in contrast to the several spiraling turns in the dust cyclone. On the other hand, the radius of this 180-degree turn, a fraction of an inch, subjects the dust particles to much greater centrifugal forces.

The efficiency of this type of dust collector, having pressure drops of 2 to 3 inches w.g., is probably of the same order of magnitude as that of the small-diameter multiple-tube cyclone dust collector.

Dynamic Precipitators. Another category of inertial dust collector are those which subject the dusty gas stream to centrifugal force in devices resembling centrifugal fans and which serve a dual function of moving the gas stream while separating out the larger particles in the dust mixture. For performance of one device of this class, see Table 16-9.

Pebble filter. Inertial impaction is the particle separation mechanism also in the pebble filter which may be operated (Lynch) to condense moisture from the gases on the aggregate elements and thus minimize re-entrainment of impacted particles. Dust removal involves continuous or periodic removal of the pebbles, sieving or washing them and returning by conveyor to the top of the chamber.

Water Scrubbers

[Editor's note: The following paragraphs present an overview of typical and historical water scrubbers developed over the years.]

Spray Towers and Chambers

Spray towers in which dusty gases pass upward, counter current to the gravity flow of water droplets from spray nozzles, have relatively low efficiency and therefore have limited industrial application. An analysis of their performance, however, provides a useful introduction to various other types of scrubbers, and a simple illustration of the application of impaction principles.

There are actually two distinct dust-capture processes occurring within the space. One is in a zone adjacent to each spray nozzle, where velocity differential between the water droplets, the impaction targets, and the dust or mist particles derive from high initial velocity of water droplets issuing under pressure from the spray nozzles. The other capture process is that due to the relative velocity of water droplets falling by gravity through the dust particles being carried upward in a the gas stream.

Spray Nozzle Impaction. The most favorable condition for impaction of particles on water droplets issuing from a spray nozzle is in the zone nearest the nozzle where there is a relative velocity of particles and droplets. With increasing distance droplet velocities decay in accordance with the principles discussed in Chapter 2 and the dust collision potential is correspondingly decreased.

The significant dimension in estimating the extent of total impaction between particles and droplets is the volume of space traversed by the droplets before their velocity decays to zero, i.e., the "swept volume." It is the product of projected area, A_w, of all the droplets from a given quantity of water, their radius of travel, L, and their collision efficiency, E. The first two terms are fixed for droplets of specified size and initial velocity, but the impaction efficiency is variable, decreasing from initial high to final low values.

$$\text{Swept Volume, W} = (A_w)(L)(E)$$

Such swept volumes can be derived by calculating first the radius of travel of droplets of a given size and initial velocity. Then inertial parameters (K) for various sizes of particles, droplets and velocities are calculated and from these impaction efficiencies are derived. From graphs of efficiencies against distance, *average* collection efficiencies for the droplets throughout their travel are obtained. This laborious procedure was carried out by Walton and Woolcock for several droplet and particle sizes, results of which are given in modified form in Table 16-6.

The fractional penetration, P, of a cloud of gasborne particles through a rain or mist of water droplets can be roughly estimated by

$$P = e^{-\frac{W}{G}}$$

where W is effective volume of space swept by the droplets, and G is the total dust-laden gas volume flowrate, or

$$\log P = \frac{-W}{2.3G}$$

[Notes for Table 16-6: The initial drop velocity of 5,900 fpm corresponds to a water feed pressure of 80 psi for a plain atomizer with a coefficient of discharge of 0.9, or 260 lbs per square inch for a swirl atomizer with a coefficient of discharge of 0.5.]

Table 16-6
Dust Collection Efficiency of Water Spray Nozzles
Pressure Atomization; Dust SG = 1.37
(after Walton and Woolcock)

Initial drop velocity, fpm*	Drop diameter, mm	Range, feet	Dust diameter, micron	Average collection efficiency %	Effective swept volume, cu ft/gal
5900	0.5	10.8	10	78	1020
			5	57	740
			3	39	500
			2	24	320
			1	5	65
5900	0.3	5.4	10	85	910
			5	66	700
			3	49	515
			2	34	360
			1	10	105
5900	0.2	2.9	10	93	820
			5	73	650
			3	57	500
			2	41	370
			1	17	160
5900	0.1	1	10	110	660
			5	87	530
			3	69	410
			2	55	330
			1	28	170
5900	0.5	9.1	10	73	790
			5	50	540
			3	32	360
			2	17	180
			1	3	26
5900	0.3	4.4	10	81	710
			5	59	530
			3	41	360
			2	26	220
			1	6	50

The following example illustrates an application of this last relationship, using the data of Table 16-6.

Example 2. An air stream, Q = 5,000 cfm, containing 5 micron dust particles (sp. gr = 1.37) is scrubbed in the zone adjacent to a spray nozzle which generates droplets of 0.3 mm diameter at a rate of 3 gal per min. Initial nozzle velocity of water is about 5,900 fpm.

Estimate the percent penetration of dust particles.

Solution.

$$\log P = \frac{-W}{2.3G}$$

W = 700 x 3 = 2100 ft^3/min (from Table 16-6)

G = 5,000 ft^3/min

$$\log P = \frac{-2,100}{2.3 \text{ x } 5,000}$$

$$= -0.1825$$

Therefore

$$P = 0.66$$

Gravity Water Spray. In the main space of a spray tower, beyond the zone of spray nozzle impaction, water droplets fall toward the bottom due only to force of gravity, counter current to upward flow of dust laden gases. Settling velocity of dust particles relative to that of water droplets is insignificantly small and their effective velocity, therefore, can be taken to be that of the gas stream itself.

Thus the relative velocity of the target water droplets and of the dust particles which determines impaction efficiency is the terminal velocity of the water droplets.

Impaction efficiencies of individual droplets falling at their terminal velocity are given in Table 16-7 based on the prevailing theory of impaction. The figures provide striking indication of the vast difference in "collectibility" of particles in such "small" size ranges as 2, 3, and 5 microns.

The data of Table 16-7 were employed by Walton and Woolcock to calculate the values in Table 16-8: the effective, capture cross-section in impaction of various droplets and dust particle sizes.

Table 16-7
Impaction Efficiency of Individual Water Droplets,
Falling at Terminal Velocity, on Dust Particles (SG = 1.37)
(after Walton and Woolcock)

Droplet			Particle	Size	Micron	
Size	2	3	5	10	15	20
Micron			Percent	Eff.		
100	0.1	0.3	2.1	31	-	-
200	0.3	1.6	6.8	41	69	89
300	0.4	3.6	12	49	73	88
500	1	7	23	60	80	90
1000	1.1	10	34	70	85	92
1500	0.5	7.8	34	71	86	92
2000	0.3	5.9	31	70	85	91
3000	0.2	4.6	27	67	83	90

Table 16-8
Dust Collection Potential of Water Droplets,
Falling at Terminal Velocity, in Terms of S, ft^2, Effective Capture
Cross-Section (particle SG = 1.37) (after Walton and Woolcock)

Droplet			Particle	Size	Micron	
Size	2	3	5	10	15	20
Micron			S, sq. ft.	per gal.	water	
100	0.8	2	13	190	-	-
200	0.8	5	-	-	-	28
300	0.8	7	25	100	150	180
500	1.2	9	29	74	98	110
1000	0.8	6	21	43	53	57
1500	0.2	3	14	29	35	37
2000	0.1	2	9	21	26	28
3000	0.0	1	5	13	17	19

Example 3. Estimate the percent penetration of 3 micron dust particles (sp. gr = 1.37) contained in a 25,000 cfm gas stream subject to water sprays in a scrubbing tower 20 feet tall. Water is sprayed at a rate of 50 gallons per minute and produces droplets one millimeter diameter.

Solution. Effective capture area, or swept area of droplets, from Table 16-8

$$S = 6 \times 50 = 300 \text{ ft}^2/\text{min}$$

The swept volume is the product of swept area and tower height

$$W = 300 \times 20 = 6,000 \text{ ft}^3/\text{min}$$

$$\log P = \frac{-W}{2.3G}$$

$$= -6,000/(2.3)25,000$$

$$= -0.104 \quad \text{and} \quad P = 0.79$$

Mechanisms of Particle Separation in Some Historical Commercial Scrubbers   [Not all are still commercially available.]

It is not possible in the present state of knowledge [circa 1963], to predict efficiencies of particle separation in scrubbers but it is feasible to make good estimates of their relative performance by reference to the principles of impaction as illustrated in the preceding section on spray towers and chambers.

It has been shown that the size of water droplets in a scrubber and their velocity relative to the gasborne particulate matter are primary factors in determining the effectiveness with which droplets and particulates will collide.

The size of water droplets evolved from hydraulic spray nozzles of various types is given by the following equation (Nukiyama and Tanasawa):

$$D_m = \frac{16,000}{v} + 1.4\left(\frac{\text{gals}}{1,000 \text{ ft}^3}\right)^{1.5}$$

where D_m is the average volume-surface diameter of the water droplets, microns and v is initial velocity of the atomizing water stream, ft per sec.

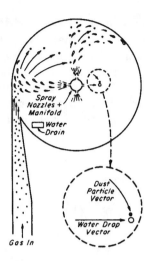

Figure 16-7. Dust scrubber in which water spray droplets are generated by spray nozzles at the axis of a cyclone chamber. Centrifugal force accelerates motion of droplets toward the wall; enroute, the particles in cyclonic motion collide with them.

The principal significance of this relationship in the present context is in the indication it affords of the fact that high relative velocity of the water and atomizing air results in smaller and smaller droplets of water, which are more effective in the impaction process. The reader will see repeated application of this principle in the following discussion of particle separation mechanisms in the various types of commercial scrubbers.

Centrifugally Accelerated Mists

The dusty gas stream in the scrubber illustrated in Figure 16-7 enters the cylindrical tower tangentially as in the conventional dry cyclone dust collector. There is a central manifold on which numerous water spray nozzles are mounted. Immediately adjacent to the spray nozzles, the process of impaction between dust particles and water droplets occurs as discussed in the earlier section on spray towers and chambers. The unique mechanism in this apparatus, however, is in the fact that the droplets are subjected to centrifugal force, accelerating their momentum toward the walls of the collector. Spray droplets smaller than those in spray towers, for example, with their favorable particle collision characteristics, can therefore be employed in the particle separation process.

The scrubber illustrated in Figure 16-8 belongs in this same category, although with some obvious differences. The gas stream passing outward through angular louvers to the outer annular space

encounters streams of water flowing down the top side of the cover cone of each stage and there will therefore tend to be a little atomization at this point, due to interaction of the air and water streams. Dust and dust-impacted particles are subject to centrifugal force and tend to be flung to the wet outer wall. One could reasonably infer that the principal dust separation mechanism is centrifugal force rather than that of water atomization and impaction.

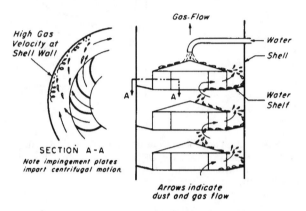

Figure 16-8. Scrubber in which the principle separation mechanism is centrifugal force, due to the angular louvers through which gases pass into the annular space. Wetted wall aids in retention.

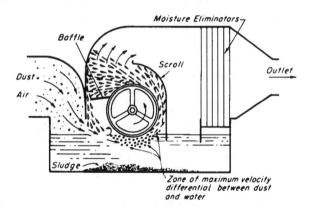

Figure 16-9. The power-driven cylinder dipping into a pool of water generates water droplets with which particles collide in the passage under and around the rotor.

Mechanically Induced Atomization

The apparatus illustrated in Figure 16-9 involves the atomization of water by a power-driven cylinder rotating at high

speed, horizontally, in a pool of water and the dust-laden gas stream is caused to pass through the water spray so generated. Judgement as to the relative particle separating ability of such a device would depend upon information of the extent to which the rotor atomizes the water supply, both as to droplets size and water droplet population, i.e., gallonage.

Air-Induced Atomization (Low Velocity)

Another form of commercial apparatus has arrangements as illustrated in Figure 16-10 whereby the dusty gas stream is caused to flow underneath a barrier which would otherwise be submerged in a pool of water, whereby the high gas velocity in relation to that of the water supply in the pool causes atomization and simultaneous impaction. The downstream, curved baffles encourage the centrifugal deposition of the larger water and mud droplets, and the finer mist is removed in the mist eliminator.

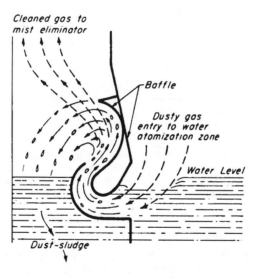

Figure 16-10. Atomization of water from the pool is induced by the flow of dusty gases under the "submerged" baffle, where impaction between particles and droplets occurs.

Although the equation for water spray droplet atomization may not be quantitatively applicable to such an atomization mechanism as this, it indicates the significant relationships and draws one's attention to the importance of gas velocity as it passes under the barrier and effective water atomization rate, as was true in the case of the previously described scrubber.

Air-Induced Atomization (High Velocity)

In the venturi scrubber (Figure 16-11) water is supplied at the throat of a venturi, through which dusty gas is forced at high velocity. Water is supplied through nozzles at low pressure. The principal atomizing mechanism is the high velocity of the main dusty gas stream itself. As illustrated in the diagram, the nature of activity at the throat is two fold: atomization of water and simultaneous impaction of particulate matter on the resulting droplets in the same zone.

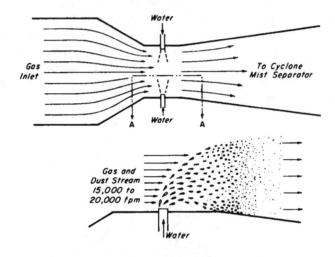

Figure 16-11. Venturi scrubber wherein high gas velocity at the throat atomizes water from low pressure nozzles. Simultaneously, dust particles moving at high velocity collide with water droplets being accelerated from low initial velocity.

As in all scrubbers, the length of the effective zone extends only as far as the point where water droplets have become accelerated to a velocity nearly that of the dust particles, i.e., gas stream. The venturi shape of this device does not contribute to particle separation; its function is to reduce unnecessary pressure drop losses and, consequently, conserve energy for gas movement. In this device, one would expect the water-spray droplet size equation to indicate the water droplet size. In fact, industrial experience with this scrubber clearly demonstrates the increasing efficiency with increasing pressure drop, i.e., gas throat velocity.

The device illustrated in Figure 16-12 operates on the same principle although its shape is quite different. The high gas velocity is developed by a restriction at the entrance, formed by an adjustable

plate and the outer wall of the tangential entry piece. The following wet cyclone corresponds to the mist eliminator that is employed in all water scrubber devices of the high velocity type.

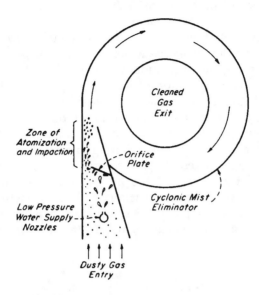

Figure 16-12. Scrubber which operates on a principle similar to that of a venturi wherein high gas velocity at the rectangular inlet atomizes water supplied at that point from low pressure nozzles. Cyclone chamber downstream functions as a mist eliminator.

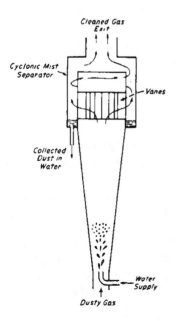

Somewhat similar to the preceding scrubber is the one illustrated in Figure 16-13 where the zone of maximum atomization and consequent particle separation is at the point of high dust velocity, i.e., high gas velocity at the bottom throat. The following vertical diverging section will recover some velocity pressure and the vane cylindrical piece at the top, essentially a wet cyclone, eliminates water mist and mud mist from the gas stream.

Figure 16-13. Scrubber type similar to the one shown in Figure 16-12. Atomization and impaction occur at the lower end of the diverging section and mist is eliminated in the cyclonic arrangement at the top.

Air-Induced Atomization and Mechanical Impaction

Another type of scrubber consists of horizontal plates with hundreds of small holes above each of which a small metal plate target is located at approximately one hole diameter above each such orifice. A pool of water lies on top of the perforated plate, with provision for overflow of excess water to a lower stage (Figure 16-14). Dusty gas passing upward underneath the plate is subjected to a preliminary separating action of water sprays directed toward the under side of the plate. As the gas passes up through the holes at high velocity, atomization of water from the pool occurs around the gas jets as they issue from the orifices and impaction of particles of the resulting mud spray and of larger unwet particulate matter against metal targets fixed above each orifice. A consideration of the dimensions of the hole and the metal target in relation to the quantitative relations given earlier in the impaction equation indicate that the principal particle separation mechanism is that due to the water atomization and simultaneous impaction.

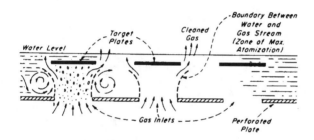

Figure 16-14. Impingement-type scrubber in which dusty gases pass through scores of small holes in a horizontal plate on which a circulating pool of water rests. Fixed metal plates above each orifice function as impaction targets and as mist eliminators.

Air-Inducing Spray Scrubber

The device illustrated in Figure 16-15, in addition to its particle separation function, also serves to induce a flow of gas and hence, where applicable, avoids the necessity for an exhaust fan.

It comprises a high pressure hydraulic spray nozzle in a housing the gas exit of which is a tapering enlargement, followed by the usual mist eliminator.

The particle separation zone is that space adjacent to the spray nozzle where the velocity of water droplets is high relative to that of the gasborne particulates entering the space from the side. The previous discussion "Spray Nozzle Impaction" and the data of Table 16-6 are also applicable to this device.

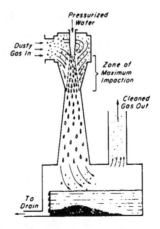

Figure 16-15. Water spray device which has a dual function. Water droplets issuing from nozzles induce a flow of gases through the unit and simultaneously effect particle separation by impaction due to high velocity of droplets relative to particulate matter in the entering gas stream.

The air induction mechanism can be better understood by reference to the principles discussed in Chapter 7. The equations developed there may be applied to a study of air induction of such a device.

This type of equipment consumes large quantities of water and this usually limits its application to special process situations.

Comparative Performance Data

As is indicated elsewhere in this chapter, no reliable conclusions concerning the relative performance of different types of dust collectors can be made on the basis of measurements made by different workers using different test dusts and techniques. Conversely, where the data are obtained from a single study by techniques standardized for these purposes, useful comparisons can be made.

Table 16-9
Percent Penetration by Particle Size in Three Water Scrubbers

Scrubber Type	Percent Penetration by Weight Particle Size, micrometer		
	5	2	1
Wet impingement scrubber	3	8	20
Self-induced spray scrubber	7	25	60
Venturi scrubber, 6" throat, 18,000 fpm	0.4	1	3

Stairmand found the comparative efficiencies given in Table 16-9 using a standardized dust whose particle size analysis enabled him

to report separation in relation to particle size. The self-induced spray scrubber is probably a Rotoclone, or one of that class. The wet impingement scrubber resembles that illustrated in Figure 16-14.

Cloth Bag Filters

Cloth filter dust collectors are typically the most efficient of all common types of gas or air cleaners, not excepting electrical precipitators. Their collection efficiency can be so high that rating them on the basis of percentage collected is almost meaningless. For example, on a weight basis it may exceed 99.99% and even on a basis of number of particles, the cloth bag efficiency is often more than 98%. If the effluent air is to be discharged outdoors, efficiencies of this order can be regarded, for all practical purposes, as near 100% except for certain especially hazardous materials like radioactive dusts. If the filtered air is to be recirculated to an occupied space, the efficiency should be described in terms of the quantity rate of flow of the dust in the effluent air stream, e.g., millions of particle per minute, or milligrams of dust per minute. This will be discussed further under "Recirculation."

The cloth fibers of ordinary filter cloth serve as a framework for retention of a permanent "cake" of dust within its interstices and the high efficiency of cloth filters results from filtration by the mass of dust rather than by any straining action by the fibers of the cloth. New, clean cloth is markedly lower in efficiency than cloth which has become saturated with some of the same dust it is to filter.

To further illustrate this point, a filter identical in principle could be formed using a background grid of ordinary window screening upon which a mat of finer fibers would be built by blowing them in air suspension through the screen. Cotton, wool, asbestos, glass fibers, mineral wool and the like would accomplish the purpose. Following this, the dusty air to be filtered would be passed through the fibrous framework so formed. At first a portion of the dust would pass through the pores, but gradually a larger proportion would be retained, progressively bridging the space around and between the fibers until a filter cake of dust is formed. Then the efficiency will be as high as in the conventional bag filter. The filtering action is identical—use of a thin bed of the dust itself as the filter element, held in place by a fiber framework.

The obvious difference in the two arrangements is in the renewing process. As dust accumulates in and on the filter, the pressure drop for a fixed rate of gas flow increases continually until a point is reached when the filter must be renewed. This is done in bag filters by vibrating or gently shaking the dust-laden cloth, dislodging the loosely adherent dust but leaving behind within the pores of the

cloth a permanent load of dust which comprises a permanent filtering element of high efficiency (Figure 16-16). Thus, after the shaking operation, the filtering is resumed and unlike the initial period with new cloth, the efficiency is high at the beginning of each cycle.

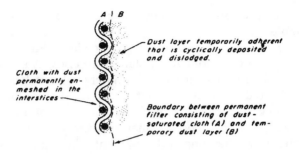

Figure 16-16. Representation of a cloth filter depicting the two elements comprising it.

It will now be apparent that the engineering problems in the application of cloth bag filters are not concerned with efficiency as is the case with practically all other types of dust collectors for its efficiency in most common applications can be regarded as "100%" for all practical purposes. The major design problem is concerned with pressure drop because unlike all other types of collectors where pressure drop is constant at all times for any fixed rate of gas flow, in the bag filter pressure drop fluctuates with the dust load accumulated.

Pressure Drop in Cloth Filters

We may disregard the characteristic pressure drop through clean filter cloth since it represents an evanescent condition when the collector is new, to which it never returns once it is placed in service and a permanent load of dust has become enmeshed within its fibers. (Pressure drop through common types of new filter cloth amounts to 0.01 to 0.1 inches w.g. at an airflow rate of 1 cfm per sq ft of cloth area.)

The pressure drop through a filter of fine pore size generally varies directly with the rate of gas flow through it. This suggests that flow through the small diameter interstices is in streamline motion. It is convenient to describe the pressure drop through a filter at a standard flow rate of 1 cfm per square foot of filter area which reduces to units of velocity, feet per minute, referred to as filtration velocity. Then the pressure drop at any other flowrate is readily obtained. For example, if the pressure is 0.2 inches w.g. at 1 cfm per sq ft, i.e., a filtration velocity of 1 fpm, it will be 1.0 inch w.g. at a

filtration velocity of 5 fpm. Calling pressure drop, R, and filtration velocity, V_f,

$$R = KV_f$$

where K, the proportionality constant, applies to a specific invariable filter.

The resistance of a cloth bag dust collector may be regarded as consisting of the sum of two resistances in series. One resistance is that of the dust-saturated cloth immediately after it has been cleaned (e.g., vibrated or shaken), that is, at the beginning of the dust collection cycle (Element A in Figure 16-16). Second is the additional resistance due to the loosely adherent layer of dust which is variable throughout the operating cycle (Element B in Figure 16-16). That is

$$R = R_o + R_d \qquad \text{and} \qquad R = K_o V_f + K_d V_f$$

where R_o designates the initial resistance and R_d that due to the detachable dust layer.

[Editor's note: At this point in the text, Hemeon presents theoretical and empirical approaches to determining the cloth and dust resistance factors, K_o and K_d, in bag filters. Those pages were moved to the Appendix. Both cloth and dust resistance factors range widely (0.25 to 4 and higher) but are sometimes assumed to be K = 0.3 for conceptual design purposes using cotton cloth. Always check supplier's data before design or selection of equipment.]

Fan Performance with Filters

The variation of airflow through a centrifugal fan due to different resistance was discussed in a previous chapter. The resistance of a duct system is, of course, fixed at any definite airflow, and, therefore, when serviced by a particular fan operating at a definite speed, the resultant airflow is a definite and predictable quantity. Now when a dust filter having a resistance that is variable from hour to hour is combined with a fan in which capacity varies with resistance, the relationships are a little less simple, although readily perceived by reference to the basic principles of each.

Consider, first, the simpler situation in which a fan draws air through a bag filter with no duct work other than connection between it and the fan inlet. One can plot on a single graph the pressure-volume characteristic curve of the fan and the pressure-volume curve of the system, i.e., the filter, in accordance with the discussion in a

previous chapter. This is done in Figure 16-17. Instead of a single system resistance curve there are, of course, an infinite number, one for each instant of time in any period. In practice, only two are of interest, one at the beginning of the dust collection cycle (just after dislodgement, by bag shaking, of the dust accumulated in the previous operating cycle) and one at the end of the period.

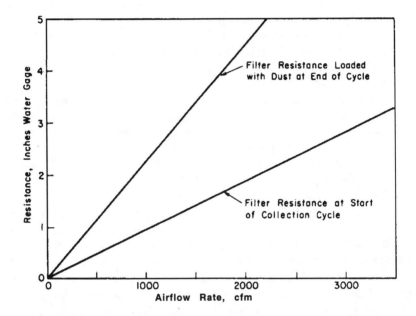

Figure 16-17. System resistance curves consisting of bag filter alone, with no other system resistance, are essentially straight lines (in contrast to duct system resistance curves).

In the design of a bag filtration system, it is seldom necessary to make such a graphical analysis. The resistance of the filter is computed at the *end* of the operating cycle which is the least favorable condition, i.e., lowest airflow, the resistance due to the duct system is added, and the total used in specifying fan performance. The increased airflow at the beginning of the dust collection cycle is ignored.

Because of the inherent uncertainty in resistance calculations for systems with bag filters, it is especially important to employ fans with a steep pressure-volume characteristic curve, such as the paddle wheel fan or the backward-curved blade fan, for they change their exhaust capacity least for a given change in resistance. The forward-curve blade fan is, for this reason, never used in this application.

Bag Resistance Effect on Fan Laws. The linear relationship between resistance and airflow in a bag filter causes an important exception to the fan laws as usually stated, which to our knowledge is not set forth in texts dealing with fan performance. [Circa 1963.]

The first law which states that in any particular system the airflow varies directly with changes in speed of the fan is unchanged.

The second law, as generally expressed, states that the pressure (against which the fan is called upon to operate when connected to a particular system) varies with the square of the fan speed (i.e., with the square of airflow rate, since the first law establishes the direct relationship between flowrate and fan speed). This pressure and its mode of change, however, is a function of the *system* resistance, and we know that the resistance of a duct system does vary with the square of velocity, i.e., the rate of airflow. However, if the system consists only of a bag filter, then the resistance varies linearly with airflow rate, i.e., fan speed. In this particular application, therefore, a special fan law can be phrased:

(a) *Where a fan is connected to a bag filter alone, the pressure will vary directly with the fan speed.*

The third law describes the change in power required with changes in fan speed. In the ordinary duct system, power varies with the cube of fan speed. This law stems from the fact that power is proportional to Q multiplied by h_L (the rate of airflow and system resistance, respectively), and since Q varies directly with speed, and h_L as the square of speed, the product will vary as the cube of fan speed. However, in the particular case of a bag filter, h_L varies directly with the fan speed, hence the product of the two results in the factor, square of fan speed. So a special law can be phrased as follows:

(b) *Where a fan is connected to a bag filter alone, the power demand will vary with the square of fan speed.*

In the usual application where the system resistance is the sum of bag filter resistance and duct resistance, total resistance will vary with power intermediately between unity and two, and could be readily estimated for any particular combination.

Capacity Rating for Small Unit Filters

Small capacity filters have been placed on the market [for many] years complete with exhaust fan, in standardized sizes as to cloth area and attached fan. As "package units," they are useful for small applications that do not warrant expenditure of engineering time. A useful method of rating their capacity can be derived from the principles that have been set forth in preceding sections, i.e., a

description of the dust holding capacity consistent with a specified minimum air flow.

For example, suppose a particular unit could be described as having a dust holding capacity of 10 pounds with a minimum airflow of 400 cfm. Suppose it was estimated that the process will produce air-borne dust at a rate of 1 to 3 pounds per hour. That would suggest that the unit would have a useful operating time of 3 to 10 hours before it needed to be renewed by shaking the cloth. On the other hand, for an operation judged to produce dust at a rate of 20 to 30 pounds per hour, the useful operating cycle would be only 20 to 30 minutes and hence of inadequate size.

The data required for deriving such a rating are as follows:

(1) Characteristic curve of the fan, pressure vs. volume.
(2) Cloth area.
(3) An estimate of basic cloth resistance, K_o.
(4) An estimate of basic dust resistance, K_d.

The last two require data from the manufacturer, or, in the case of conceptual design, some convenient estimate. One method for developing the capacity rating is outlined in the following example.

Example 4. Derive the approximate dust holding capacity for a unit cloth collector from the following data:

Cloth area, A = 80 ft

Pressure-volume characteristic curve of fan—see Figure 16-18.

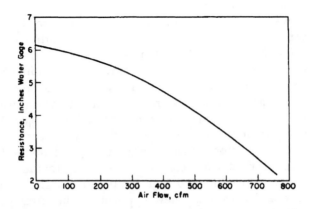

Figure 16-18. Fan characteristic curve for a unit filter.
Cloth area = 80 sq ft.

Solution. For the particular dust of interest, assume

$K_o = 0.3 \qquad K_d = 0.3$

Substitute these values in the following equation. [Please see the derivation of a similar equation in the Appendix.]

$$R = h_{SP} = K_o V_c + \frac{K_d V_c (16M)}{A} = 0.3 V_c + \frac{0.3 V_c (16M)}{A}$$

where R, the filter resistance, is set equal to h_{SP} (inches w.g.), the *available static pressure of the fan* at the indicated flowrate. M is total collected weight of dust in pounds. Substituting $V_f = Q/80$ and rearranging gives

$$M = \frac{1350}{Q} \left(h_{SP} - \frac{Q}{267} \right)$$

Using this formula, a series of calculations can be made for different values of Q, obtaining the corresponding available static pressure, h_{SP} from the curve, Figure 16-18. Thus for a Q of 300 cfm, the available static pressure is 5.2 inches which gives a value M = 18 lb.

High Temperature Filtration

Historical Overview. The development of synthetic fibers [e.g., early versions of Nylon, Orlon, Dacron, Teflon] extended the allowable temperature for bag filters from the limit of about 180°F for cotton up to approximately 500°F.

Technological developments in the finishing of glass cloth [circa 1950] overcame in large measure previous limitations in the application of glass cloth to high temperature gas filtration. Gas fibers of ordinary types can withstand temperatures well in excess of 500 to 600°F but the usual sizing materials which lubricate the glass fibers in woven glass cloth are destroyed at such temperatures and the cloth then assumes a harsh texture and, when subject to the periodic flexing required to dislodge accumulated dust, rapidly deteriorates due to friction between individual glass fibers. It was discovered that certain silicone compounds would provide effective lubrication and would withstand high temperatures. The advent of these new finishes led to many new applications beginning in the 1950s and 1960s.

The best of the glass cloth filter bags are still relatively fragile objects and successful application demands utmost care in handling during installation and in the design of mechanisms for cleaning them at the end of the filtration cycle. Realization of this fact in various bag cleaning arrangements has contributed in a major degree to the success of high temperature filtration installations.

The maximum practical operating temperature for glass cloth bags has been variously reported as 500°F to 650°F.

One common dust dislodging mechanism has been one in which the dusty gas stream is interrupted and a reverse flow of air induced sufficient to cause relaxation and gentle collapse of the cylindrical bag. This causes the bulk of the dust cake to fall away from the cloth surface and drop into the collection hopper below.

The K_0 resistance factor of a cloth filter, it has been previously noted, is an indication of the quantity of dust remaining adherent to the cloth after shaking, as well as its permeability. For glass cloth fibers, K_0 factors are higher than those usually observed in low temperature cotton cloth bag filters (0.25 to 0.80; average 0.3), previously cited. Herrick reported K_0 factors of 3 to 4 inches per fpm in pilot studies filtering fumes from open-hearth furnaces.

These high factors are a reflection partly of the high basic dust resistance and partly to the quantity of particles that remains adherent to the cloth after operating of the dislodgement mechanism. Herrick determined residual weight of fume ranging from 1/10 to 1/4 pound per square foot of cloth with one particular mode of bag shaking; and less than 0.02 lb per sq. ft with another mechanism.

In early years there were many failures in glass cloth bag applications related to excessive cloth flexing mechanisms, condensate of volatile salts in the cloth, and corrosive properties of certain dusts and gases. For comprehensive and current discussions of recent industrial experiences with glass cloth bag filters, see the McIlvaine citation in References and Bibliography.

[Editors's note: Four pages describing fiber-bed filters were moved to the Appendix.]

Electrostatic Precipitators (ESP)

The high voltage electrostatic precipitator (ESP) is typically used for particle separation in large gas streams at elevated temperatures and where costs for electricity are low.

In one of its common forms (after Cottrell) the gas flows horizontally in numerous vertical parallel channels formed by electrically-grounded plates spaced 6 to 12 inches apart, with plate heights ranging up to 24 feet. In the channel space, electrically

charged wires (25,000 to 100,000 volts) are suspended at frequent intervals along the direction of gas flow. The resulting corona discharge charges the gas-suspended particles which under the influence of the intense electric field causes them to migrate toward the grounded plates. Dust collected on the plates is dislodged by rapping or washing the plates, falling into collecting hoppers below.

Theoretical precipitating efficiency (distinguished from overall operating efficiency) is determined by the trajectory of particles in their motion toward collecting plates in relation to length of the electrical passage. The suspending gas carries them horizontally at velocities of 100 to 450 fpm, or greater, while they migrate laterally toward the plates under the influence of the electric field. The result of their forward velocity and lateral velocity in relation to the length of the charged plates theoretically determines whether or not they are deposited before being carried out of the precipitator.

The velocity of migration is the result of the coulomb force (analogous to force of gravity in sedimentation) and the frictional resistance of the gas opposing its motion (as in sedimentation of particles). The velocity for spherical conducting particles is approximated by

$$u = \frac{D \bullet E_o \bullet E_p}{4\pi\mu}$$

where D is particle diameter, centimeters, E_o is the charging field, and E_p the collecting field, kilovolts per centimeter; and μ is the gas viscosity, poise.

Table 16-11 shows calculated values for migration velocities of various particle sizes in representative precipitator conditions.

Table 16-11
Calculated Migration Velocities for Various Particle Sizes
in Indicated Precipitator Conditions, and where Gas Flow
is in Streamline Motion (after White)

Charging field = 5Kv /cm; collection field = 4Kv / cm; gas temp = 660°F.

Particle Size, micrometer	Migration Velocity, fps
0.5	0.1
1	0.2
10	1.9
20	3.9
100	19

Particle trajectories estimated by reference to calculated migration velocities and velocity of the gas would be applicable only to the circumstance that the gas flow was stream line. Because it is turbulent in actual practice such simple calculations are not valid. This was the case also, it will be recalled, in the case of the shelf dust collector (see Figure 16-3).

The following approximating expression for precipitating efficiency takes account of the probability of particle capture and escape due to gas turbulence, assuming infinitely perfect mixing. According to this expression, (re: White) the fraction, theoretically, penetrating a precipitator would be

$$P = e^{-\frac{A_p u}{Q}}$$

or

$$\log P = -\frac{A_p u}{2.3Q}$$

where A_p is the total area of collecting electrode (e.g., plates), u is the particle drift velocity, and Q is the gas flow rate.

It will have been noted that migration velocity is directly proportional to particle size and one might therefore conclude that larger particles would be more effectively precipitated than smaller ones. In practice this is found not to be the case, because these larger particles are less adherent to the collecting plates and are more readily re-entrained in the gas stream.

Rapping of electrodes to dislodge collected dust results in some inevitable entrainment into the gas stream of a fraction of the dust as it falls toward the collecting hoppers, so that precipitator manufacturers have designed a variety of arrangements attempting to minimize this loss.

One of the most troublesome factors encountered in the application of precipitators relates to the high resistivity of most of the common industrial dusts. The corona current must flow through the layer of collected dust and, if the resistance of the layer is excessive, electrical breakdown or puncture of the layer occurs resulting in a condition known as "back ionization." Its effect is to neutralize the charges on dust particles and, equally serious, greatly reduce the sparkover voltage. Since higher voltages favor higher efficiency, sparkover voltage limitations also mean limitation on realizable efficiency, all other factors being equal.

Resistivity of the precipitated dust layer is determined partly by the inherent electrical properties of the particulate matter itself and partly to the presence or absence of absorbed moisture from the gas or some other agent which would tend to lower resistivity.

Higher temperatures lower the resistivity of particulate matter and, hence, back ionization, thus favoring the precipitation process. On the other hand, lower temperatures favor the retention of adsorbed moisture and, consequently, lower resistivity, favorable to precipitation.

Higher gas temperatures are unfavorable to precipitation on another count. The lower gas density lowers the sparkover voltage and this limitation on applicable voltage gradient limits the charge on particles and results in lower efficiency.

Finally, higher gas temperatures require larger and hence more costly precipitators, for two reasons. The larger gas volume in relation to design gas velocity is one factor. The other is the potentially increased viscosity of the gas with increasing temperature which results in higher frictional resistance to the motion of particles and hence lower migration velocities.

The latter effect of high temperature is of course similarly operative in gravity sedimentation and in the impaction mechanisms previously discussed.

Electric Precipitator (after Penney). The two-stage precipitator invented by Penney primarily for cleaning of ventilation air is quite different from the Cottrell precipitator. Gas-borne particles pass first through a charging stage comprising very fine charged wires opposite grounded electrodes, to which potentials of 6,000 to 8,000 volts are applied.

The charged particles then pass through numerous parallel passages formed by closely spaced (1/2 inch and less) metal plates charged at 12,000 to 15,000 volts to which particles are attracted.

Particles reaching the plates lose their charge and are retained only by the process of ordinary adhesion. In the Cottrell precipitator a precipitated particle which becomes detached from the plate is still subject to precipitating forces and tends to be re-precipitated.

Commercial applications of this type of precipitator have been principally for cleaning air for ventilation of buildings. The bulk of the precipitated solids in this application is small which makes it practical to typically operate the equipment for periods of days before removal of accumulated solids by water washing.

Industrial applications are quite limited because of its low dust holding capacity. One application to which it is well adapted is in removal of oil mist from the air of exhaust systems serving various

machine tool operations because precipitated oil drains continuously from the plates by gravity.

Considerations in the Selection of a Dust Collector

In preceding sections we have attempted to provide the reader with an understanding of the particle separation mechanisms that are operative in the various types of devices, with particular reference to their relative efficiency in particle collection.

Table 16-12
Qualities of Dust Collectors
in Relation to Various Considerations

A. Cost installed.

B. Efficiency: hourly, monthly, yearly

C. Daily operating requirements
   Inspection; labor time

D. Maintenance requirements: daily, weekly, monthly

E. Application limitations:

   (1) Characteristics of the gas stream
      (a) Temperature: 70-200°F; 200-300°F, 300-600°F, >600°F
      (b) Presence of corrosive gases, e.g., SO_2 in fuel burning
      (c) Presence of water mist, or dew problems

   (2) Characteristics of particulate materials
      (a) Fineness (re: collectiblity)
      (b) Adhesive qualities due to dominance of fine materials
      (c) Abrasive qualities due to dominance of coarse materials
      (d) Plugging potential of liquid particulates

For the purpose of further understanding the many factors that enter into the problem of matching particle collection requirements in particular circumstances to the selection of the most appropriate device, Table 16-12 describes operating characteristics.

The question of collection efficiency in relation to the operating period considered has already been alluded to in the beginning of this chapter. An electrostatic precipitator may operate at high efficiency

during relatively brief hourly periods, whereas losses of dust due to re-entrainment when plates are vibrated to dislodge collected dust result in markedly increased dust emissions and consequently lower average efficiencies during that period.

A cloth dust collector will have a consistently high efficiency not only hourly but monthly, but its yearly efficiency may be lessened if maintenance practices in relation to torn, leaky bags are inadequate.

This is similar to various other types of particle separators wherein plugging, leakage due to wear, over-filling of hoppers, and the like may impair the efficiency over long time periods, unless an effective program of periodic checking and repair is followed.

Daily operating requirements, depending on the characteristics of the device, may involve no more than daily inspection of indication or recording instruments or they may be required in the case of some actual expenditure of labor time inherent in the operating characteristics. In all cases, of course, disposition of the collected dust involves an expenditure of labor time.

Application Limitations

These limits may be considered in relation to characteristics of the gas stream and also in terms of characteristics of the particulates in the gas stream. Gas stream temperatures are listed in four separate temperature intervals which have logic in terms of basic design problems. For example, in the application of cloth filters, ordinary cotton cloth can be employed at ordinary temperatures (70°F). Synthetic filament cloth bags are available at materially higher cost for the 200 to 300°F temperature range. For gases at 500 to 600°F, glass cloth bags may be applied, and this temperature range also represents the area above which special problems respecting strength of steel structural elements arise. Finally, extremely high temperatures present a number of obvious problems and usually suggest the need for a preliminary stage of gas cooling.

Corrosive Gases. The burning of sulfur-containing fuels results in the generation of sulfur dioxide in the flue gases which has a corrosion potential if liquid water is present simultaneously, and is also related to the problem of the formation, at high temperatures, of SO_3 and sulfur acid mist.

Liquid Mist. The presence of water mist droplets emanating from a process or formed by collecting gas containing high proportions of water vapor below its dew point produces a condition presenting obvious difficulties and limitations in applicability of some types of particle separators.

Table 16-13
Maintenance Problems in Application of Various Classes of Dust Collector to Different Types of Dust and Gas Streams

Type of particle	Particle quality	Cyclone, etc.*	Scrubber	Cloth bag filter	ESP
Abrasive dust due to large (>10μm) particle content	Non-sticky (dry)	Wear due to high velocity. Requires metal construction	No problem	No problem	No problem
Adhesive (sticky) particles; due to lack of large particles	Sticky (dry)	May adhere to metal, gradually plugging equip.; bridging in collector hoppers	No problem (usually)	May require excess size (cloth area) due to non-dislodged dust cake; bridging in hoppers	May cause difficulty in dislodging precipitated dust to hoppers; bridging in hoppers
Wet particles, due to liquid (mist) content; water spray carryover; cooling to dew point	Usually aqueous (tar mist, etc., special case)	Not applicable	No problem except where insoluble, like tar mist, etc.	Requires water be vaporized before; and avoid dew point. Not applicable to tar mist, etc.	No problem, except where precipitated liquids will not drain from plates.
Gas at high temperatures		No problem	Requires attention to wet-dry boundary where buildup and plugging can occur	Limitations of fabric	No problem except dust resistivity at high temps, and structural limits
Corrosive gases due to SO₂, etc.		No problem if above dew point	Requires stainless steel, or not applicable	No problem, above dew point	No problem.

*Includes all dry inertial devices involving change of direction at high velocities, e.g., dust concentrating devices, used often in conjunction with cyclones.

Characteristics of Particulates

The four characteristics of particulates listed in the outline complete the check list. The particle size characteristic which has a bearing on the efficiency of a dust collector has already been discussed.

All industrial dusts and fumes are heterogenous mixtures of particles of various sizes ranging from fine to coarse, the relative proportions of each being highly variable. Dusts and fumes that are lean in coarse particulates have adhesive qualities that are often the cause of operating problems.

The presence of large particles, when present in sufficient proportions, is advantageous in their ability to cancel or neutralize the adhesive properties of the co-existing fine material due to the sand-blasting action. We noted this same phenomenon in the discussion of duct transport velocities for dusts in an earlier chapter.

Some dusts containing a high proportion of coarse particulates have marked ability to abrade metal surface and this may constitute a problem of wear and tear in dry type dust collectors which depend upon high velocities as, for example, in cyclones.

Finally, the plugging potential of liquid particulates generated in some industrial processes would result in serious problems if the wrong type of particle separator were selected for their removal.

A view of the problems of application and maintenance of various types of dust collectors in relation to these characteristics of gas streams and of particulates is described in concise form in Table 16-17.

Required Dust Collector Efficiency

We alluded earlier to the common tendency to compare the relative merits of dust collectors in terms of their efficiency figure. We have remarked elsewhere that "if the collection efficiency is in the 90s, some quirk of the adult mind automatically concludes that this is good, perhaps a subconscious reaction resulting from school-day experience with scales of scholastic performance." (Hemeon, 1962.) This is of course an unsound basis and, now reminded, every reader can contrive examples to demonstrate this.

Engineers will do well to acquire the habit of referring instead to percent penetration in a comparison of dust collectors and for evaluating his requirements in particular circumstances. It will guide him/her well in realizing for example that the difference between 97 percent and 99 percent efficiency is a three-fold difference; and he or she will be even more conscious of total quantity of particulate to be treated and raise intelligent questions as to permissible rates of discharge to the atmosphere.

There are a number of criteria relative to the question as to what is a tolerable dust or fume discharge rate. The simplest situation is where EPA or local government regulations specify maximum permissible stack dust concentrations in the discharged gases. For example, 0.85 lb solids per 1,000 lbs stack gases (referred to as 50 percent excess air) [was] a widely promulgated standard regulating emissions from coal burning furnaces. In those days, Los Angeles [specified] maximum permissible weight rates of industrial emissions on a sliding scale, with a maximum of 40 lbs per hour. These [were] tangible and provided a concrete basis for deriving collector performance specifications.

In circumstances and locations where there are no EPA or other official regulations, one or more of the following criteria can be referred to:

(1) Prior experience with the particular process emission and with the performance of known dust separating apparatus.
(2) Good industrial practice, referring to the same process elsewhere and commonly accepted control measures.
(3) The dustfall emission rate.
(4) Visual appearance of the residual emission of dust or fume in relation to the character of the locality.

[Editor's Note: The visual criterion in the case of smoke from fuel burning, for example, traditionally referred to the use of the "Ringlemann chart," but measurements on a quantitative basis were also made, employing apparatus for aspiration of a stack gas sample through white filter paper and evaluation of its intensity by measurement of light transmission.]

Atmospheric Diffusion

[Editor's note: Much work has been accomplished in the science and practice of dispersion modeling during the past three decades that was not available to Hemeon. A summary of dispersion mechanics relating to industrial ventilation exhaust systems can be found in the latest version of ASHRAE's *Fundamental Handbook* and in the Editor's *Workbooks*. The following discussion is retained here for its historical and background value.]

It is possible in some [rare] circumstances to avoid the necessity for any air cleaning equipment, utilizing the natural dilution capacities of the atmosphere.

Where the quantity of dust is very small and the establishment remote from neighbors, one can visualize the magnitude of dilution in the short distances that affect neighboring windows and other

openings of one's own establishment according to the laws of behavior governing air jets. In terms of practical realities, these principles are more commonly applicable to a consideration of the dilution of gas and vapor emissions than to dust and fumes.

A stream of air issuing from a duct is a simple air jet, and this can provide a basis for initially judging the degree of dilution for short distances. These laws apply to the discharge of an air stream into a still air space where the degree of mixing is determined by the turbulence imparted to the stream by its own flow characteristics. This is in contrast to the dilution that occurs in natural wind currents, discussed later. The jet relations can be useful only in a consideration of dilution rates within relatively short distances not exceeding, say, 50 to 100 feet.

The jet expansion equation applicable is

$$\frac{q_x}{q_o} = \frac{0.45X}{\sqrt{A_o}}$$

Note that q_x/q_o (the ratio of total air in motion at distance, X, to the initial flowrate of air issuing from the stack) is inversely proportional to the ratio of concentrations of contaminant, C_o/C_x. We may also consider $A_o^{1/2}$ to be sensibly the same as stack or nozzle diameter, D, which is the point of issuance of the air stream into the open. The jet relationships then becomes

$$\frac{C_o}{C_x} = \frac{X}{2D}$$

Let X/D, the number of diameters, be designated by N. Then

$$\frac{C_o}{C_x} = \frac{N}{2}$$

and

$$C_x = \frac{2C_o}{N}$$

That is, the concentration of contaminant in an open air stream will decrease approximately with the distance. Its magnitude can be approximated by this equation.

Example 5. The concentration of dust in a 1-ft diameter duct discharging from an exhaust system into the open air is 50 million particles per cubic feet. Estimate the average concentration at a distance of 100 ft.

Solution. Note: $C_0 = 50$; $N = X/D = 100/1 = 100$. Therefore, concentration, $C_x = (2) (50)/100 = 1$ million particles per cubic foot. This will be the approximate concentration at 100 ft from the point of discharge, or lower.

It is because of turbulent mixing of contaminated air with fresh air, and the subsequent dilution, that the inevitable re-entry of contaminated exhaust air through nearby windows may often avoid major recontamination. (Figure 16-19.) Even though such short-circuiting may not result in sensible contamination of the interior air when time-average concentrations are considered, every effort should be expended in designing the system to arrange the point of discharge away from window openings.

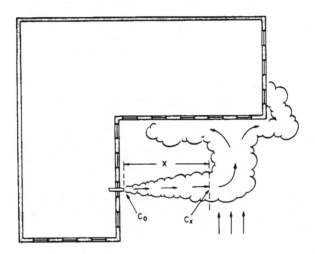

Figure 16-19. While ideally exhaust air would always be cleaned before discharge outdoors, in practice the quantity of contaminant is sometimes insignificant and natural dilution operates effectively to avoid nuisances downstream of the building. Jet formulas applied in Example 5 provide a basis for visualizing the magnitude of the problem.

Long Distance Dilution. When the distance between the point of emission and location of possible offense is great, one refers to the quantitative relationships of micrometeriology. For a source located at or near ground level as illustrated in Figure 16-20(a), the concentration near the ground at a distance is approximately

proportional to the quantity rate of emission of the dust or other contaminant, as follows:

$$C = \frac{2B}{\pi C_y \bullet C_z \bullet u \bullet x^{2-n}}$$

where B is the mass rate of emission, u is wind velocity, x is distance from the source to the point of interest, and terms n C_y and C_z are parameters related to the diffusing power of the atmosphere. The magnitude of these values and their significance is beyond the proper scope of this text and will therefore not be further elaborated. The relationship itself is of interest, however, indicating the relationship between the engineering parameters shown in the equation and, more particularly, to indicate the superior dilution capacity of elevated emissions discussed below.

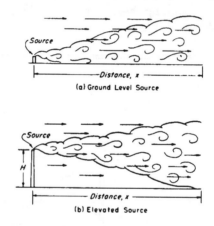

Figure 16-20. Dilution of contaminants due to atmospheric diffusion in the open atmosphere. (a) Contaminated gas is discharged at or near ground level. (b) The discharge is at an elevation.

Elevated Sources. The discharge of contaminants from exhaust system or other equipment can often be effectively diluted by discharging the contaminated gases through a stack well elevated above ground level and roof tops and other adjoining structures (Figure 16-20b). The equation describing the rate of dilution is similar to the last equation in terms of concentrations at ground level directly below the axis of the "smoke" plume except that it also takes into account the two-sided dispersion and stack height above the ground. A stack provides an important advantage to dilution, where the emission is elevated as compared with those discharged near the ground.

It will be noted than when distance X is very large in relation to stack height H, the equation reduces to the former one for sources at ground level. In other words, the advantage of an elevated source is realized in ground level locations relatively closer to the source.

Recirculation

[Editor's note: NIOSH established recirculation criteria in the 1970s which are still current. Some OSHA regulations (e.g., that for inorganic lead) and other codes actually prohibit the recirculation of some toxic materials in air. The following text is retained because of its historical and practical nature, and because many recirculating "unit collectors" are widely used in medium and small plants.]

Discharge of air that has been cleaned by passage through a dust collector or other air cleaner back into the ventilated space comes up properly for consideration in two circumstances: (1) where the quantity of air is so large as to have economic importance in relation to heating of replacement air in cold weather, and (2) where small unit dust collectors are used.

In the first instance, if the exhaust system is handling a nuisance dust such as that from woodworking there is little question as to feasibility of this practice [when NIOSH criteria are met]. A final high-efficiency cloth or HEPA-type filter has adequate efficiency to provide an effluent air stream suitably free of excessive dust.

If the dust being handled is of a toxic nature, such as mineral dusts containing more than 5 to 10% of quartz or dusts containing heavy metals, considerable caution is indicated to ensure that the initial high efficiency of the collecting unit is constantly maintained. Practically, the only dust collecting units having potentially adequate dust collecting efficiency for toxic dusts are HEPA or ULPA filters. Such filters may, however, leak slightly at the point of contact, between filter media and metal/wood framework, or holes may develop in the media due to wear.

These considerations make it imperative that special provisions be incorporated into any system providing for recirculation of air from a dust collector system where the dust is of toxic character. Initially, testing by dust counts or similar measurement is necessary to insure adequate operation. Then periodic or constant measurements or other suitable means of inspection are needed to make sure that initial conditions are always maintained. Alarms are required to notify exposed employees of malfunctions. Automatic instruments that constantly check operation are desirable.

It is rather obvious that the economic argument for recirculation of air from a cleaner in these circumstances would have to be a strong

one to warrant the trouble and expense involved in planning and execution of necessary safety measures, not to mention long-term O&M costs.

Recirculation of air from unit dust collectors for either nuisance or toxic dusts is a considerably simpler problem because of the small scale of the operation. The quantities of air flowing are small—a few hundred cubic feet per minute—and therefore the total quantity of dust escaping may also be correspondingly small. The criterion of permissibility is defined by two factors: the total rate of efflux of the dust and the total rate of dilution ventilation of the room in which it operates. This is illustrated in the following example.

Example 6. A small, portable, unit dust collector with integral exhaust fan handling 500 cfm is employed to control dust at an operation. The maximum permissible dust concentration is 5 million particles per cubic foot. A dust count in the effluent air from the collector shows a concentration of 1.5 million particles per cubic foot. Total ventilation from other exhaust systems and general ventilation arrangements in the room amount to 5,000 cfm. Appraise the adequacy of the dust collector. (Assume good mixing of dilution air in the plant.)

Solution. The rate of dust escape is the product of the dust concentration in escaping air and the airflow rate— $q = 1.5 \times 500 = 750$ million particles per minute. This quantity escaping into the room is diluted continuously by the room ventilating capacity of 5,000 cfm. The net effect is then a diluted dust concentration as follows:

$$C = q/Q = 750/5,000 = 0.15 \text{ million particles per cubic foot.}$$

The calculation suggests that the net effect of the operation is to raise the general dust count in the room by about 0.15 million particles per cubic foot. It also demonstrates the fallacy of judging the performance of such a unit in terms of the effluent dust count alone, as is sometimes done.

Calculations of the kind illustrated above can demonstrate in some circumstances that even though local concentrations of dust nearest the operation are excessive in the absence of a local exhaust arrangement, the total situation may be corrected by a simple unit exhaust system without any dust collector at all. (The unit exhaust system does not exhaust to the outside.) This can be the case where the quantity of contaminant is small in relation to the room ventilation although not diluted rapidly enough by natural air

movement to be a satisfactory situation in the absence of some simple exhaust system near the source. Such an arrangement is comparable, in effect, to merely blowing the contaminant out of the breathing zone into the general room air; indeed, a simple blowing air jet or circulating fan can sometimes be employed to function in that manner.

Importance of Large Particle Collection

If particle separation is inadequate, the effect will be an offensive situation in the vicinity due to the sedimentation of dust particles onto surfaces, causing an annoyance either to neighboring portions of the given plant or to residents in the neighborhood, usually referred to as dustfall nuisance.

The rate of gravity fallout is the product of the dust particle concentration and particle setting velocity. For a given concentration of dust in a cloud near the ground, the rate of fallout is directly proportional to the settling velocity of the particles comprising the cloud and, more particularly, that the larger particles having the higher settling velocity contribute most to fallout. The concentration of particulates in the open atmosphere is also proportional to the rate of emission, and this rule may therefore be re-stated in these terms: a small percentage of large particles in a dust mixture passing out into the open atmosphere can contribute to a major degree to any fallout nuisance in the vicinity of the source.

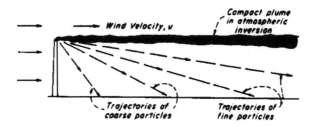

Figure 16-21. Trajectories of coarse and fine particles from an elevated source in conditions of atmospheric inversion are the result of horizontal wind velocity and particle settling velocity. The figure illustrates the importance of high efficiency removal of the largest particles to avoid nuisance dustfall.

The picture of dust fallout illustrated in Figure 16-21 is sometimes employed to demonstrate the greater importance of large particles in contributing to dust fallout nuisances. This particular description is valid for atmospheric conditions of *inversion* and atmospheric wind flow condition analogous to streamline flow in engineering hydraulics. It is not, however, valid for the other common atmospheric condition where gases and particles, whether large or

small, are transported in turbulent vortices naturally present in the atmosphere.

In the atmospheric conditions illustrated in Figure 16-20(b), all particles regardless of size will be transported rapidly to ground level. The sedimentation of the larger particles will nevertheless occur preferentially.

In both cases, the larger particles are of dominant importance to nuisance dustfall. In one case (inversion), dustfall is due entirely to the heavier particles according to the distance pattern indicated in Figure 16-21. In the second case (turbulent atmosphere), the smaller particles also contribute to dustfall although at the proportionately reduced rate illustrated in the example.

References and Bibliography

[For this Edition, 1998]

--, ACGIH, *Industrial Ventilation Manual*, latest Edition, contact ACGIH, 1330 Kemper Meadow Drive, Cincinnati, Ohio 45240; about 400 pages.

--, AIHA, *The Occupational Environment, Its Evaluation and Control*, (the "White Book"), AIHA, Fairfax VA, 1997, 1,300 pages; see chapters on Control.

--, AMCA, *Fan Application Manual*, 1997, 30 West University Drive, Arlington Heights, Illinois 60004.

--, ANSI Z9 (Various industrial ventilation standards; various dates.) Contact AIHA, Z9 Secretariat, 2700 Prosperity Ave, Suite 250, Fairfax, VA 22031.

--, ASHRAE, *Handbook of Fundamentals*, latest edition, about 600 pages; one of four basic handbooks.

--, British Occupational Hygiene Society, *Controlling Airborne Contaminants in the Workplace*, Science Reviews Ltd., 1987, 175 pages; contact BOHS, 1 St. Andrews Place, Regents Park, London, NW1 4LB.

--, McIlvaine Company (various scrubber and filter manuals); 2970 Maria Ave., Northbrook, IL 60062.

--, SMACNA, *HVAC Systems -- Testing, Adjusting, and Balancing*, 1983, about 250 pages.

Burton, D. J., IVE, Inc., *Industrial Ventilation Workbook*, 4th Edition, 1997; *Lab, Ventilation Workbook*, 1998; contact IVE, Inc., 2974 South Oakwood, Bountiful, Utah 84010; each about 320 pages; also available from ACGIH, AIHA, ASHRAE, ASSE, NSC, CRC, and others.

Burton, D.J. IVE, Inc., *IAQ & HVAC Workbook*, 1997; contact IVE, Inc.; 300 pages.

Goodfellow, Howard D., *Advanced Design of Ventilation Systems for Contaminant Control*, Elsevier Publishers, NY, NY; 1985, 750 pages.

Guffey, S. "Estimation of Junction Losses," unpublished paper, 1990, UW, SPH&CM, DEH, SC-34, Seattle, WA 98195

Jorgensen, Robert, *Fan Engineering*, 8th Edition, Buffalo Forge Company, Buffalo, NY, 1983; about 800 pages.

[For the 2nd Edition, 1963]

Caplan, K.J., and Bush, P.D., American Petroleum Institute, *"Cyclone Dust Collectors. (Report On Removal of Particulate Matter from Gaseous Wastes.)"* Stern, A.C.; (1955)

Caplan, Knowlton J. *Air Pollution*, Vol. II (1962), pp. 247-291. Stern, A.C., Ed. Academic Press, New York, NY.

First, Melvin W. and Silverman, Leslie. *"Factors in the Design of Cyclone Dust Collectors."* Heating and Ventilating (July 1948), pp. 80-86.

Golovin, M.N., and Putnam, A.A. *"Inertial Impaction on Single Elements."* Industrial and Engineering Chemistry Fundamentals, Vol. 1, No. 4 (Nov. 1962), pp. 264-273.

Haines, George F., Jr.; Hemeon, Wesley, C.L.; and Puntureri, Samuel D. *"Rating of Dust Collectors According to Dust Settling Velocities."* Journal of the Air Pollution Control Association, Vol. 11, No. 6 (June 1961), pp. 264-267.

Hemeon, W.C.L., Haines, George F. Jr. and Ide, Harold, M., *"Determination of Haze and Smoke Concentrations by Filter Paper Samplers."* Air Repair (Aug. 1953), pp. 22-28.

Hemeon, W.C.L. *"Dust Particle Inertial and Various Consequences."* Heating, Piping & Air Conditioning, ASHRAE Journal Section; (Feb. 1957), pp. 247-250.

360

Hemeon, Wesley C.L. *"Characteristics of Equipment for the Removal Of Particulate Matter from Gas Streams."* Monograph No. 4, Journal Of The Franklin Institute Series (March 1958), pp. 61-73.

Hemeon, Wesley C.L. *"Gas Cleaning Efficiency Requirements for Different Pollutants."* Journal of the Air Pollution Control Association, Vol. 12, No. 3 (March 1962), pp. 105-108.

Herrick, Robert A. *"A Baghouse Test Program for Oxygen Lanced Open Hearth Fume Control."* Journal of the Air Pollution Control Association, Vol. 13, No. 1 (Jan. 1963).

Hofman and Hayward. *"Metallurgy Of Copper."* 2nd Ed. (1924), McGraw-Hill Book Co., New York, NY.

Johnson, Glenn A., Friedlander, Sheldon K., Dennis, Richard, First, Melvin W., and Silverman, Leslie. *"Performance Characteristics of Centrifugal Scrubbers."* Chemical Engineering Progress, Vol. 51, No. 4 (Apr. 1955), pp. 176-188.

Johnstone, H.F. and Roberts, M.H. *"Deposition of Aerosol Particles From Moving Gas Streams."* Industrial and Engineering Chemistry, Vol. 41 (Nov. 1949), pp. 2417-2423.

Kleinschmidt, R.V. *"Factors in Spray Scrubber Design."* Chemical and Metallurgical Engineering (Aug. 1939).

Lapple, C.E. *"Dust And Mist Collection."* Chemical Engineers' Handbook; Perry, J.H., Ed., CRC, (1940).

Lynch. *Fuel Economist*, Vol. 12, No. 47 (Oct. 1936).

Nukiyama, S. and Tanasawa, Y. *Trans. Soc. Mech. Eng. (Japan)*, 5, No. 18 63 (1939).

O'Mara, Richard F. and Flodin, Carl R., *"Filters and Filter Media for the Cement Industry."* Journal of the Air Pollution Control Association, Vol. 9, No. 2 (Aug. 1959), pp. 96-101.

Penney, G.W. *"A New Electrostatic Precipitator."* Electrical Engineering (Jan. 1937), P. 159.

Peters, E.D. *"The Practice Of Copper Smelting."* (1911) McGraw-Hill Book Co., New York, NY.

Ranz, W.E. and Wong, J.B. *"Impaction of Dust and Smoke Particles on Surface and Body Collectors."* Industrial and Engineering Chemistry, Vol. 44, No. 6 (June 1952), pp. 1371-1381.

Richardson, E.G., Ed. *"Aerodynamic Capture of Particles."* (1960) Pergamon Press, New York, NY.

Rietema, K. and Verver, C.B., Eds. *"Cyclones in Industry."* (1961) Elsevier Publishing Co., New York, NY.

Spaite, Paul W., Stephan, David G., and Rose, Andrew H., Jr. *"High Temperature Fabric Filtration of Industrial Gases."* Journal of the Air Pollution Association, Vol. 11, No. 5 (May 1961), pp. 243-247.

Stairmand, C.J. *"The Design And Performance of Modern Gas Cleaning Equipment."* Journal of the Institute of Fuel (Feb. 1956).

Vajda, Stephen. *"Open Hearth Dust Control."* Iron and Steel Engineer (July 1952), pp. 111-120.

Walton, W.H. and Woolcock, A. *"Aerodynamic Capture of Particles,"* pp. 129-153. Richardson, E.G., Ed. (1960) Pergamon Press, New York, NY.

White, Harry J. *"Modern Electrical Precipitation."* Industrial and Engineering Chemistry, Vol. 47, No. 5 (May 1955), pp. 932-939.

Index

Appendix

The Appendix contains pages from the second edition which were deleted from the main text of this edition. These provide background, context, and historical value to the current edition (and may generate interest in experimentally confirming Hemeon's theoretical work).

The following pages are included in the Appendix:

Third Edition Pages	Second Edition Chapter	Second Edition Pages	Topics
366-374	2	28-45	Particle behavior under various conditions.
375-379	7	122-131	Behavior of particles in motion.
380	8	166, 168	Figures 8-5 and 8-7.
381-385	16	440-448	Behavior of cloth in baghouses.
386-388	16	454-458	Fiber and fiberbed filters.

Coefficient of Resistance

The magnitude of the force tending to retard the motion of a particle in air, regardless of the type of relative motion prevailing, includes a term, C, the drag coefficient, or coefficient of resistance

$$F_R = \frac{C\rho u^2 A}{2g} \quad (1)$$

where F_R is the resisting force, pounds; ρ is the density of the air (0.075 lb per cu ft at standard temperature and pressure); u is the particle velocity relative to the air, feet per second; A is the projected area of the particle, square feet; and C is a dimensionless number.

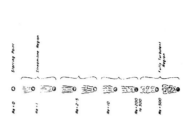

FIG. 2-3. Representation of changing character of the wake, as to development of turbulence, in a falling particle at various stages in its descent. Starting from rest.

The coefficient, C, has a value variable with Reynolds number, Re, as shown in Fig. 2-4. The values, from Lapple and Shepherd, on which the chart is based, are given in Table 2-4. The three types of motion represented therein are discussed in the following sections.

Rate of Change of Velocity

To derive velocity-time and velocity-distance relationships, reference is made to the basic equations which follow, ignoring the factor of buoyancy.

$$F = Mg - F_R$$
$$Ma = Mg - \frac{C\rho A u^2}{2g}$$
$$a = \frac{du}{dt} = g - \frac{C\rho A u^2}{2W}$$
$$dt = \frac{du}{g - \dfrac{C\rho A u^2}{2W}} \quad (2)$$

where F is net accelerating force, pounds; M is mass or W/g; W is

TABLE 2-4

Values of Drag Coefficient, C, for Various Values of Reynolds Number Re (from Lapple and Shepherd[1])

Re	C	Re	C
0.1	240	700	0.50
0.2	120	1,000	0.46
0.3	80		
0.5	49.5	2,000	0.42
0.7	36.5	3,000	0.40
1.0	26.5	5,000	0.385
2	14.4	7,000	0.390
3	10.4	10,000	0.405
5	6.9	20,000	0.45
7	5.4	30,000	0.47
10	4.1		
		50,000	0.49
20	2.55	70,000	0.50
30	2.00	100,000	0.48
50	1.50	200,000	0.42
70	1.27	300,000	0.20
100	1.07	400,000	0.084
200	0.77	600,000	0.10
300	0.65	1,000,000	0.13
500	0.55		
		3,000,000	0.20

[1] Lapple, C. E., and Shepherd, C. B., Calculation of Particle Trajectories, *Ind. and Eng. Chem.*, **32**, No. 5 (1940).

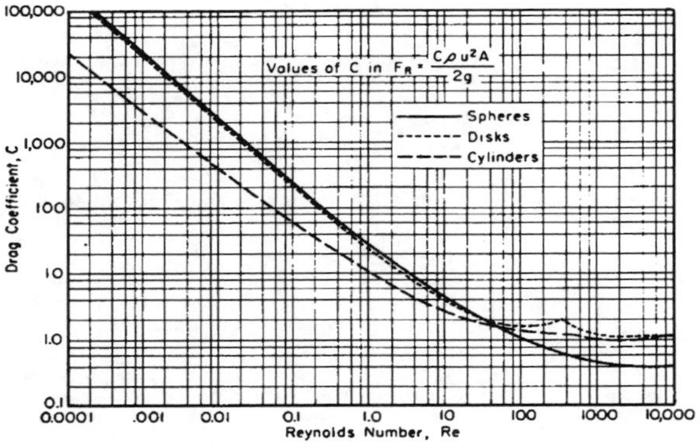

Fig. 2-4. Variation of coefficient C with Reynolds number.

weight, pounds; a is net acceleration, and other notation as before. The distance of fall corresponding to various velocities is represented by

$$\Delta s = \int_{u_o}^{u} u\,dt$$
$$= \int_{u_o}^{u} \frac{u\,du}{g - \dfrac{C\rho A u^2}{2W}} \quad (3)$$

Integration of this equation between suitable limits of u_o and u, and corresponding values of C, results in expressions by which velocity-distance relationship can be computed for each region of flow.

Streamline Motion in Fall by Gravity

It may be seen from Fig. 2-4 that the slope is constant for values of Re up to, approximately, one; and that in this region—

$$C = \frac{24}{Re}$$

This is the region of streamline flow. When this value of C is substituted in general equation (1), noting that $Re = Du\rho/\mu$ and $A = \pi D^2/4$, where D is particle diameter in feet, one obtains the well known expression, Stokes law—

$$F_R = \frac{3\pi\mu Du}{g} \quad (4)$$

Terminal Velocity. The terminal velocity, u_t, of a particle subject to a positive force is attained when that force is in equilibrium with the resistance force. For a falling body, the force is that of gravity, F_g,

$$F_g = \frac{\pi}{6} D^3 \rho_s$$

where ρ_s is density of the solid particle.
When F_g is equated to the resistance force one obtains

$$\frac{\pi}{6} D^3 \rho_s = \frac{3\pi\mu Du}{g} \quad (5)$$

whence

$$u = u_t = \frac{g\rho_s D^2}{18\mu} \quad (6)$$

Substituting $g = 32.2$; $\rho_s = 62.4z$ (z is specific gravity, with water = 1); $D = d_m/(3.05 \times 10^5)$ (d_m is particle diameter in microns); and for air viscosity $\mu = 1.3 \times 10^{-5}$, the expression, in more convenient form, is

$$u_t = 9.23 \times 10^{-5} z d_m^2 \quad (7)$$

Rate of Velocity Increase. By integration of equation (3) with $C = 24/Re$ the equation is derived that describes the velocity—distance relationship, during the acceleration phase, before the particle has attained its terminal velocity within the region of *streamline* flow

$$s = \frac{u_t^2}{g}\left[2.3 \log \frac{u_t}{u_t - u} + \frac{u}{u_t}\right] \quad (8)$$

where u_t is given by (7), and g is 32.2.

Intermediate Flow in Fall by Gravity

The relation between C and Re in this region is not the simple one such as prevails in the zone of streamline flow. Simple and accurate derivations of resistance, and of terminal velocity are therefore not possible. However, it has been shown by Shepherd[*] that a straight line

[*]Shepherd, C. B., quoted by Walker, Lewis, McAdams, and Gilliland, *Principles of Chemical Engineering.*

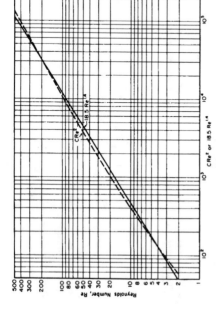

Fig. 2-5. Chart for finding terminal velocity (in intermediate motion), and for use in step-wise integration to find $\Delta u/\Delta t$, and u vs. s relation. Also compares $18.5\,Re^{1.4}$ with true values.

in the C vs. Re graph can approximate the true relation over this region and gives this equation.

$$C = \frac{18.5}{Re^{0.6}} \qquad (9)$$

Employing this value of C in the general frictional force formula (1), and $Re = Du\rho/\mu$, we get

$$F_R = \frac{18.5}{2g} \cdot \frac{\pi}{4} \cdot \rho^{0.4} u_t^{1.4} D^{1.4} \mu^{0.6}$$

which is further simplified, by assigning values of $\rho = 0.075$ and $\mu = 1.3 \times 10^{-5}$, to

$$F_R = 0.94 \times 10^{-4} u_t^{1.4} D^{1.4} \qquad (10)$$

Terminal Velocity. When the expression for resistance force above is equated to force of gravity, as demonstrated for streamline motion, we

obtain a reasonably good approximation for terminal velocity in intermediate motion,

$$u_t^{1.4} = 0.56 \times 10^4 \rho_s D^{1.6}$$

Substituting $\rho_s = 62.4z$ and $D = d_m/(3.05 \times 10^5)$, the following equation results.

$$u_t = 0.49 \times 10^{-2} d_m^{1.14} z^{0.72} \qquad (11)$$

Basis for an exact evaluation is provided in a graph of CRe^2 against Re (Fig. 2-5) based on the data and method of Lapple and Shepherd. The limiting value of CRe^2, i.e., the value corresponding to the terminal velocity, u_t, is shown to be expressed in the following.

$$CRe_t^2 = \frac{4g\rho(\rho_s)D^3}{3\mu^2} = 41.9 \times 10^{-6} z d_m^3 \qquad (11a)$$

Its use is illustrated in the following.

Problem. Find the terminal velocity of a 100-micron spherical particle of specific gravity, $z = 3$.

Solution. The value of CRe^2 at the terminal velocity is—

$$CRe_t^2 = 41.9 \times 10^{-6}(3)(10^2)^3$$
$$= 126$$

Reading from Fig. 2-5, the value of Re corresponding to $CRe^2 = 126$ is 4. Thus

$$Re = 4 = 0.0188 d_m u_t;$$

and where $d_m = 100$,

$$u_t = 2.1 \text{ fps}$$

In Fig. 2-6, terminal velocities, calculated according to the exact method illustrated in the preceding problem, are given for various particle sizes and specific gravities and will be found convenient for reference.

Accuracy. The relation $C = 18.5/Re^{0.6}$ is a good approximation for the entire region from $Re = 2$ to $Re = 500$. The corresponding values of CRe^2, i.e., $18.5Re^{1.4}$, are plotted on Fig. 2-5 and demonstrate the degree of agreement with the more exact values. In accordance with these considerations we shall herein arbitrarily define the upper limit of intermediate flow to correspond to Reynolds number of 500, although in actuality there is a gradual merging of the two regions.

where $\phi = 4gp \cdot \rho_s D^3 / 3\mu^2$, and other terms as before. This equation reduces to

$$\frac{\Delta Re}{\Delta t} = \frac{0.15 \times 10^5}{z d_m^2} \left(41.9 \times 10^{-6} z d_m^3 - CRe^2\right) \quad (12)$$

Construction of a curve of u vs. s may be effected by calculation of successive points, using (12). The steps to be taken and their order are illustrated in the following example.

Problem. Construct an approximate curve of velocity vs. distance for gravity fall, for a spherical particle of diameter, $d_m = 500$ microns; specific gravity, $z = 1$.

$$Re = 0.0188 d_m u = 9.4u$$

From (12),

$$\Delta t = \frac{17.3 \Delta Re}{5240 - CRe^2}$$

		I	II	III
(a)	Let u =	0.3	1.2	2.4
(b)	Δu =	0.3	0.9	1.2
(c)	$Re = 9.4u$ =	2.82	10.8	21.6
(d)	ΔRe =	2.82	8.0	10.8
(e)	$17.3 \Delta Re$ =	48.5	138	187
(f)	CRe^2 (from Fig. 2-5) =	85	450	1200
(g)	$5240 - CRe^2$ =	5155	4790	4040
(h)	Δt = item (e)/item (g) =	0.0094	0.029	0.0465
(i)	u_{av} =	0.15	0.75	1.8
(j)	$\Delta s = u_{av} \cdot \Delta t$ =	0.0014	0.022	0.084
(k)	Cumulative s =	0.0014	0.023	0.107

Curves of Fig. 2-7 illustrate the results of calculations of u vs. s obtained by the method illustrated in the preceding problem. The curve for frictionless falling motion, according to $u = \sqrt{2gs}$, and its intersection with the terminal velocity line of the particle having $z = 1$, is also given for comparison. In the early stages of fall the frictional effect is seen to be negligible. The particle of specific gravity, $z = 1$, has attained 80% of its terminal velocity in 2.75 feet.

Turbulent Flow in Fall by Gravity

The region of turbulent motion has been defined for present purposes as the one starting at $Re = 500$. While the value of C ranges from about 0.40 to 0.55, it may be regarded as essentially constant; the

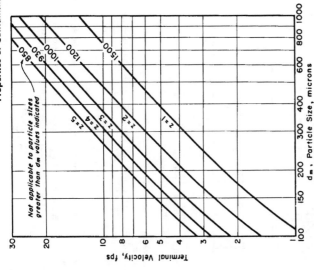

FIG. 2-6. Terminal velocities for particles in intermediate motion from plot of CRe^2 vs Re. Approximate $u_t = 0.49 \times 10^{-2} d_m^{1.14} z^{0.72}$.

Rate of Velocity Increase. Integration of equation (3) with the approximate C value given by equation (9), as was done for streamline and turbulent motion, results in an unduly cumbersome expression. It is simpler to develop the u, vs. s relationship by a step-wise integration based on a basic equation due to Lapple and Shepherd.

A rearrangement of equation (2), incorporating the velocity term, u, into Re (noting that $Re = Du\rho/\mu$) gives

$$\frac{\Delta Re}{\Delta t} = \frac{\mu A (\phi - CRe^2)}{2DW}.$$

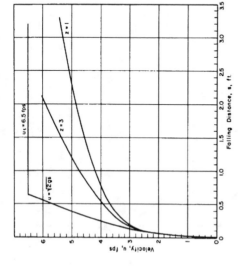

FIG. 2-7. Comparison of u vs. s relation of falling particle 500 microns in diameter. Starting from rest, entire motion is within intermediate zone. Curves for $s = 1$ and $z = 3$, compared with $u = \sqrt{2gs}$.

average value is conventionally taken at 0.44 and that value will be used herein.

Using equation (1) with $C = 0.44$ results in an expression for force of air resistance, F_R.

$$F_R = \frac{0.44\rho u^2 A}{2g} \qquad (13)$$

where F_R = resistance, in pounds; ρ = air density, and other symbols as before.

Terminal Velocity. When F_R is equated to force of gravity acting on the particle, $F_g = (\pi/6)D^3 \rho_s$, one obtains the terminal velocity

$$u_t = 9.9 \sqrt{\frac{\rho_s D}{\rho}}$$

Transforming D (ft) to d_m (microns) by $d_m = .305 \times 10^5 D$;

$\rho_s = 62.4z$; $\rho = 0.075$ at normal temperature, a convenient form of the equation results.

$$u_t = 0.54 \sqrt{z \cdot d_m} \qquad (14)$$

Rate of Velocity Increase. Integration of equation (3) to obtain the relationship between s and u, is straightforward because C is constant at 0.44.

$$\Delta s = s - s_{500} = \frac{-2.3}{2K} \log_{10} \frac{g - Ku^2}{g - Ku_o^2}$$

where $K = \rho CA/2W = 120/d_m \cdot z$; $g = 32.2$; s is distance (ft), corresponding to u_t, and s_{500} is the distance at which Re has attained a value of 500. $u_o = u_{500}$, corresponding to s_{500}.

In a simpler form, the above equation becomes

$$\Delta s = s - s_{500} = - \frac{d_m \cdot z}{105} \log_{10} \left(\frac{32.2d_m z - 120u^2}{32.2d_m z - 120u_o^2} \right) \qquad (15)$$

CHRONICLE OF A FALLING PARTICLE

Every particle starting from rest, and falling by gravity, passes successively through the three flow regions: streamline, intermediate and turbulent motion, as depicted in Fig. 2-3. Where one type predominates over the range of interest, others may sometimes be ignored without serious impairment in accuracy of the calculated result, but this is not a safe assumption, in general.

Velocity and Distance Limits

The boundaries of the three regions, in terms of particle velocity are defined by

$$Re = 2 = \frac{Du_2 \rho}{\mu}$$

$$u_2 = \frac{106}{d_m}$$

This indicates the velocity where motion changes from streamline to intermediate and

$$Re = 500 = \frac{Du_p}{\mu}$$

$$u_{500} = \frac{26,400}{d_m}$$

the velocity where motion changes from intermediate to turbulent.

38   Properties of Contaminants

For a preliminary survey, the preceding equations may be converted to distance units by use of the relation $u = \sqrt{2gs}$ to obtain the distance (s) which represents the *earliest possible* point of transition from one zone to the next.

$$s_2 = \frac{u_2^2}{2g}$$
$$= \frac{106^2}{64.4 d_m^2} \qquad (16)$$

This distance marks the earliest possible transition point between streamline and intermediate motion.

And similarly,

$$s_{500} = \frac{11.1 \times 10^6}{d_m^2} \qquad (17)$$

which marks the earliest possible transition point between intermediate and turbulent motion.

Particle Size Limits

Particles below a certain weight and size cannot attain velocities high enough to carry them beyond streamline motion; others, similarly, cannot attain turbulent flow conditions. These limits are fixed by the terminal velocity of the particle and by the arbitrary Re limits of 2 and 500, respectively.

The limits are derived as follows: For *streamline motion,* $u_t = 9.23 \times 10^{-5} z d_m^2$. When combined with $Re = 2 = 0.0188 d_m u$, one obtains

Particles Falling in Streamline Motion

Motion of particles smaller than values given is confined to streamline region, i.e., Re always less than 2 when falling from initial state of rest. From $d_{m(t)} = 103/\sqrt[3]{z}$.

Specific gravity, z	Maximum particle size, $d_{m(t)}$, microns
1	100
2	80
3	70
4–5	60

Properties of Contaminants   39

$$d_{m(t)} = \frac{103}{\sqrt[3]{z}} \qquad (18)$$

Size limits for particles of several specific gravities in streamline motion ($Re < 2$) calculated by equation (18) are listed in Table 2-5.

The particle size limits *between* intermediate and turbulent motion are obtained by reference to the following: at $Re = 500$, $C = 0.55$, and therefore $CRe^2 = 1.38 \times 10^5$. Since, at the terminal velocity, $CRe^2 = 41.9 \times 10^{-8} d_m^3 z$, equating these two expressions gives the values of $d_{m(t)}$ and z that correspond to the upper limit ($Re = 500$) of intermediate motion,

$$d_{m(t)} = \frac{1.48 \times 10^3}{\sqrt[3]{z}} \qquad (18a)$$

Values calculated by solving equation (18a) for several specific gravities show maximum particle sizes with terminal velocities in the zone of intermediate motion ($Re < 500$) as shown in Table 2-6.

Method of Analysis

The method of analyzing the falling behavior of a particle which attains velocities in the turbulent region is illustrated by Fig. 2-8, which presents the velocity-distance relationship for a 2000-micron particle of specific gravity = 3, and includes, for comparison, the theoretical velocity-distance curve, according to $u = \sqrt{2gs}$, that would result in the absence of air resistance, and its intersection with the terminal velocity line of the actual particle in air.

It was clear from a preliminary examination that the streamline region would be negligible. Therefore, the motion curve from the origin to

TABLE 2-6

Particles Falling in Intermediate Motion

Motion of particles smaller than listed values is confined to intermediate flow region, i.e.. Re always less than 500 (when falling from initial state of rest). From equation (18a).

Specific gravity, z	Maximum particle size, $d_{m(t)}$, microns
1	1500
2	1200
3	1000
4	930
5	850

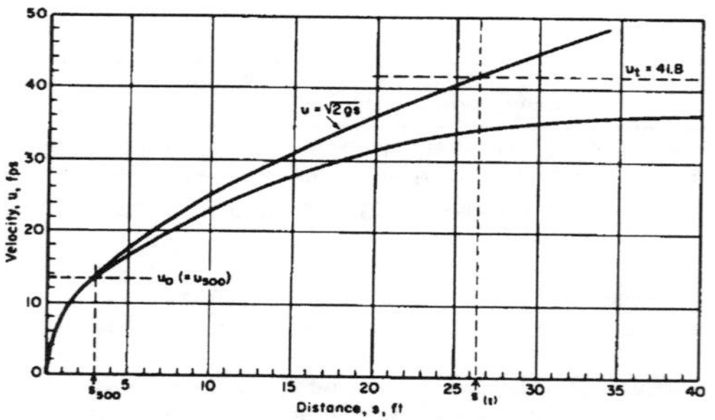

40

Properties of Contaminants

FIG. 2-8. Comparison of actual u vs. s relation, in both intermediate and turbulent zones, with $u = \sqrt{2gs}$. Particles 2000 microns, $z = 3$, starting from rest.

$s = 3$ ft is taken to be entirely in the intermediate zone, and u vs. s figures for it were obtained by the incremental integration method illustrated previously. It is of interest that within this zone, air resistance proved negligible and, therefore, the curve coincides practically with that given by $u = \sqrt{2gs}$.

At u_{500} and beyond, the points were obtained by equation (15) wherein $u_o = u_{500} = 26,400/d_m$. The degree of divergence between this portion of the curve and the one by $u = \sqrt{2gs}$ will be of interest in the discussion of Chapter 7. Notice that $s_{(t)}$ is an odd sort of theoretical distance which would be attained in frictionless fall by the actual particle at its actual terminal velocity. It has practical utility in dealing with air induction by falling materials discussed in Chapter 7.

PROJECTION OF PARTICLES

Where particles are projected, bullet-like, from a source with a high initial velocity, analysis of the motion may be made on the basis of negligible gravitational effect.

A particle may start its travel in the turbulent flow region because of high velocity (i.e., high Reynolds number), pass through the next zone, intermediate motion, and finally into streamline flow. By the time its velocity has attained low enough velocities for the last stage, however, it is unlikely that gravity effects could be neglected.

Smaller particles, or those with lower initial velocities, may be pro-

jected entirely within the region of intermediate motion where Re is always less than 500.

Rate of Change of Velocity

A basic equation corresponding to (2), from which velocity-distance relation may be derived is as follows

$$dt = \frac{du}{-\dfrac{CpAu^2}{2W}}$$ (19)

and, as before

$$\Delta s = \int_{u_o}^{u} u\,dt = \int_{u_o}^{u} \frac{u\,du}{-\dfrac{CpAu^2}{2W}}$$ (20)

Projection Through Turbulent Flow Region

Where the initial stages of motion are turbulent, C is taken, as before, equal to 0.44. Integration of (20) then gives

$$\Delta s = 7 \frac{\rho_s}{\rho} \cdot D \log_{10} \frac{u_o}{u}$$

which is transformed to a more convenient equation in which ρ is assigned the value 0.075, $\rho_s = 62.4z$ and diameter is expressed in microns, d_m.

$$\Delta s = 19 \times 10^{-3} z d_m \log_{10} \frac{u_o}{u}$$ (21)

Approximate Formula

The velocity-distance relation for particles of several weights is plotted in Fig. 2-9. Study of the curves discloses that for a considerable portion of the travel, they are approximately straight lines with equations of the form $\Delta u/\Delta s$ = a constant. Determination of the slope enables development of a simpler and approximate substitute for equation (21) of the form given by (22). This formula gives results departing relatively little from equation (21).

$$\Delta s = \frac{(u_o - u)zd_m}{100u_o}$$ (22)

also

$$u = u_o\left(1 - \frac{100s}{zd_m}\right)$$ (22a)

Properties of Contaminants

Substitution in equation (20) and expansion of $Re^{0.6}$ into its elements $(Du\rho/\mu)^{0.6}$, produces the form,

$$\Delta s = -0.93 \times 10^4 z D^{1.6} \int_{u_0}^{u} u^{-0.4} du$$

where $u_0 = u_{500}$ or below. Integration results in

$$\Delta s = 1.8 \times 10^4 \cdot z \cdot D^{1.6}(u_0^{0.6} - u^{0.6}) \tag{23}$$

and in terms of d_m.

also

$$\Delta s = 3.05 \times 10^{-5} z d_m^{1.6}(u_0^{0.6} - u^{0.6}) \tag{23a}$$

$$u = \left[u_0^{0.6} - \frac{\Delta s \times 10^5}{3.05 \cdot z d_m^{1.6}} \right]^{1.67}$$

Projection Through Stream-Line Flow

Probably of little practical importance is particle motion in the streamline region where Re is less than about 2, and where, as before, gravity is negligible.

The appropriate relation is obtained by integration of (20) after substituting $C = 24$ Re.

$$\Delta s = -\frac{W}{3\pi\mu D} \int_{u_0}^{u} \frac{u\,du}{u}$$
$$= \frac{W(u_0 - u)}{3\pi\mu D}$$

which becomes

$$\Delta s = 0.29 \times 10^{-5} d_m^2 z(u_0 - u) \tag{24}$$

The principal equations are summarized in Table 2-8, while the symbols are as follows:

A = cross-sectional area of particle, sq ft
a = net acceleration of particle, ft per sec²
C = coefficient of resistance, equation (1), and Fig. 2-4
c = maximum % concentration in air of solvent vapor with liquid phase at p_f
D = particle diameter, feet
d_m = particle diameter, microns, = $3.05 \times 10^5 D$
$d_{m(t)}$ = particle diameter, microns, corresponding to terminal velocity
F = net accelerating force, pounds
F_g = force due to gravity = W, pounds

Properties of Contaminants

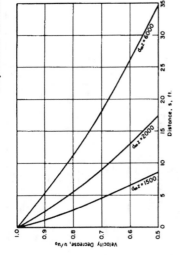

FIG. 2-9. Velocity-distance relationship in turbulent projection according to $\Delta s = 19 \times 10^{-3} z d_m \log_{10} u_0/u$.

Values of slope, K_t, in simplified expression applicable to turbulent projection $K_t = (u_0 - u)/s$, are shown in Table 2-7 derived from log $u_0/u = (s/z d_m) \cdot 10^3/19$. Note that in the u/u_0 range from 0.9 to 0.25 the coefficient in K_t is sensibly 100 within about 15%.

TABLE 2-7
Values of $K_t = \Delta u/\Delta s$ for Turbulent Motion

u/u_0	Δs	K_t
0.9	$0.86 \times 10^{-3} z d_m$	$116 u_0/z d_m$
0.8	$1.84 \times 10^{-3} z d_m$	$109 u_0/z d_m$
0.7	$2.95 \times 10^{-3} z d_m$	$102 u_0/z d_m$
0.5	$5.72 \times 10^{-3} z d_m$	$88 u_0/z d_m$
0.25	$9.1 \times 10^{-3} z d_m$	$83 u_0/z d_m$
0.1	$19 \times 10^{-3} z d_m$	$47 u_0/z d_m$

Using round value of $K_t = 100(u_0/z d_m)$: $u = u_0 - K_t s$ or $u = u_0(1 - 100 s/z d_m)$.

Projection Through Intermediate Flow Region

For particle motion in which Re is below 500, either following a period of turbulent motion or initially, the velocity-distance relation is obtained by another integration of the basic equation (20). In this case C is taken as $18.5/Re^{0.6}$, the approximate value described earlier in this chapter.

K = group of constants, $\rho CA/2W$, for equation (15)

k_1 = slope of lines in Fig. 2-9, $\Delta u/\Delta s$

M = mass

p_l = Vapor pressure of liquid

Re = Reynolds number

s = distance traveled by particle

$s_{(t)}$ = distance traveled by particle at frictionless rate to attain terminal velocity

s_2 = distance traveled by particle to point where $Re = 2$

s_{500} = distance traveled by particle to point where velocity results in $Re = 500$, i.e., start of turbulent motion

t = time, seconds

u = particle velocity, ft per sec (fps)

u_o = particle velocity, fps, at beginning of any particular region. In turbulent flow, $u_o = u_{500}$; in intermediate flow, $u_o = u_2$

u_s = particle velocity, fps, after travel of s feet

u_t = terminal velocity of particle, fps

u_2 = particle velocity, fps, corresponding to $Re = 2$, end of streamline motion

u_{500} = particle velocity, fps, corresponding to $Re = 500$, end of intermediate motion

v = gas velocity, fps

W = particle weight, pounds

z = specific gravity of particle, water = 1

ρ = density of air, = 0.075 lb per cu ft at 70F

ρ_s = density of solid particle, = 62.4z lb per cu ft

μ = viscosity of the air, 1.3×10^{-5}, lb per (sec) (ft)

ϕ = group of terms in equation (12)

σ = specific gravity of vapor, air = 1

F_R = force due to air resistance, pounds

g = acceleration due to gravity, 32.2 ft per sec²

TABLE 2-8
Summary of Equations

$$F_R = -\frac{C\rho u^2 A}{2g}$$

Streamline	Intermediate	Turbulent
Terminal Velocity: $u_t = 9.23 \times 10^{-6} z d_m^2$	Terminal Velocity: $CRe^2 = 41.9 \times 10^{-6} z d_m^3$ From Fig. 2-5, find Re corresponding to calculated CRe^2. Then $u_t = \dfrac{Re}{0.0188\, d_m}$ Alternate Method/(approx.) $u_t = 0.49 \times 10^{-2} d_m^{1.14} z^{0.72}$	Terminal Velocity: $u_t = 0.54\sqrt{z d_m}$

General, for u vs. s:

u vs. s: $s = \dfrac{u_t^2}{g}\left[2.3 \log \dfrac{u_t}{u_t - u} + \dfrac{u}{u_t} \right]$	u vs. s: Incremental Integration of $\dfrac{z d_m^2 \Delta u}{0.145\times10^9(41.9\times10^{-6} z d_m^3 - CRe^2)} = \Delta t$ Then $a = \Delta u/\Delta t$, whence $\Delta s = u^2/2a$	u vs. s: $\Delta s = s - s_{500}$: $= \dfrac{d_m^2}{105} \log_{10}\left(\dfrac{32.2 d_m z - 120 u^2}{32.2 d_m z - 120 u_o^2} \right)$

$$\Delta s = \int_{u_s}^{u} \frac{u\,du}{g - \dfrac{C\rho A u^2}{2W}}$$

Boundaries of Regions

Velocity Limits:

$$u_t = \frac{106}{d_m} \qquad u_{400} = \frac{26,400}{d_m}$$

Distance Limits (Earliest values of s): by $u = \sqrt{2gs}$:

$$s_2 = \frac{175}{d_m^2} \qquad s_{500} = \frac{11.1 \times 10^6}{d_m^2}$$

Particle Size Limits:

$$d_{m(t)} = \frac{103}{\sqrt{z}} \qquad d_{m(t)} = \frac{1480}{\sqrt{z}}$$

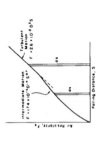

Fig. 7-1. Variations of air resistance, F_R, with falling distance, s, wherein area under the curve is work done on air by a single particle. (Equations for F_R are based on $n^2 = 2\rho/s$; implications of which are discussed later.)

Energy and Power of Particles in Motion

A particle moving through air is subject to the frictional force of air resistance tending to reduce its velocity, or to maintain it at its constant terminal velocity when falling by gravity.

The work represented by the force of friction, F_R, acting for the differential distance, ds, on a single particle (see Figure 7-1) is given by

$$\text{Work} = \int F_R \, ds$$

and the power developed by a stream of particles, by

$$P = \frac{R}{W} \int F_R \, ds \qquad (1)$$

where R is material flow rate, weight per unit time, and W is weight of one particle. Note that $\frac{R}{W}$ is number of particles per unit time. The force, of course, represents that exerted by the particles on surrounding air, as well as the reverse. Therefore the power expression, above, is a measure of the rate of energy transfer to the air which will appear as air motion and as heat. The air motion can be expected to be partly vertical, typical of turbulent motion, and partly linear in the direction of the path of the particle. A primary assumption necessary in the present development is that the energy transformed into heat is negligible, and that the resulting air motion is totally linear.

Expressing power and its elements in engineering units: horsepower, foot-pounds per minute, and pounds per square foot, the expression for power required to maintain flow of air in any system is

$$\text{H.P.} = \left(\frac{ft^3}{min}\right)\left(\frac{lb}{ft^2}\right)\left(\frac{1}{33,000}\right)$$

i.e., horsepower is dependent on the product of volume rate of flow and pressure.

Expressing the pressure or head in units commonly employed in air flow, i.e., inches of water column, h, and using q to designate air flow in cubic feet per minute, the horsepower becomes

$$\text{H.P.} = \frac{qh}{6360}$$

The head, h, is in this case that required to accelerate the air from rest to some velocity, V, i.e., the velocity head, h_v. It is, at ordinary temperatures, numerically equal to

$$h_v = \left(\frac{V}{4005}\right)^2$$

where V is air velocity in feet per minute.
Substitution of h_v in the horsepower equation gives

$$\text{H.P.} = \frac{q}{6360}\left(\frac{V}{4005}\right)^2$$

Consider now a continuous stream of moving particles, trajectories of which, taken in the aggregate, comprise a definite cross-sectional flow area, A (Fig. 7-2). This area may be defined solely by the pathways of the outermost particles in an unenclosed stream, or the stream may be enclosed by solid barriers consisting in a chute. The symbol, V, then represents the *average* velocity of induced air in any selected plane area. Noting that $V = q/A$, the expression for horsepower becomes

$$\text{H.P.} = \frac{q}{6360}\left(\frac{q}{A \cdot 4005}\right)^2$$

$$= \frac{q^3}{(1.02 \times 10^{11}) \cdot A^2}$$

whence, in round numbers, one obtains

$$q^3 = (\text{H.P.})A^2 \cdot 10^{11} \qquad (2)$$

The significance of these relationships is as follows. By reference to relations developed in following sections, from equation (1) it is possible to compute the power exerted on the air through which particles are

moving, by gravity or by projection. Assuming that all this energy is devoted to the linear movement of air within the particle stream, one can calculate the maximum possible induced air flow rate by equation (2) which was developed on the basis that all the power is devoted to the acceleration of air from a state of rest to an average velocity, without frictional losses.

Power in Streams of Particles Falling by Gravity

In the following sections, an expression for horsepower developed by a stream of falling spheres is derived as a function of falling height, particle size and weight. In the first case, the particles have not attained their terminal velocity within the distances considered. It will be necessary to consider, separately, turbulent motion and intermediate motion. Streamline motion is of little significance.

Turbulent Fall—Starting from Rest. The horsepower generated by $\frac{R}{W}$ particles per second falling in turbulent motion a distance, $s - s_{500}$, is given by

$$\text{H.P.} = \frac{R}{550 \cdot W} \int_{s_{500}}^{s} F_R \cdot ds$$

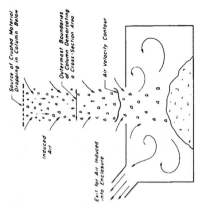

Fig. 7-2. Particles falling into an enclosure inducing a flow of air.

Fig. 7-3. Illustrating nature and magnitude of difference, in turbulent fall, between work based on actual velocity (Area A) and that calculated according to $u_r^2 = 2gs$ (Area A + B). u_r = velocity in vacuum.

The value of F_R for turbulent motion (Equation 13, Chapter 2) is substituted. Then

$$\text{H.P.} = \left(\frac{R}{550 \cdot W}\right)\left(\frac{0.44\rho A}{2g}\right)\int_{s_{500}}^{s} u^2 \cdot ds$$

The relation between u and s is given by Equation 15 of Chapter 2, but use of it would result in an unwieldy equation. A simple expression can be obtained by using the relation $u^2 = 2gs$. The nature and magnitude of the resulting error can be visualized by reference to Figure 7-3. The error is not important in the final result because of the cube root relation between power and air flow potential. Moreover it is on the conservative side.

Substituting $u^2 = 2gs$ in the preceding equation gives

$$\text{H.P.} = \frac{R\,(0.44\rho A)}{550 \cdot W} \int_{s_{500}}^{s} s\,ds$$
$$= \frac{R\,(0.44\rho A)}{2(550)W}\,(s^2 - s^2_{500})$$

For practical purposes, s_{500} may be considered to be negligibly small.

Substitute $\rho = 0.075$; $A = \frac{\pi}{4}D^2$; $W = \frac{\pi}{6}\cdot D^3\cdot(62.4)z$;

$$D = \frac{d_m}{3.05 \times 10^{-5}}$$

The resulting equation is

$$\underset{\text{(Turbulent zone)}}{\text{H.P.}} = 0.22\,\frac{R \cdot s^2}{z \cdot d_m} \qquad (3)$$

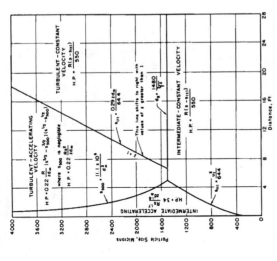

Fig. 7-4. Chart showing zones within which indicated power formulas for falling particles are applicable.

Fall in Intermediate Motion—Starting from Heat. Equations for the power generated by particles falling in the zone of intermediate motion are derived in a manner similar to that demonstrated for turbulent motion.

The value of F_R for intermediate motion (Equation 10, Chapter 2) is employed in Equation 1 to give

$$\text{H.P.} = \frac{R}{550 \cdot W} \cdot (0.94 \times 10^{-4} \cdot D^{1.4}) \int_0^s u^{1.4} \, ds$$

The initial period of streamline fall is neglected and this is justified on practical grounds.

In Chapter 2 it was shown that it was not practical to derive a useful equation for the relation between u and s; a step-wise integration was employed instead. As in the corresponding development for turbulent fall we elect to use the expression $u^2 = 2gs$, recognizing and accepting the error which, again, is on the high side.

Integration, and substitutions for D and W, result in the final form:

$$\text{H.P.} \atop \text{(Intermediate zone)} = 34 \frac{R \cdot s^{1.7}}{s \cdot d_m^{1.6}} \qquad (4)$$

Falling at Terminal Velocity. After attainment of the terminal velocity, u_t, regardless of the type of motion prevailing, the force of resistance, F_R, is equivalent to the force of gravity, W, i.e., the weight of the particle. The following simple relations are therefore applicable:

$$\text{Work} = \frac{R}{W} \int F_R \, ds$$

$$= \frac{R}{W} \cdot W \cdot s$$

$$= Rs$$

$$\text{H.P.} = \frac{Rs}{550} \qquad (5)$$

This holds for all types of motion.

Distance, Velocity, and Particle Size Boundaries

From the various relations developed in Chapter 2, the limits of applicability of the three power equations are summarized in Fig. 7-4.

The height of fall is considered in a given instance, together with the size of particle, or range of sizes, whence by reference to the figure a

conception of the range of applicability is provided. It is clear from this diagram that intermediate motion is confined to falling distances of 4 to 5 feet and less for particles of sizes around 1500 microns (1.5 millimeters, or 12 to 14 mesh) and to 1 to 2 feet for sizes of 500 to 1000 microns or 2000 to 2500 microns and greater.

Exact adherence to the indicated boundaries is, of course, not necessary in the calculation of practical problems; reference to assumptions that led to their derivation will remind one of their approximate nature. Thus a mass of material in the particle size range from 1500 to 4000 microns, falling a distance of 10 feet, could properly be considered to be within the "turbulent-accelerating" zone, even though the fraction

378

between 1500 to 2200 is seen to range into the "constant velocity" zone. If the total mass were to consist of particles lying between 1500 and 2200 microns, it would be proper to divide the travel into three zones: the first, 3 feet (approximately) in the "intermediate accelerating" zone; from 3 feet to 8 feet in the "turbulent accelerating" zone; and the remainder in the "constant velocity" zone.

Some practical difficulties arise in applying these relationships, foremost of which stems from the fact that particle size, d_m, is never a single value and one must, therefore, make a number of calculations, one for each segment of particle size provided by a screen analysis. To obviate this laborious procedure, Figs. 7-5 and 7-6 which make it possible to read power quantities directly have been prepared. Figure 7-7, based on equation (2), then permits a final estimate of induced air flow. Examples of the use of these charts are given in later sections.

TABLE 7-1

Induced-Air-Flow Equivalent of the Energy in Falling, Unenclosed Streams of Particles

(Solids flow rate, $R = 1$; specific gravity, $s = 1$)

Falling Distance, $s = 2\,ft$

Stream area, sq ft	Particle size, millimeters	
	1	2.5
	Air flow equivalent, cfm (multiply by $(R/z)^{1/3}$)	
½	200	100
1	300	200
2	500	300
4	800	500
8	1200	700
15	2000	1000
25	2500	1500
50	4000	2500

Falling Distance, $s = 3\,ft$

Stream area, sq ft	Particle size, millimeters						
	1	2	5	10	20	50	100
	Air flow equivalent, cfm (multiply by $(R/z)^{1/3}$)						
½	450	350	220	180	150	80	65
1	750	550	350	300	250	150	100
2	1200	850	550	450	350	200	170
4	1900	1400	850	700	600	350	250
8	3000	2200	1400	1100	900	500	400
15	4600	3300	2000	1700	1400	800	650
25	6400	4600	3000	2500	2000	1000	900
50	10000	7500	4700	4000	3200	1800	1400

1 Millimeters $\times$ 0.04 = inches.

(Table continued on next page)

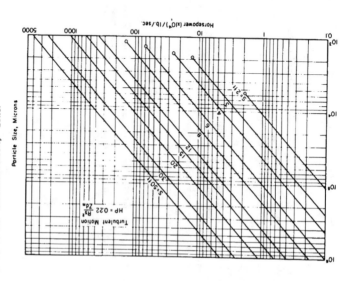

Fig. 7-5. Power induced by particles of $z = 1$ falling in turbulent motion. In practice, multiply chart values by R/z.

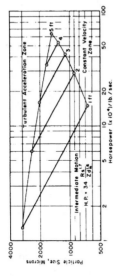

FIG. 7-6. Power induced by particles of $z = 1$ falling in intermediate motion from initial state of rest. In practice, multiply chart values by R/z.

Nomenclature for Chapter 7

A' = cross-sectional area of particle, sq ft.
A = cross-sectional area of stream of particles, sq ft
D = particle diameter, feet
d_m = particle diameter, microns, = $3.05 \times 10^5 D$
$d_{m(t)}$ = particle diameter, microns, corresponding to terminal velocity
F_R = force due to air resistance, pounds
g = acceleration due to gravity, 32.2 ft per sec²
h = pressure head, inches w.g.
K = $10^5/3.05 s d_m^{1.6}$
k_1 = $100 \, u_t/z d_m$
N = total weight of material in falling column
q = air flow, cfm
R = solid material flow rate, pounds per second
Re = Reynolds number
s = distance traveled by particle, ft.
$s_{(t)}$ = theoretical distance traveled by particle in frictionless fall to attain actual terminal velocity, ft.
s_{500} = distance traveled by particle to point where velocity results in $Re = 500$, i.e., start of turbulent motion, ft.
u = particle velocity, ft per sec (fps)
u_o = particle velocity, fps, at beginning of any particular region. In turbulent flow, $u_o = u_{500}$; in intermediate flow, $u_o = u_2$
u_t = terminal velocity of particle, fps
u_{500} = particle velocity, fps, corresponding to $Re = 500$, end of intermediate motion
W = particle weight, pounds
z = specific gravity of particle, water = 1
ρ = density of air, = 0.075 lb per cu ft at 70F
ρ_s = density of solid particle, = 62.4 z lb per cu ft

TABLE 7-1—Continued

Falling Distance, s = 6 ft

Stream area, sq ft	Particle size, millimeters					
	Less than 2	5	10	20	50	100
	Air flow equivalent, cfm					
½	650¹	350	300	250	200	150
1	1000¹	600	500	350	300	200
2	1500¹	900	700	600	400	350
4	2000¹	1500	1200	900	700	500
8	4000¹	2300	1800	1500	1000	900
15	6500¹	3500	2800	2200	1500	1300
25	9000¹	5000	4000	3200	2300	1800
50	—	8000	6000	5000	3700	3000

Falling Distance, s = 12 ft

Stream area, sq ft	Less than 2	5	10	20	50	100
½	800¹	560	450	350	270	220
1	1300¹	900	700	600	420	350
2	2000¹	1500	1100	900	700	550
4	3300¹	2300	1800	1500	1000	850
8	5100¹	3600	2900	2300	1700	1300
15	8000¹	5300	4200	3500	2600	2100
25	—	7800	6000	5000	3600	2900
50	—	—	10000	8000	6000	4700

Falling Distance, s = 20 ft

Stream area, sq ft	Less than 2	5	10	20	50	100
½	950¹	800	650	500	370	300
1	1500¹	1300	1000	800	600	500
2	2400¹	2000	1600	1300	960	760
4	4000¹	3300	2600	2000	1500	1200
8	6000¹	5000	4100	3200	2400	1900
15	9500¹	8000	6400	5000	3700	2900
25	—	—	9000	7000	5000	4000

Falling Distance, s = 30 ft

Stream area, sq ft	Less than 2	5	10	20	50	100
½	1100¹	1000	850	650	500	400
1	1800¹	1700	1400	1100	800	600
2	2800¹	2700	2100	1700	1300	1000
4	4500¹	4400	3500	2700	2000	1600
8	7000¹	6800	5400	4200	3200	2500
15	—	10000	8200	6600	4800	4000
25	—	—	—	9000	6800	5400

¹ Multiply by $\sqrt{R}$; all others, multiply by $\sqrt{R/z}$.

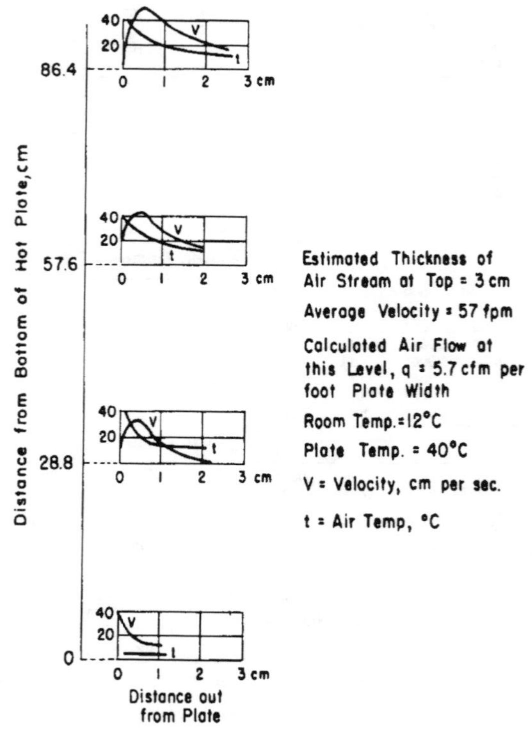

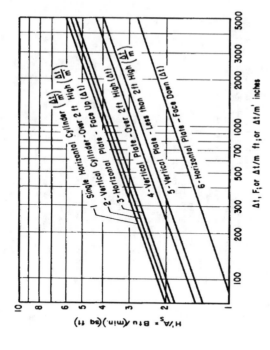

40
20
V
t
86.4
0 1 2 3 cm
57.6
28.8
0
Distance from Bottom of Hot Plate, cm
Distance out from Plate
Estimated Thickness of Air Stream at Top = 3 cm
Average Velocity = 57 fpm
Calculated Air Flow at this Level, q = 5.7 cfm per foot Plate Width
Room Temp. = 12°C
Plate Temp. = 40°C
V = Velocity, cm per sec.
t = Air Temp, °C

FIG. 8-5. Air velocity and air temperatures in warm air stream adjacent to a heated vertical plate (after Griffiths & Davis).

H'/A_s = Btu/(min.)(sq ft)
10
8
6
5
4
3
2
1
1 Single Horizontal Cylinder (Δt/m')
2 - Vertical Cylinder - Over 2 ft High (Δt/m')
3 - Horizontal Plate - Face Up (Δt)
4 - Vertical Plate - Over 2ft High (Δt)
5 - Vertical Plate - Less than 2ft High (Δt/m)
6 - Horizontal Plate - Face Down (Δt)
100 200 300 500 700 1000 2000 3000 5000
Δt, F; or Δt/m ft; or Δt/m' inches

FIG. 8-7. Heat transfer from various hot bodies to air. Where vertical cylinder is less than 2 ft high, multiply H'/A from curve 2 by appropriate factor:

Height, Ft	*Factor*	*Height, Ft*	*Factor*
0.1	3.5	0.4	1.7
0.2	2.5	0.5	1.5
0.3	2.0	1.0	1.1

Basic Cloth Resistance

The initial resistance, R_0 is proportional to the filtration velocity, V_f.

$$R_0 = K_0 V_f$$

where K_0 is a particular factor which we call the *basic cloth resistance*. When V_f equals one, $R_0 = K_0$, hence it is seen that K_0 can be defined as the pressure drop, in inches of water, at a filtration velocity of 1 fpm.

The magnitude of K_0 is determined partly by the fineness of the dust that is lodged in the pores of the cloth and partly by the quantity, after the cloth has been shaken. The residual quantity of dust is largely dependent on the violence with which the cloth shaking mechanism operates to dislodge the dust. In one instance in our experience, the basic cloth resistance of a flat bag collector handling crushed stone dust was 0.83, and after installation of an auxiliary pneumatic vibrating mechanism for more vigorous shaking of the cloth, the value was reduced to 0.49.

A number of observations of the magnitude of basic cloth resistance made by the author are given in Table 16-11. They indicate materially higher values in the case of flat bag collectors than for cloth tube collectors. This is not surprising when one compares the shaking action characteristic of each type of collector. In the flat bag type, each cloth bag is stretched over a metal framework, and dust on the cloth is dislodged by vibrating the framework, a relatively gentle action. The bags of the cloth tube or stockingleg type are suspended at the upper end to a framework which, when vibrated, subjects the cloth tubes to a whipping action that is relatively vigorous. Evidently, then, less dust remains adherent to the cloth tubes than to the surfaces of flat bags, hence the basic cloth resistance of the latter tends to be higher.

The size of collector might influence the value of basic cloth resistance if the shaking mechanism be such as to subject the cloth in small units to more vigorous action than in larger units. There is some suggestion in the data of Table 16-11, that this is sometimes true, especially if one regards the dust of stone cutting to be essentially the same as that from stone crushing. However, other data in the table indicate no difference in the value over wide ranges of cloth area.

The type of dust being handled does affect the value of K_0, the finer

TABLE 16-11

Values of the Basic Cloth Resistance Factor, K_0, Observed in Specified Applications

No.	Type of dust	Cloth area	K_0	Remarks
	Flat Bag Collectors			
1.	Stone crushing operations	250	0.83	
2.	Stone crushing operations	250	0.49	Same as No. 1 but following installation of pneumatic vibrator
3.	Stone crushing operations	500	0.83 0.78 0.75 0.74	
4.	Stone crushing operations	2250	0.79	
5.	Stone crushing operations	9000	1.01	
6.	Synthetic abrasive crushing operations	—	0.80	
7.	Clay crushing in dry pan	500	1.6	
	Cloth Tube Collectors			
7a.	Stone crushing operations	2150	0.47 0.45 0.60	
8.	Stone crushing operations	4300	0.45 0.37	
9.	Stone crushing operations	1500	0.40	
10.	Stone chiseling	400–1000	0.17–0.27	
11.	Electric welding fume	10	0.7	
12.	Iron cupola fumes	—	2.5	
13.	Foundry dust	5200	0.28	
	Core knockout		0.25 0.58	
	Shot blast-room ventilation	2350	0.63 0.39 0.39	
14.	Same—pneumatic lift	950	0.34, 0.36, 0.59	
	Clay crushing in dry pan	500	0.6	

dusts leading to higher values. This would be expected, as will be clear from the later discussion of dust resistance.

Measurement of K_0 Factor. The elements of basic cloth resistance, K_0, are given in the equation

$$K_0 = \frac{R_0}{V_0} = \frac{P_2 - P_1}{Q/A}$$

where P_2 and P_1 are pressures at exit and at entrance to the collector,

TABLE 16-12
Values of Basic Dust Resistance, K_d, Observed on Some Industrial Installations
(See also Table 16-13)

Type of dust	Cloth dust loading W, oz sq ft	K_d
Stone Crushing (Plant A)	5	0.18
	12	0.12
	14	0.08
	17	0.12
	22	0.11
	25	0.02
	28	0.07
Stone Crushing (Plant B)	7	0.16
	8	0.10
	8	0.08
Stone Crushing (Plant C)	1	0.82
Foundry, castings cleaning		
Shot blasting	0.2	0.82
	0.3	0.25
	1.3	0.25
Pneumatic shot lift	0.2	0.66
	2.4	0.40
Core knockout	0.2	0.55
	0.1	0.68
Sandblasting (scale)	7	0.20

The coefficient, K_d, we have termed *basic dust resistance*. When V_f is 1 fpm and W is 1 oz per sq ft, K_d is numerically equal to R_d from which it is seen that K_d units are inches of water per 1 fpm per 1 oz per sq ft. Some measurements made by the author in conjunction with the data of Table 16-11 are given in Table 16-12. They indicate a rather considerable spread of values within any one category. If, however, the maximum and minimum values among the stone crushing data are omitted, those remaining fall for the most part within the range 0.1 to 0.2 and one would have some confidence in calculations based on those numbers. The variations are undoubtedly due partly to differences in the quantity of very coarse particles which contribute greatly, in weight, but insignificantly to pressure drop, and also probably to differences in degree of porosity of the dust layers.

Measurement of Dust Resistance, K_d. The determination of K_d on an industrial filter unit is made by measurement of (1) pressure drop

---

Q is the air flow rate, and A the total cloth area, and this indicates the measurements needed to establish its value. The pressure readings are made in the housing of the collector not in connecting duct work. This is important as to the exit duct because the static pressures are different in housing and duct by the amount of velocity pressure in the duct plus entrance energy loss. Pressures are measured with an ordinary U-tube.

Dust Resistance

The technical literature dealing with pressure drop through filter beds as a function of the several dimensional elements of such beds is extensive.* It is known that pressure drop for air at a fixed rate of flow is a function of particle size characteristics, degree of packing, porosity of the bed and thickness of the dust layer, i.e., length of air travel. Specifically, it has been shown that pressure drop, R_d, through the dust layer varies as follows:

$$R_d = K \frac{L}{d^2} \frac{(1-f)^2}{f^3}$$

where K is a constant, L the thickness of the layer, d the particle size, and f the porosity of the bed, the fractional ratio of void volume to total volume. It can be seen that very small changes in f will cause large changes in R_d.

It is not practical in applying this knowledge to the design of dust filters to deal directly with particle size, d, nor with degree of packing, f. Consequently in the present treatment they are combined with the constant, and their magnitude therefore implied only by a description of the dust and of the process in which it originates; the constant is written K_d. Thickness of the dust layer, L, can be conveniently described in terms of weight of deposited dust per unit area of filter, thus implying that one expects all layers of the deposit to be of uniform porosity, i.e., bulk density.

Designating dust loading, ounces per square foot, as W, and filtration velocity, V_f, as before, we can describe pressure drop through the temporarily adherent layer of dust on the cloth as follows:

$$R_d = K_d V_f W = \frac{K_d Q W}{A}$$

---

* An authoritative summary is provided by Hatch, L. P., *Flow Through Granular Media, J. App. Mech. 7*, A109-12 (1940).

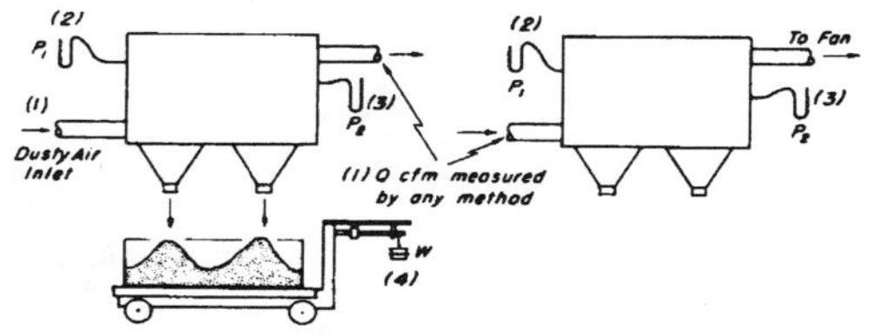

(a) Measurements (1)(2)(3) made on loaded filter. For final measurement, W, shut off fan, remove and weigh dust.

(b) Measurement for K_0. Following (a) start up fan and measure (1)(2)(3).

FIG. 16-17. Procedure for determination of K_0 and K_d.

after a suitable dust filtering time interval, (2) air flow rate, Q, and (3) weight of dust on the filter corresponding to the data of (1) and (3). After the first two items have been observed the dust is shaken from the cloth, removed from the collector hopper and its weight determined.

After removal and weighing of dust, the air flow is resumed and the "clean-filter" pressure drop determined, together with the corresponding air flow rate. Simple algebraic calculation of K_d then follows. The procedure for determining both constants is illustrated in Fig. 16-17.

Application to Design Problems

The total pressure drop through a cloth dust collector exclusive of duct entrance to and exit from the housing is the sum of "basic cloth resistance" and "dust resistance."

$$R = K_0 V_f + K_d V_f W$$

In order to apply this equation to practical problems it is necessary to accumulate a background of data on values of K_0 and K_d. It still will be necessary to estimate the quantity of dust that may be collected within any specified time interval but this is a tangible factor that can readily be visualized.

The data that have been acquired* on stone chiseling operations are summarized in Table 16-13 and are of general applicability for all stone chipping operations as illustrated in the following problems.

---

* Hemeon, W. C. L., Dust Control in the Granite Cutting Industry, *Heating and Ventilating*, pp. 41–45, August, 1940.

TABLE 16-13
Design Data for Cloth Filters
Applicable to Chiseling of Stone

Resistance Factors	
K_0, for cloth tube collectors	0.25
K_0, for flat bag collectors	0.80
K_d, for dust	0.20

Dust Production Rates, Chiseling Granite	Pounds per hour per man
Hand-held pneumatic tools (i.e., each "banker")	$\frac{1}{2}$ to 2
Pneumatic surfacing machine	15 to 20
Abrasive blasting for stone lettering	4 to 7

Example 1. A filter of the cloth tube type is desired for the exhaust of a surfacing machine, with a minimum air flow of 600 cfm. It is decided to allow for a pressure drop through the filter of about 5 inches w.g. at the end of the cycle (just before shaking). Working periods are 7:30 a.m. to 12:00 and 1:00 to 4:30 p.m. The schedule will call for shaking the bags each noon and at the end of the day.

Solution. Taking the longer morning period of $4\frac{1}{2}$ hours, we assume that there will be not over 4 hours intensive dust production. Reference to Table 16-13 indicates that a dust production rate of 20 lb per hr will not be exceeded. Therefore we calculate 80 lb of dust on the cloth at the end of the dust cycle.

The basic cloth resistance of the cloth tube collector is 0.25, the basic dust resistance, 0.20.

A cut and try method of solution is simplest. As a first try, assume a filter with 200 sq ft of cloth.

Filtration velocity, $V_f = 3$ fpm

Dust loading, $W = \dfrac{80 \times 16}{200} = 6.4$ oz per sq ft

Substituting in the equation

$$R = K_0 V_f + K_d V_f W$$
$$= 0.25(3) + 0.20(3)6.4$$
$$= 4.6 \text{ inches w.g.}$$

Which is sufficiently close to the maximum pressure drop specified and we therefore conclude that a filtration velocity of 3 fpm satisfies the stated conditions.

384

Example 2. Using the other data of Example 1 assume instead that a minimum air flow of 1200 cfm is desired. Calculate the pressure drop through a cloth tube filter of size to result in a filtration velocity of 4 fpm.

Solution.

Filtration velocity = 4 fpm

$$\text{Cloth area} = \frac{1200}{4} = 300 \text{ sq ft}$$

$$\text{Dust loading} = \frac{80 \times 16}{300} = 4.25 \text{ oz per sq ft}$$

Substituting in equation (3)

$$R = 0.25(4) + 0.20(4)4.25$$
$$= 4.4 \text{ inches w.g.}$$

Example 3. A cloth tube filter is to be used in conjunction with an exhaust system for 6 bankers. A 5-hour work cycle (before shaking the filter) is to be provided for and the air flow at the end of the cycle must not be less than 400 cfm through any exhaust inlet. It is established that this means a total minimum air flow of 2700 cfm. Calculate the pressure drop at the end of the cycle for a filtration velocity of 10 fpm.

Solution. We note by reference to Table 16-13 that the maximum amount of dust to be expected in this situation, employing hand pneumatic tools, is 1 lb per hour per man, that is, in 5 hours an aggregate of 30 lb.

$$\text{Cloth area} = \frac{2700}{10} = 270 \text{ sq ft}$$

$$\text{Dust loading, } W = \frac{30 \times 16}{270} = 1.78 \text{ oz. per sq ft}$$

Substitution in equation (3) gives

$$R = 0.25(10)1.78$$
$$= 6.1 \text{ inches w.g.}$$

Laboratory Determination of Dust Resistance. Recognizing the design principles set forth above, it will be seen that there is a need for practical techniques enabling derivation of the necessary constants for a new

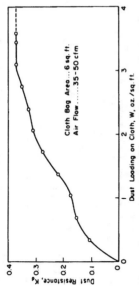

Fig. 16-18. Laboratory measurement of dust resistance using a dust taken from a large commercial collector; field tests on which gave $K_d = 0.12$ ($W = 17$ to 22 oz/sq ft)

situation, especially dust resistance. Skinner and Hemeon[*] explored the possibility of determining dust resistance in the laboratory. Dust samples acquired from a commercial collector where full scale dust constants had been determined were reprocessed on a laboratory scale to see if the full scale dust resistance constant could be reproduced. Such reproduction was not successful as seen in Fig. 16-18. The study was not continued to a final conclusion, however, and it may not be stated that a valid procedure could not be developed. With such a technique worked out, it would be possible to secure a sample of dust from the operations for which a dust collection system is being designed, process the dust sample in a laboratory filter and derive a dust resistance constant for use in estimating pressure drop corresponding to some logical time cycle.

The Y-Factor for Dust Resistance. An alternative method of developing design data for estimating pressure drop applies to a specific process in which the dust production rate and the resistance characteristics of the dust are combined in a single factor, Y, which is the rate of resistance increase in inches pressure drop per fpm per hour. It appears in the term

$$R_d = YV_f^2 t$$

where V_f is the filtration velocity, fpm, and t is the operating time in hours. If, for example, the Y-factor be measured as 0.1 inch per fpm

_____
[*] Skinner, J. B. and Hemeon, W. C. L., unpublished data.

per hour, one can then calculate the resistance increase at any filtration velocity and for any operating time. For example, for a filtration velocity, $V = 3$ fpm, and an operating cycle of 4 hours, the pressure drop increase due to dust would be

$$R_d = (0.1)(3.0)^2(4) = 3.6 \text{ inches w.g.}$$

and to this value the initial resistance due to dust-saturated cloth, at the beginning of the cycle must be added. Suppose K_0 to be 0.3, then

$$R_0 = 0.3(3) = 0.9 \text{ inches}$$

and the total resistance, R, would equal the sum of these values, 4.5 inches.

As a practical matter, we believe that development of the Y-factor as a basis for design of cloth dust collectors will be profitable although it should not be developed to the exclusion of the basis previously described since each will have useful application not, in general, overlapping.

Measurement of Y-factor. The Y-factor appears in the resistance equation as

$$R = K_0 V_f + Y V_f^2 t$$

then

$$Y = \frac{R - K_0 V_f}{V_f^2 t}$$

and from this it is clear what measurements need to be made on an existing dust collector to ascertain its value. R is total resistance, inches of water corresponding to a time period of filtering operation, t, in hours; V_f as previously set forth is filtration velocity, cfm per sq ft; K_0 is obtained as described in an earlier section of this chapter.

The value of Y, it will be observed, is not related to the characteristics of the dust collector and its shaking mechanism (as is K_0). It is, rather, dependent on the rate at which dust is deposited on the cloth and on the fineness and related characteristics of the dust. Clearly, then, if a Y-factor derived from a single cloth dust collector installation is to be applied to another projected installation, it is implied that the rate of dust production is the same and that it is of the same fineness. This means, practically, that these data cannot be transposed to a different process, although with the development of sufficient data it may eventually be possible to apply data from one situation to a somewhat different one for a rough estimation.

FIBER BED FILTERS

The filtration of ventilation air in thick high velocity fibrous filters is well known and the dust collection efficiency of some types is probably as good as that of many industrial dust scrubbers. They are not, however, applicable to industrial dust collection because there are no provisions for cyclical removal of accumulated dust.

The principle of particle separation, impaction on fibers in a thick bed, in such air filters, is illustrated, however, in two cyclically renewable filters, one of historical and scientific interest and another a recent development.

500-Foot Thick "Filter"

A notable and spectacular particle separator was applied in the copper smelting industry more than fifty years ago. Metal dust and fumes from various processes in the plant of the former Boston and Montana Smelting Company at Great Falls, Montana, were to be collected, mainly for the economic value, from a gas stream of 1,100,000 cfm. Following extensive large scale pilot experimentation with various devices, a system was designed, constructed and placed in operation for this purpose around 1910 and operated for a period of almost four years.

The dust and fume laden gases were passed through a particle separator 478 feet long, 176 feet wide, and 21 feet high inside. Below this chamber a 12-foot high basement was provided to house its 1,040 collection hoppers above a series of parallel tracks which permitted transfer of the collected particulate matter into railroad cars.

Within the chamber wires were suspended from the ceiling 1⅝ inches apart, the wires being of No. 10 and No. 12 Birmingham gauge. A total of 1,219,000 wires weighing almost one pound each, for a total weight of 608 tons, comprised the "filter." This mass of wires was arranged in two banks, the first in a section from the entrance to a point 150 feet distant and, following a vacant space 47 feet long, the second bank 150 feet in length.

The average linear gas velocity horizontally through the chamber was 300 feet per minute and gas temperature was 500°F. The wire assembly was provided with a vibration mechanism whereby they could be shaken to dislodge dust collected on the wires. This was done for a 30-minute period at intervals of 60 to 90 days.

The total cost in currency of that day was $1,100,000. In 41 months of operation the system collected 191,000 tons of dust, an average of 4660 tons per month. An account of this installation(Hofman), (Peters) states that "after being in successful operation for some time, the corrosion of the supporting network allowed the wires to drop to the bottom of the flue and they have never been replaced."

The collection efficiency of the full-scale installation is indicated by the performance of an experimental chamber which preceded construction of the full-scale unit and on which its design was based. The experimental flue was 300 feet long, 4 feet wide and 4½ feet high. Efficiency of collection was determined first with the open flue without wires, and subsequently with wires installed. Collection efficiency versus length of travel through the wire "filter" is shown in Figures 16-21.

Theoretical particle separation efficiency. Two particle separation mechanisms would be operating in this system: gravitational sedimentation and impaction.

The theoretical penetration of particulate matter through the forest of wires relative to the impaction mechanism of separation can be calculated by the same method as was employed for spray towers. In that case the impaction targets were the spherical water droplets falling by gravity through the gas stream and in the calculation we derived the swept volume rate of these targets as the product of their effective cross-sectional area for unit time, and falling distance.

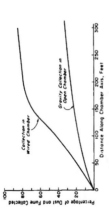

Fig. 16-21. Collection efficiency of 300-foot-long experimental wire "filter" on which design of the 500-foot "filter" of the Boston and Montana Smelting Company was based.

An analogous situation appears in the wired chamber, where the targets however are stationary and cylindrical rather than spherical. The swept volume is the product of the effective impaction area and gas velocity

$$\log P = \frac{-W}{2.3G}$$

where, as before, P is fractional penetration, by number, W is the swept volume rate, and G is the gas volume rate. This relation is applied in the following problem.

Problem. Calculate the collection efficiency of the 500-foot-thick filter for 3-micron particles having specific gravity = 5. Diameter of wire targets is 0.13 inch or 1700 microns; their length is 20 feet; and their number is 1,200,000. Gas velocity is 5 ft/sec and gas volume is 1,100,000 ft^3/min.

Solution. First calculate the impaction parameter

$$= \frac{0.88 u_o \cdot Z \cdot d_m^2}{j_m}$$

$$K = \frac{0.88\ (5)\ (5)\ (3)^2}{1.7 \times 10^3}$$

From Table 16-2, a K-value of 0.16 corresponds to a fractional efficiency of 0.03 which we shall use.

Effective area of all wires is, then

$$A_e = \frac{0.03\ (1.2 \times 10^6)\ (0.13)\ (20)}{(12)}$$

$$= 7.8 \times 10^3\ ft^2$$

Swept volume is

$$W = 7.8 \times 10^3 \times 5 \times 60$$

$$= 2.34 \times 10^6\ ft^3/min$$

$$G = 1.1 \times 10^6$$

$$\log P = \frac{-W}{2.3G}$$

$$= \frac{-2.34 \times 10^6}{2.3 \times 1.1 \times 10^6}$$

$$= -0.0925$$

$$P = 0.81$$

Efficiency = 19%, by number

Limitations of theory. The penetration equation is based on the assumption of perfect turbulent mixing following each layer of targets. In the case of the 5C0 ft wired chamber where a distance of 1⅝ inch (12 wire diameters) separates each layer from the next one downstream, it seems almost certain that this condition is realized and one would have confidence in applicability of the equation as in the preceding problem.

This is in contrast to the case with ordinary compact ventilation-air filters for which the same relationship could be considered. Because of the close spacing of fibers, in contact with each other, it is certain that the perfect mixing assumption is invalid.

It should also be noted that the calculated result assumes adhesion of all impacted particles onto the target and no blowoff.

Water-Wash Fiber Bed Filter

A recently developed particle separator having unique operating characteristics comprises an annular filter bed (e.g., glass fibers) as the outer portion of a rotatable cylinder (Figure 16-22). Dusty gas passes into the housing, through the annular filter element (1-2 inches thick) at filtration velocities of 100-300 fpm, usually, then outward axially to the clean gas duct.

At the end of an operating cycle the collected dust is washed out of the filter. The gas seal is broken to permit rotation of the cylindrical filter unit, with jets of water supplied to the exterior and, simultaneously or alternatively, a stream of water to the interior, while rotating the unit. The water entrains the dust and the resulting mud is thrown

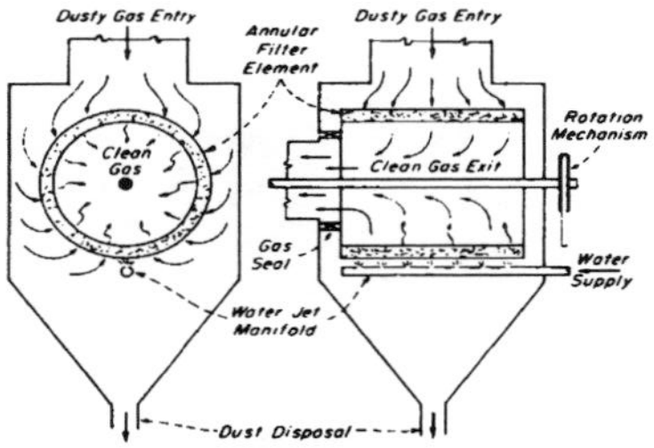

FIG. 16-22. Water wash fiber bed filter in which the filter bed is cyclically renewable by water washing and simultaneous spinning of the annular filter bed. (Patented, Hemeon, 1961)

outward by centrifugal force and the clean, palpably dry filter is then ready for another operating cycle.

Separation mechanism. The mechanism of particle separation is one of filtration, differing however from that in cloth bag filtration, the filter element of which is essentially the cake of dust as it accumulates. In this deep bed filter, impaction of dust particles occurs on the individual fibers throughout the thickness of the filter.

Filtration efficiency is a function of the design and fiber composition of a particular filter bed.

Pressure drop characteristics are best described in terms of the parameters K_o and K_d previously described in the discussion of cloth filters. Typical initial cyclical pressure drops in terms of K_o are compared below with typical values for cloth bag collectors, from Tables 16-1 and 16-3.

	K_o: inches pressure drop, at 1 fpm
Cloth bag	0.25 — 0.80
Fiber bed filter	0.010 — 0.015

Thus, at a filtration velocity of 100 fpm the pressure drops would be 25 to 80 inches, and 1.0 to 1.5 inches, respectively.

Pressure drop increase due to accumulated dust is described in terms of K_d.

	K_d: inches at 1 fpm and 1 oz/ft²	
Cloth bag	0.10 to 0.60	inches/fpm/oz/ft²
Deep bed	0.0002 to 0.0006	inches/fpm/oz/ft²

For example, at 100 fpm and after an increment of 100 ounces (6.3 lbs) dust per square foot, the *increased* pressure drop due to dust for the cloth bag filter would range from 10 to 60 inches, and for the deep bed filter, 0.02 to 0.06 inch.